Seismic Imaging

A Review of the Techniques, their Principles, Merits and Limitations

Etienne Robein

ISBN 978-90-73781-78-8

EAGE Publications bv
PO Box 59
3990 DB HOUTEN
The Netherlands

Contents

Introduction – What is a reflection seismic image exactly?

As the search for new resources means that we are forced to maximize the production of discovered reservoirs and explore new ones in domains that are increasingly complex, seismic imaging is becoming more and more important as a tool.

Seismic imaging is the ultimate stage of a time-consuming and complex data processing sequence that aims to produce accurate images of the Earth's subsurface that are suitable for interpretation by geoscientists.

This course gives an overview of today's most popular seismic imaging techniques used in the oil and gas industry. Two main classes of techniques are addressed: time-imaging and depth-imaging. Both approaches require an estimate of how fast the seismic waves travel at any given point in the Earth but with different degrees of accuracy. If time-imaging embeds velocity analysis almost naturally, though at a cost in image quality, depth-imaging requires the explicit construction of a velocity model. Recent advances in seismic acquisition, imaging technology and high-performance computing, allow us to correctly take into account a much greater complexity of subsurface model and consequently, start to image structures that were previously invisible.

The book presents in simple terms the principles of different techniques in each class of methods, while pointing out their respective merits and limitations. The relationship between acquisition and imaging on the one hand and interpretation and imaging on the other will also be discussed.

The book does not address, however, or briefly touch upon topics related to data pre-processing, such as signal processing, multiples attenuation or static corrections. For those topics, the reader may refer to classical references (Yilmaz, 1987, 2001). Multiple attenuation has been thoroughly treated in a previous EAGE Education Tour (Verschuur, 2006).

Similarly, niche or just emerging processing and imaging technologies, such as converted wave imaging, imaging with multiple reflections, borehole seismics, various applications of interferometry, micro-seismic will be at best briefly mentioned in the text.

What is a reflection seismic image exactly?

The reflection seismic method is based on sending elastic waves into the Earth and letting them propagate in all directions. We say that we send a 'seismic signal'.

In their travel into the subsurface, waves encounter heterogeneities, where they 'refract', 'reflect' and 'convert'. The reflected energy travels back to the surface, where it is eventually measured by specific receivers and recorded (fig. 1).

Seismic processing and imaging have the objective to create from the recordings an image of these heterogeneities, which are indeed mainly the geological features present in the subsurface that we are interested in. Often in the subsurface, these 'geological heterogeneities' have a lateral continuity. This is why they are called 'layers' or more often 'interfaces' or 'reflectors' in seismic jargon. They may also be more localized; we will then call them 'diffractors'.

To be more precise, refraction occurs wherever there is a change in propagation velocities (we will discuss this term further on) and reflection takes place any time when the propagating wave encounters a change in the 'acoustic impedance'. Acoustic impedance is an elastic property of any medium. It is the product of propagation velocity and density. A key point in imaging is that *the amplitude of the reflected wave is proportional to the amplitude of the incidence wave*. The proportionality coefficient is called the *reflection coefficient*, a parameter function of three (in the isotropic case) elastic parameters on each side of the heterogeneity (generally a geological layer) and incidence angle, as given in fig. 2. The 'imaging condition' that we will use many times in our discussions later on has the objective of computing this reflection coefficient. As a consequence, we can claim that the seismic image is *an image of the 'elastic reflectivity' at each point of the subsurface*.

Here again, we need to be more accurate. For that, we note that another phenomenon occurs at interfaces: part of the incident energy is 'converted' from one type of waves into another type, as schematized in fig. 1. The waves that seismic reflection imaging deals with are mainly 'P-waves' (see below). At elastic interfaces, part of the wave energy is converted into 'S-waves'. The amount of conversion depends on the angle at which the elastic wave hits heterogeneity in the ground. This angle, measured from the normal to the interface, is called the *incidence angle*. The reflection coefficient of the P-wave depends on this angle. The basic image we want to produce corresponds to the reflectivity for a zero incidence angle, often called 'normal incidence', but the way reflectivity varies as a function of incidence angle is also of great importance to the geophysicist. This is because it provides more information on the elastic properties of the Earth, which can possibly be related in the interpretation step to geological ones. Analysis or display of how reflection coefficients vary as a function of incidence angle is another product of seismic imaging, called 'amplitude versus angle' ('AVA') analysis.

We can note already at this point that the elastic signal that is sent into the ground is of limited bandwidth, i.e., it does not include all frequencies and it is of a given duration. Imagine two layers so close to each other (in a vertical sense) that the corresponding reflected waves arrive at the surface with a time delay much shorter that the signal duration: the two 'reflections' will be mixed up and will not appear as two distinct layers on the seismic image. We say that the band-limitation of the seismic signal limits the resolution of the seismic image, which is consequently a 'band-limited version of the elastic reflectivity of the Earth'. Note that band-limitation of the seismic signal comes both from the intrinsic physical limitations in sending very low frequencies into the ground and the 'natural' attenuation of the high-frequency components of waves as they propagate into the Earth.

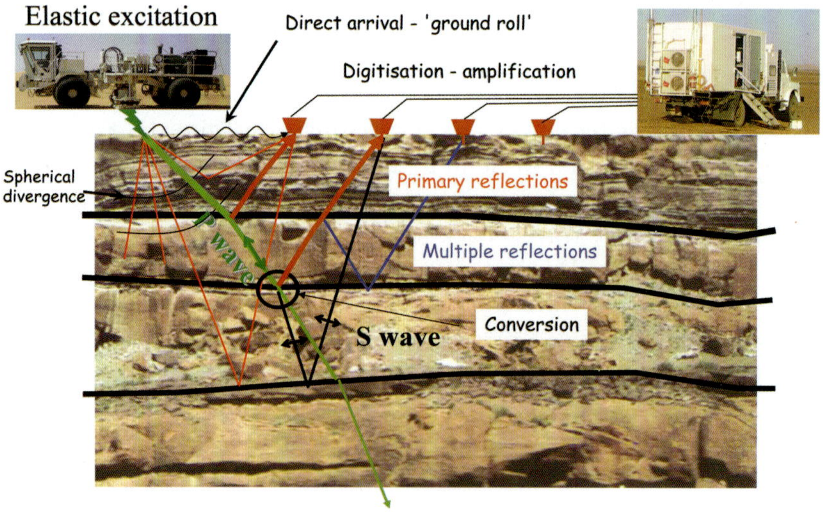

Figure 1: The Seismic Experiment. Various types of waves are generated. Some of them are recorded at the surface.

In order to create the image of the subsurface, we need to estimate in some way the velocity field of the Earth in which waves propagate. This will be discussed extensively all through the book. This velocity field is a third piece of information derived from seismic processing and imaging.

Hence, the results we expect from P-wave seismic imaging are:
- an image, which is a band-limited version of the normal incidence elastic reflection coefficient characterizing heterogeneities present in the subsurface;
- information on the variation with incidence angle of this reflection coefficient;
- information on the P-wave propagation velocity field in the Earth.

We note from its definition that the reflection coefficient is the derivative of the logarithm of acoustic impedance. It is the role of 'stratigraphic inversion' (not to be confused with other types of inversion that we will present later) to integrate, in a mathematical sense, the reflection coefficient series resulting from imaging, in order to reconstruct an acoustic impedance log at each trace location. However, since seismics provide us with a band limited version of reflection coefficient only, it must be combined with other pieces of information that may include velocities derived from the imaging process itself but also further information from measurements in the wells and other *a priori* information if one wants to reconstruct the 'true' impedance. This process goes beyond our scope here.

Transmission, attenuation, refraction and other primary reflections

We said that when the elastic wave hits heterogeneity in the Earth, part of the energy, controlled by the reflection coefficient, is reflected. Since the reflection coefficient is in most cases much smaller than one (an interface with a reflection coefficient of 0.01 may be considered as a strong reflector), the other part of the wave will carry on travelling down into the Earth, will encounter other heterogeneities, reflect there, sending part of its remaining energy back to the surface, and so on ... This means that we can record successively '*primary reflections*' of deeper and deeper layers. This cannot

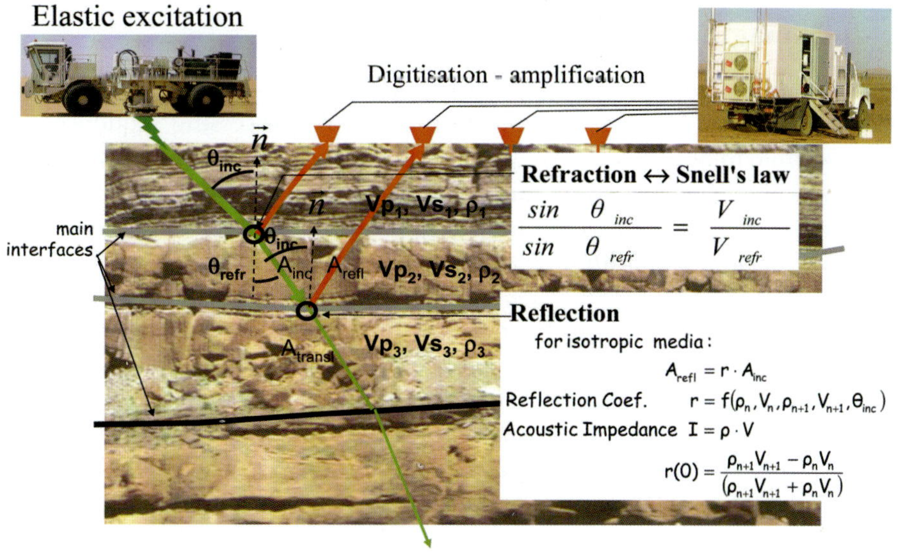

Figure 2: The two main phenomena we deal with in seismic: reflection and refraction.

go on indefinitely however, as the amplitude of wave decreases up to a point where it goes below the ambient noise level and cannot be recorded. There are at least three reasons for this:

- since part of the energy is both reflected and converted at the interface, only part of it travels further down. The ratio between transmitted and incident amplitude is called the '*transmission coefficient*' and is always smaller than 1;
- *spherical divergence*, a term that we develop later;
- *attenuation*, a concept that includes non-elastic loss of energy (friction, rock-fluid interactions, etc.).

Finally, when the travelling wave crosses heterogeneity, yet another phenomenon occurs : refraction. This is the fact that travelling direction is modified across the interface, a well-known phenomenon in optics and quantified by Snell's law.

All these different effects are important and need to be compensated for in some way in imaging. We will therefore come back to them throughout the book.

Multiple reflections

Heterogeneity in the elastic parameters at geological layers are seen not only by the downgoing waves, but also by the reflected waves as they travel back upwards toward the surface. Hence, the phenomena of refraction, reflection and transmission take place there too. It means that part of the energy travels back downwards again, though with a lower amplitude. They are called 'multiple reflections' or 'multiple scattering'. They are considered undesired waves because they do not follow the propagation 'rules' that we use in imaging. They consequently appear as misleading artefacts on the seismic image. The current way of dealing with multiple reflections in 'classical' imaging techniques is to try and remove them from the records at the processing stage. Multiple reflections however bear information on the reflectivity field of the Earth that we are after. It is hence a new and important route to use them directly in imaging rather than trying to get rid of them beforehand. We come back to this point in chapter 6.

Processing and imaging

This book is principally concerned with the issue of seismic imaging, which is different from seismic processing. We call 'imaging' all the steps within the seismic processing sequence concerned with constructing the seismic image as such. We will consider two very different routes (fig. 3):

- *time imaging* also called pre-stack time migration, with its corresponding 'velocity analysis';
- *depth imaging* with different approaches to 'pre-stack depth migration', 'full waveform inversion' and explicit velocity model building.

This means that we will not be discussing in depth issues concerned with preconditioning the data, called 'pre-processing'. Pre-processing addresses topics such as source signature processing, noise elimination, static corrections or multiples removal. However, since in modern seismic surveying all the various steps in the acquisition – processing – imaging – interpretation sequence are much more interrelated than in the past, we shall examine why the design of the acquisition parameters cannot be accomplished independently from the choice of a seismic processing sequence and discuss how interpretation interferes with imaging.

Objective of seismic imaging

We can conclude this introductory note in saying that the objective of seismic processing and imaging is to *relocate reflections recorded at the surface at their right place in the subsurface, with an 'amplitude' faithfully related to the local reflection coefficient*. In time imaging, as we shall see, the vertical axis of the image is 'in vertical time'; in depth imaging we aim at relocating reflection coefficients directly at their true depth.

At the *exploration* stage, the seismic image will be used mainly for structural interpretation, rock volumes estimation or optimal exploration of well locations.

At the *field development* stage, we will be even more demanding as the objective is to characterize the reservoir using seismic information: further structural interpretation including compartments detection, accurate faults location, volumes estimation updates, development of well trajectory optimization and even reservoir properties through seismic attributes (AVA, acoustic or elastic impedance to map porosity variations, etc.). This is the 'seismic reservoir characterization' process.

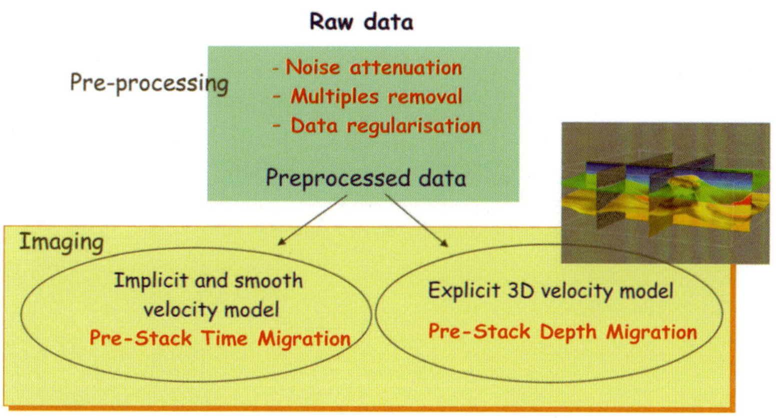

Figure 3: Seismic Processing and Imaging. The book deals mainly with seismic imaging.

1 Wave propagation: some reminders

We have seen in the introductory chapter that seismic imaging is concerned with elastic wave propagation. This first chapter will give reminders on how waves propagate, the wave equation, wavefront and ray theory, with an emphasis on P-wave velocity and velocity anisotropy.

P-wave propagation

What are elastic waves and why do they 'propagate' into the Earth? From the moment in time and the point in space of an emission from a seismic source (e.g., the soil under the vibrating plate of a vibrator truck, water surrounding the bubble created by an air gun, the walls of the cavity created by dynamite sources, etc.), particles within the surrounding medium are displaced, which results in a disequilibrium in the local pressure regime. The particles displaced by the seismic source undergo compression because they are unable to move freely and, in turn, exert a pressure on their neighbours. These particles then go on to be displaced and undergo compression themselves, exerting a further push on surrounding particles before they also once again come to rest, and so on. Subsequently, the motion is reversed, as particles in contact with the source undergo a secondary motion in the opposite direction (as the vibrator plate rises or the air bubble collapses under hydrostatic pressure). The seismic source in this way generates an oscillatory movement or *disturbance* within the neighbouring particles, a signal known as the *near-field signature*.

This particle motion and change in pressure regime created by the source will in this fashion propagate throughout the subsurface, even after the source itself has ceased emitting. In the ideal case of a perfectly elastic medium, this 'elastic wave' would propagate indefinitely.

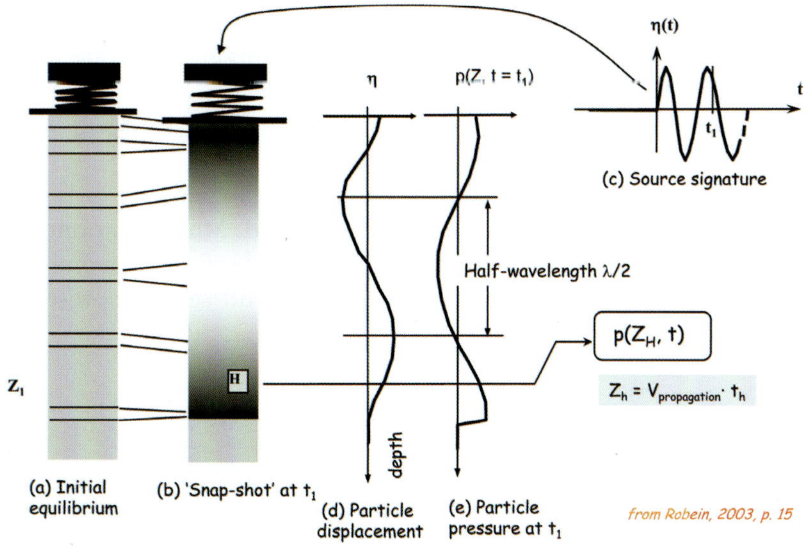

Figure 1.01: Particle velocity, pressure and wave propagation.

The phenomenon is illustrated in fig. 1.01, with fig. 1.01-a showing a representation of the subsurface in its initial equilibrium position. At time $t = 0$, a signal is emitted by the vibrator. This signal or signature s(t), is measured as the variation as a function of time in the vibrator plate's position about its equilibrium state (fig. 1.01-c). The plate acts upon the particle immediately adjacent to it in the subsurface, this particle in turn acting upon the particle immediately below it, and so on; the particle motion propagating in an elastic manner, creating a field of pressure variations within the medium. At any given instant t, the subsurface state can be represented in terms of one of two variables:

- the displacement η relative to the subsurface original equilibrium position;
- a pressure variation, p, with respect to the equilibrium pressure.

This oscillatory state, expressed by either the particle motion or local pressure variation, as a function of time, is known as the *wavefield*, here one-dimensional. When dealing with pressure, it is denoted by p(z, t), and $\eta(z, t)$ with the particle motion. Fig. 1.01-b illustrates this in the form of a *snapshot*, using a grey scale to display the pressure state at each point in the subsurface at the particular instant, t_1. In fig. 1.01 the variables η and p are sketched as a function of depth for this *snapshot* t_1.

Measuring the elastic wave with hydrophones and geophones

In marine seismic surveys, these *pressure differentials p are measured using hydrophones*, whilst in land or in borehole seismic experiments, geophones or accelerometers are used. These tools give an electric signal proportional to their velocity of displacement or *particle velocity* v, the time derivative of the displacement η, or to their acceleration.

As an example, a hydrophone positioned at H at a depth Z_h in fig. 1.01-b produces at instant t_1, an electric signal proportional to the pressure variation $p(Z_h , t_1)$, and during a time t a variable signal $p(Z_h, t)$. Equally, a geophone positioned at H gives at instant t_1 an electric current proportional to the particle velocity v, equal to the rate of change with time of the particle displacement η at this depth.

However, while pressure is a scalar quantity, displacement and particle velocity have a direction and are hence *vectors* called 'polarization'. This means a single geophone is only able to measure one particular component of the particle velocity at a time:

- In the case of a geophone in a conventional land seismic survey it will be the motion's vertical component;
- This is in contrast to ocean bottom cable and nodes on the sea floor or borehole seismics, where the objective is to is to record the entire vector wavefield. Multi-component receivers are therefore used, constructed of three geophones orientated orthogonally to one another, completed by hydrophones in the marine case. These tools enable a full vector representation of the particle motion (polarization) to be gathered.

Seismic amplitudes

The values of the electrical current coming out of geophones or hydrophones are digitized at a sampling rate of typically 2 milliseconds before being transmitted to tapes. Theoretically, the value recorded on the tape will either be proportional to the particle velocity or the pressure variation. Normally however, the coefficient of proportionality is disregarded and the seismic 'amplitudes' recorded on the tape are simply relative.

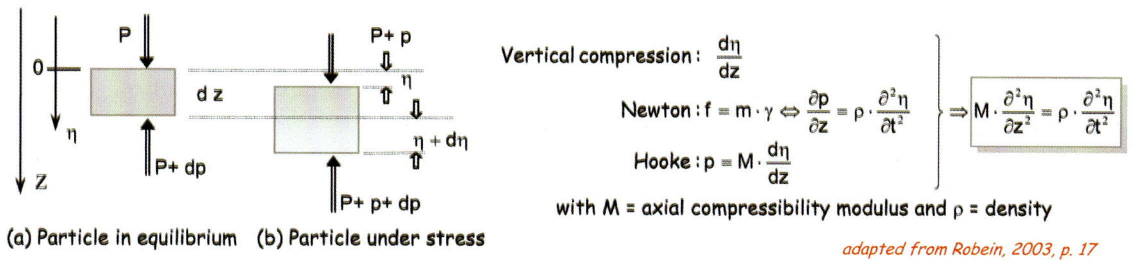

Figure 1.02 : The '1-D constant density wave equation'. When an horizontal plane elastic wave hits a particle, this one is under stress. Newton's and Hookes' laws together allow us to write a differential equation, for instance in displacement η and time, which controls the particle motion as a function of time.

The 'wave equation' and propagation velocity

It might be a counter-intuitive idea, but *rocks are elastic*. This is why waves propagate in the subsurface. If they were not and an infinitely rigid medium lay beneath a vibrator truck, the subsurface would thus move *en masse*! It is thanks to the subsurface elasticity that any movement induced by the truck is therefore not propagated instantaneously.

The wave equation

Elastic propagation of waves can be represented mathematically.

The concept is illustrated in fig. 1.02, where we show a particle positioned at a depth z and instant t. Analysing the forces acting on this particle, namely elasticity and inertia, physicists have established the 'elastic wave equation', here in its most simple one-dimensional (1D), acoustic form assuming constant density.

The rock's elasticity means that when we apply a force (stress) to a rock, it is deformed (strain). Hooke's law is an empirical set of linear relationships between the stress applied to an elastic material and the strain that results from it.

In this 1D case, the specific co-efficient of proportionality is called the *axial modulus* and denoted M. It describes one elastic property of the subsurface in which the wave is moving. Inertia is governed by Newton's 2nd law, which implies a mass. This is why the density ρ of the particle comes into the equation.

Combining both laws leads eventually to the differential equation that governs elastic wave behaviour: the *Elastic Wave Propagation Equation*, or as it is more simply known, the *Wave Equation*, which relates the derivatives of particle displacement with depth (how the wave evolves in space), derivative of particle displacement with time (how the wave evolves as time unfolds), local density ρ of the medium (assumed to be constant with z) and local elastic property. Its acoustic 1D version reads:

$$\frac{\partial^2 \eta}{\partial z^2} = \frac{\rho}{M} \frac{\partial^2 \eta}{\partial t^2} \qquad (1.1)$$

We will now use it simply to show the link between the wave equation and the phenomenon of propagation.

Propagation velocity

If we are aiming for the full wavefield $\eta(z, t)$ for every depth z and time t, beginning from an *initial condition*, we have to *integrate the differential equation* (1.1). As this is a standard equation, its general solutions are well known.

Firstly, noting M/ρ has the dimension of velocity squared, we can write:

$$V = \sqrt{\frac{M}{\rho}} \tag{1.2}$$

It can be easily shown that every function of the type:

$$\eta(z,t) = s(t \pm z/V)) \tag{1.3}$$

is a general solution for the differential equation (1.1). Suppose therefore that we know the particle motion of the source at the surface, i.e., $\eta(z = 0, t) = s(t)$, which is the initial condition of our problem. Then, functions of type:

$$\eta(z,t) = s(t \pm z/V)) \qquad (1.4))$$

will constitute solutions of the differential equation.

This means for instance, that the particle motion of the geophone H positioned at depth Z_h in fig. 1.02-b *will be exactly the same as that at the source, except for a delay equal to* Z_h/V : the signal has travelled from the surface to H at a velocity V!

The value of the V at which the waves propagate is defined by equation (1.2). It is linked directly to two major properties characteristic of the medium in which they travel:
- the *density* of the rocks;
- and the incompressibility or inversely, the *elasticity* of the rocks.

Note the + and – signs in equation (1.4) above; they say that we can 'propagate' the wave in time, but also 'back-propagate' it. Back-propagation is an important concept to which we will return. Choosing one sign only means that the equation simulates propagation in one direction only. It is called the '*one-way wave equation*'.

Our simple 'demonstration' does not take account of the fact that the source is at the limit between two media with different properties. We will not allow this point to bother us at this stage.

In our simple example however, we are able to integrate the Wave Equation between depths zero and z using an *analytical solution*. This is possible because the medium is assumed to be 1D homogeneous: the propagation velocity was constant between the two depths concerned.

For a heterogeneous medium, we would have to integrate the equation between two points sufficiently close as to suppose the propagation velocity to be locally homogeneous and is hence, only linked to the local properties of the medium. We will therefore use the term *local velocity*, or (as it is more commonly known) the *instantaneous or* (more improperly) *interval velocity field* V(x, y, z).

Finally, note that the propagation in this example takes place vertically, i.e., parallel to the particle motion. It therefore constitutes an example of *P-wave* or *compressional wave* motion, which is characterized by wave propagation parallel to particle motion.

Other types of waves commonly used in seismics are *S-waves*, where particle motion is perpendicular to propagation direction.

Simulation of wave propagation – numerical modelling

The purpose of the example above is to show that through the 'integration' of the wave equation beginning from an 'initial condition', we can simulate the propagation of acoustic waves in a medium of which we know the velocity.

Seismic imaging is very much concerned with this *propagation simulation exercise*, as we will see. We need however to simulate waves propagating in media representing as realistically as possibly the true propagation velocity field in the Earth. We call them *'Velocity models'*, noted V(x,y,z), an object that we will encounter and use a lot later!

These models are firstly three-dimensional and secondly often very complex, which means that we need a 3D wave equation on the one hand and more sophisticated ways than the analytical solution to 'solve' or 'integrate' the differential wave equation on the other.

Extension to 3D of the acoustic wave equation is straightforward:

$$\frac{\partial^2 p}{\partial^2 x} + \frac{\partial^2 p}{\partial^2 y} + \frac{\partial^2 p}{\partial^2 z} = \frac{\rho}{M}\frac{\partial^2 p}{\partial t^2} \quad \text{if we write it now for pressure changes} \qquad (1.5)$$

If we were dealing with a homogeneous Earth, the integration of equation (1.5) would remain simple, as there exists an analytical solution to it. Indeed, we can show in these conditions that a function of the form:

$$p\left(x, y, z, t \pm \frac{r}{V}\right) = \frac{1}{r} \cdot p_{source}(x_s, y_s, z_s = 0, t) \qquad (1.6)$$

where r is the distance between source S at co-ordinates (x_s, y_s, z_s = 0) and the recording point (x, y, z) is a solution of the 3-D wave equation. The initial condition is the pressure at the surface $p_{source}(t)$.

As in 1D propagation, the signal at position (x, z) is recorded with a *delay equal to the propagation time t = r/V*, where r is the propagation distance and V the P-wave propagation velocity. Note however the difference in the amplitude term. The amplitude of the recorded signal is now scaled by a factor of 1/r: this is the well-known *spherical divergence* term. Such a decrease is explained by the necessity to conserve energy during propagation in a medium in which no energy loss occurs.

Three-dimensional integration from an initial condition on a computer in a complex velocity model is however another story. Two classes of solutions exist for this integration:

- numerical integration;
- wavefront and rays.

The first one is used in the so-called 'Wavefield Extrapolation' imaging techniques of Chapters 6 and 7. The second one is the heart of Kirchhoff and other ray-based methods addressed in Chapters 3, 4 and 5.

Numerical integration means that the wavefield is computed on a *discrete grid in space and at discrete steps in time*. The differential equation (1.6) needs to be transformed into *'finite difference equations'* or sets of equations, leading to very large systems of equations to be solved in the computer. Examples are discussed in chapters 6 and 7, but the reader may refer to (Berkhout, 1982; Claerbout, 1985; or Biondi B, 2006) for more detailed and rigorous mathematical developments.

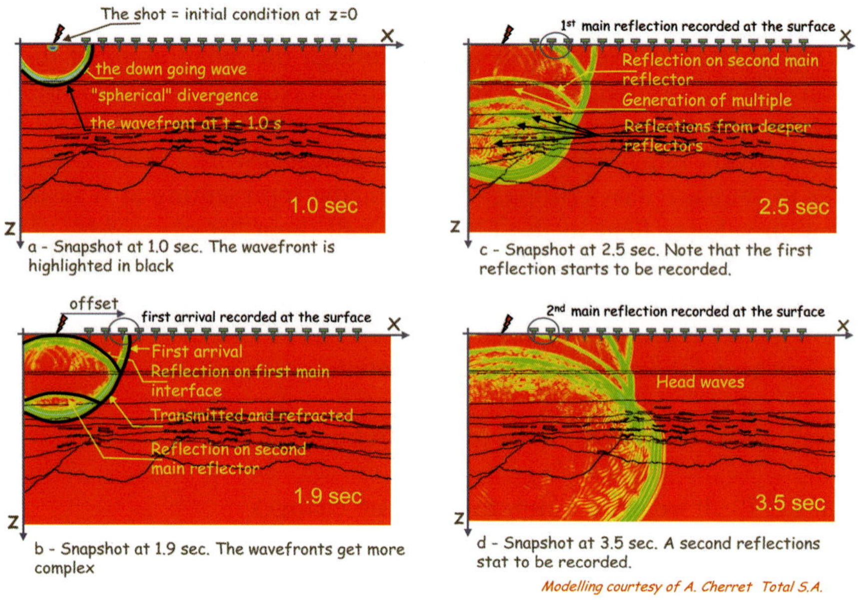

Figure 1.03: Acoustic (p-wave) modelling in 3D. Snapshots are computed by integration of the wave equation in the propagation velocity model.

Fig. 1.03 illustrates the simulation of waves propagating in velocity models computed using a numerical integration scheme. Four snapshots are shown at different times after the source has been triggered at the surface (the initial condition, also shown in fig. 1.03-a). They are a cross-section through a 3D model as the simulation was performed in 3D. These snapshots represent the pressure field changes due to the propagating wave in the Earth at that time. Pressure changes are colour-coded. 'Spherical' expansion of the initial shot, first arrivals, reflection of the first important interfaces (velocity and density change), transmitted refracted wave, reflections at deeper horizons, upgoing waves eventually hitting the surface where they are recorded by receivers are clearly visible on this 3D simulation.

A set of receivers is displayed at the surface. The pressure (or equivalently the particle velocity) appearing on the display for a given snapshot at time t is what is measured by the receiver at that very moment. This is the way the *'synthetic shot record'* is computed, taking the pressure or the particle velocity at the surface from snapshots computed one after the other, as time unfolds.
The simulation of propagation and computation of synthetic records is known as *'numerical seismic modelling'*.

How do these different features appear on a real shot record? Firstly, fig. 1.04 is a scheme of a typical marine multi-streamer acquisition. Note an important parameter, *the offset*, which is the distance between the source and the receiver. Typical seismic boats today may tow up to twelve streamers, each 8 km long. This would represent more than 1500 hydrophones, recording a total of $6{\cdot}10^6$ samples for each shot.

Fig. 1.05 is a real example of the top part (first 4 s. out of 8 s.) of a shot record, which would be recorded on one streamer of such an acquisition system. The various events described (first arrivals, head waves, reflected waves) in the modelling exercise are also visible in the real data.

Direct arrivals and head waves are generally undesired and simply *'muted'* (green window in fig. 1.05).

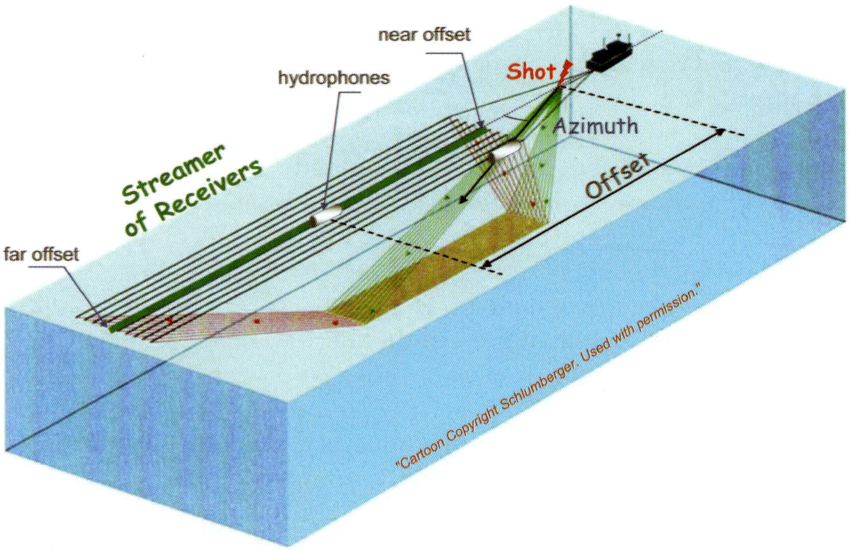

Figure 1.04: A typical multi streamer marine shooting ('narrow azimuth').

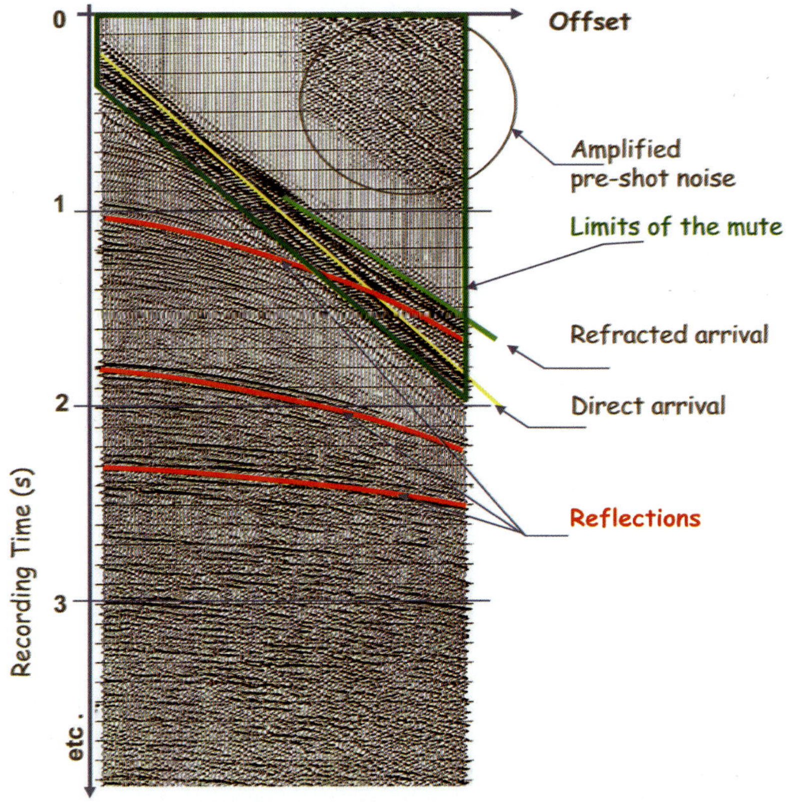

Figure 1.05: A typical marine shot (after application of agc).

Wavefront and rays

Dealing with wavefronts and rays may be more intuitive than thinking in terms of equations. We will hence use these concepts regularly.

Looking at fig. 1.03-a, we note that all particles beyond the line in black have not seen the wave yet at t = 1.0s. At a given instant t, we can define a threshold marking the limit between the particles still at rest and those just at the point of being displaced by an oncoming wave. This surface is known as the *wavefront*. In our example, this wavefront will be locally approximately spherical, as long as the velocity model is homogeneous.

Note that if the source was an extremely short impulse, i.e., composed of all frequencies up to infinity, the wavefront would actually represent the full wavefield. We sometimes use the term 'high-frequency asymptotic approximation of wave propagation' for the wavefront and ray theory. Fig. 1.06 is a schematic 2D description of how wavefront propagation may be simulated using Huygens' principle. This principle states that each point of a wavefront, say at time t, may be seen as a new virtual source (the particle just hit by the wavefront and displaced by it in turn pushes its neighbours!).

Each of these virtual sources creates its own 'local wavefront', which we can draw at time t+dt, dt being a small interval of time. The size of this local wavefront depends on the *local propagation velocity* V(x,z). The actual wavefront at time t+dt is the superposition of these small local virtual ones. This superposition or sum, of local wavefronts constructs positively at their geometrical envelope.

Note point P in the figure and its corresponding virtual wavefront. The latter concurs constructively with the summation, where it is tangent to the envelope: this is point Q in the figure. We can say that the 'packet of energy' in P travels to Q between snapshot t and t+dt. Snapshot after snapshot, we can follow a set of successive points P, Q, etc. These points define the trajectory of a given packet of energy as time unfolds: this trajectory is the well-known *'ray path'*. In a heterogeneous but isotropic medium, the ray is normal to the wave front in each point by a geometrical definition of envelope.

If the medium is heterogeneous, as schematized in fig. 1.06, the radius of the virtual local wavefront changes with location (x, z) as propagation velocity changes too. The wavefront at time t+dt is consequently not parallel to wavefront at time t and the *ray 'bends'* as schematized in the figure. This is the already mentioned phenomenon of *refraction*.

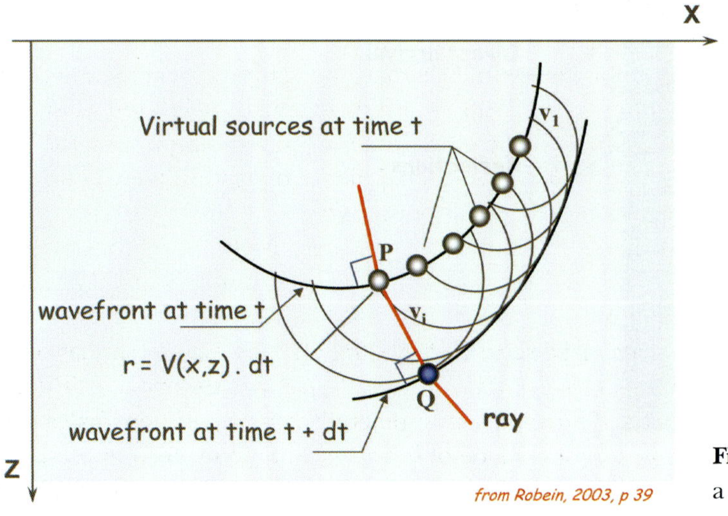

Figure 1.06: Wavefront and rays in a heterogeneous isotropic medium.

Figure 1.07: Anisotropy at macroscopic and microscopic scale: seismic propagation velocities depend on propagation direction. Anisotropy may be seen as a manifestation of layering at all scales.

A 2D illustration of wavefront and related rays is given in fig. 4.03 in Chapter 4 where we discuss Kirchhoff depth migrations. It shows how complicated wavefronts and rays may become in complex media, with the possible existence of 'caustics' (multivalued wavefront and crossing rays), also known as *'multipathing'*. Note also that wavefronts and rays are computed in a smoothed model: this is because the computation of rays may become very inaccurate in the presence of sharp velocity contrasts and high angles of incidence.

The subsurface is anisotropic

So far, we have considered the Earth as heterogeneous but isotropic. It has been shown however that it is not only heterogeneous, which is obvious when looking at any geological outcrop, but also *'anisotropic'*.

Anisotropy is the physical phenomenon that makes waves propagate at a velocity that depends on its direction of propagation. Fig. 1.07 illustrates the fact that one cause of velocity anisotropy is layering at various scales. For example, take a superposition of thin alternate shaley and sandy layers. Waves with a long wavelength (10 to 100s of meters) such as the ones we are dealing with in the seismic experiment, 'see' this layered material as an *'effective medium'* for which propagation velocity is different whether waves propagate vertically, horizontally or in any given direction. We can indeed imagine that the overall compressibility of the 'rock package' is not the same if measured vertically (compressibility will be closer to the 'soft' rock) or horizontally ('the 'hard' rocks will control compression). The medium is said to be 'anisotropic' for long wavelengths.

Looking at rocks at a much smaller scale, especially at shales (fig. 1.07-b), we see similar 'layered' microscopic features, which also produce anisotropic rock behaviour. Geophysicists tend to call 'intrinsic' anisotropy the one induced by such small-scale features. It is generally considered that shales present this kind of intrinsic anisotropy more than any other type of lithology commonly encountered.

Other physical phenomena than small-scale heterogeneity induce 'effective' anisotropy. We may mention for instance *tectonic stress or rock fractures* (Grechka, 2008).

Wavefront, phase and group angles and velocities in anisotropic media

One way to understand propagation in anisotropic medium is to consider plane waves. A 'plane wave' is a wave of which the wavefront is planar. If the medium is homogeneous and isotropic, the wavefront remains planar as the wave travels and velocity propagation will be the same whatever the direction of propagation. If the medium is still homogeneous but anisotropic, the wavefront remains planar as the wave propagates but its propagation velocity now depends on its direction. The velocity of a plane wave in a particular direction θ is known as the *phase velocity* denoted $V_{ph}(\theta)$.

To define the group velocity, we consider the case of a point source S in fig. 1.08-a. Making use of the principle of superposition, we show that a point source can be decomposed into an infinite number of constituent planes, propagating between –90° and +90°. The circular wavefront at instant t generated by the source S can therefore be considered as the envelope of the planar wavefront at this instant t. In a 2D homogeneous medium of velocity V, this wavefront at instant t would be a circle of radius $r = V \cdot t$.

For an anisotropic homogeneous medium, *each plane wave propagates at its own phase velocity* $V_{ph}(\theta)$. At instant t, each one will have travelled a different distance $r(\theta) = V_{ph}(\theta) \cdot t$. As for the homogeneous case, the envelope of the planar wavefront will represent the actual wavefront of the propagating energy.

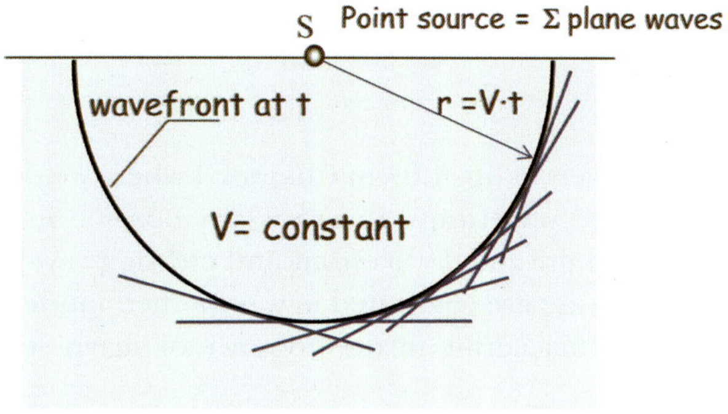

from Robein, 2003, p. 48

Figure 1.08 - a: The wave front of a point source in the isotropic case. The wavefront can be seen as the envelope of the propagation of an infinity of plane waves.

P-wave phase velocity defined by 4 parameters :

- V_{phase_vert} , ε , δ
- axis of symmetry (vertical for VTI)

The VTI model *for phase velocity (Thomsen, 1986) :*

$$V_p(\theta) \approx V_{vert} \cdot \left(1 + \delta \cdot \sin^2\theta \cdot \cos^2\theta + \varepsilon \cdot \sin^4\theta\right)$$

$$V_{hor} = V_{vert} \cdot (1 + \varepsilon)$$

$$V_p(45^\circ) = V_{vert} \cdot (1 + 0.25 \cdot \delta + 0.25 \cdot \varepsilon)$$

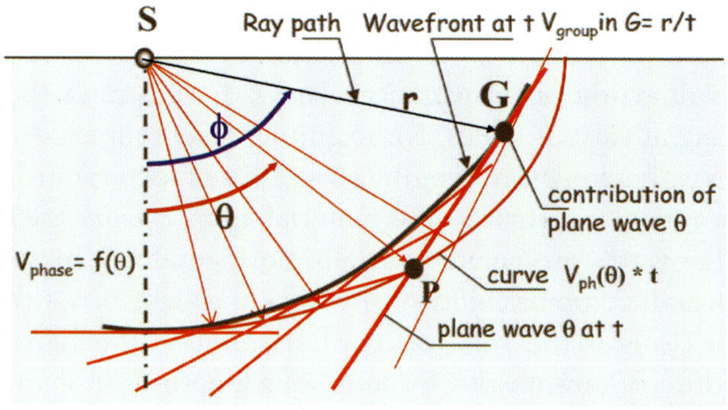

from Robein, 2003, p. 15

Figure 1.08 - b: The wave front of a point source in the homogeneous anisotropic case. The wavefront is still the envelope of plane waves, but now plane waves propagate at phase velocity that depends on their direction.

It will no longer be a circle as schematized in Fig. 1.08-b! A receiver placed at point G in this figure will record a particle motion for time t and, since we know the energy has been emitted from the source S, we deduce that the energy has travelled along ray path SG, at an angle ϕ with the vertical. This angle is known as the *group angle*.

We deduce also that the propagation velocity of the energy will be SG / t. This is the *group velocity, denoted $V_{group}(\phi)$,* as it clearly depends on the group angle ϕ.

One important point to note is that V_{ph}, V_{group}, θ and ϕ are not independent. The relationships between these quantities are derived geometrically from the triangle SPG in fig. 1.08-b. Similar to the isotropic case, the trajectory along which the energy propagates from the source to a given point is the ray path. As the figure shows, the ray is rectilinear due to the medium's homogeneity but *is not perpendicular to the wavefront*, because of anisotropy.

To be more complete, we should note that the receiver in G will record a particle motion in space, or *polarization*, which is the vector sum of the contribution from all the plane waves, yet this polarization will be neither perpendicular to the wavefront nor to the plane phase wave. For this reason, the wave is not strictly a compressional P-wave. It is consequently referred to as a *'quasi-P' wave*, often denoted *qP-wave*.

The distinction between P-wave and qP-wave is not made in ray-based imaging, which we will encounter later.

Analytical parameterization of velocity

Both today's time- and depth-imaging methodologies require a more and more accurate description of the propagation velocity field in the Earth. In this chapter we recapitulate some standard ways to describe possibly very complicated heterogeneous and anisotropic velocity fields, using only a limited number of parameters.

A first-order effect that we want to take into account is the vertical heterogeneity that comes from the natural compaction of rocks with depth. The standard way to handle it in models is to use a *'velocity gradient'* often denoted k, as shown in fig. 1.09.

The local velocity in this case depends on two parameters: a reference velocity noted V_0, and a vertical gradient k. The local propagation velocity at depth z is then modelled as $V(z) = V_0 + k \cdot z$.

This simple model takes the reference velocity V_0 at the surface, which is not appropriate in many real cases. It can therefore be naturally modified by taking the 'reference' velocity at the top of a specific layer.

Another limitation of this model is that it assumes no changes in the velocity field laterally. This can easily be improved by allowing V_0 vary laterally. We then define a 'map' of velocity at the top of the layer $V_{0\_top}(x, y)$.

Estimation of $V_{0\_top}(x, y)$ comes generally from velocity analysis or tomography, techniques that we will discuss in chapter 4. A first guess may however be the result of a thorough analysis of time-depth curves in wells drilled and logged in the area under investigation. This analysis is also very useful to estimate k. In some instances it may even lead interpreters to let k vary laterally too, as k(x,y).

The parameterization for local or instantaneous velocity within a given layer is presented on the right of fig. 1.09. It is controlled by the following 'parameters':

- $V_{0\_top}(x,y)$, the map of local velocity at the top of the layer, that may be interpreted as a representation of the burial effect;
- $Z_{top}(x,y)$, the depth map of the layer;
- k(x,y), the map of velocity gradient within the layer, embedding the compaction effect.

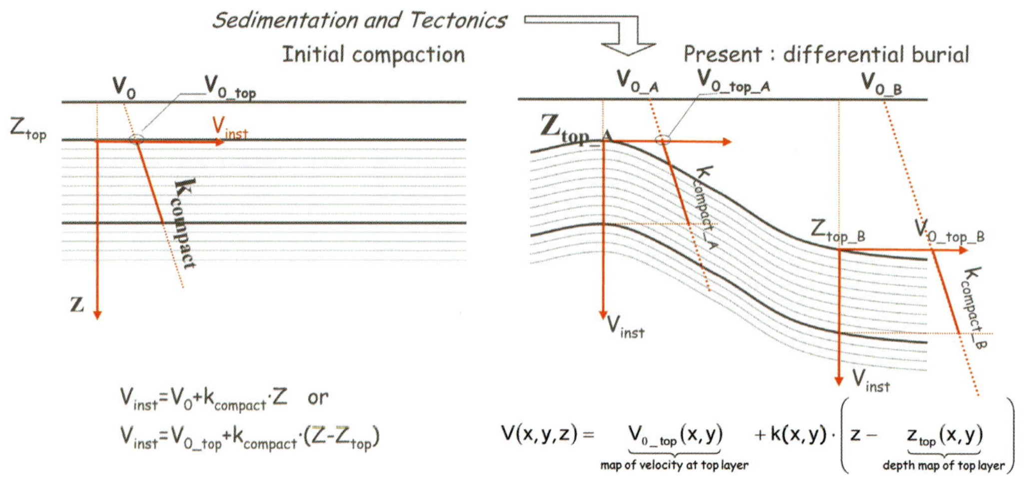

Figure 1.09: The V_0 + kZ model. Various versions of this popular model allow to account for both compaction and burial.

Parameterization of velocity anisotropy

Next step is to 'parameterize' anisotropy in order to apply it in computers. The by far most popular way to do it is to use Thomsen's set of parameters (Thomsen 1986), presented in fig. 1.08-b. He derived his model with a weak anisotropy assumption but his model is now used for sometimes very strong anisotropy all the same.

The first point to note is that anisotropy is *parameterized in terms of phase velocity*. Secondly, we deal here with qP-waves only. Thirdly, we assume that anisotropy has an axis of symmetry that for the time being, is vertical.

For this case, we speak of 'Vertical Transverse Isotropy', or VTI and Thomsen's relationship for P-wave phase velocity as a function of phase angle θ appears in fig. 1.08-b.

There are consequently four parameters to characterize anisotropy in a given location (x,y,z) for P-wave propagation in the subsurface:

- Phase velocity for plane waves propagating vertically $V_{\text{phase\_vert}}(x,y,z)$;
- two *Thomsen's parameters denoted* ε *and* δ;
- and the direction of the axis of symmetry that is vertical for VTI.

The next question is how to combine heterogeneity and anisotropy. For that purpose, we use the '*factorized anisotropy model*' which again is an approximation of the real behaviour of anisotropic solids.

This model says that phase velocity bears the heterogeneity described for instance by a V_0+k.Z or a gridded (chapter 4) model and anisotropy multiplies the now variable vertical phase velocity according to the Thomsen relationship. An example of the parameterized model *within a layer* is given in fig. 1.10.

Note that although in current practice we allow velocity to vary within a given layer, we still use *constant* ε *and* δ *values per layer*, although this is changing rapidly. This is not driven by any physical or geological considerations but rather, by practical ones. We simply lack the data to estimate these

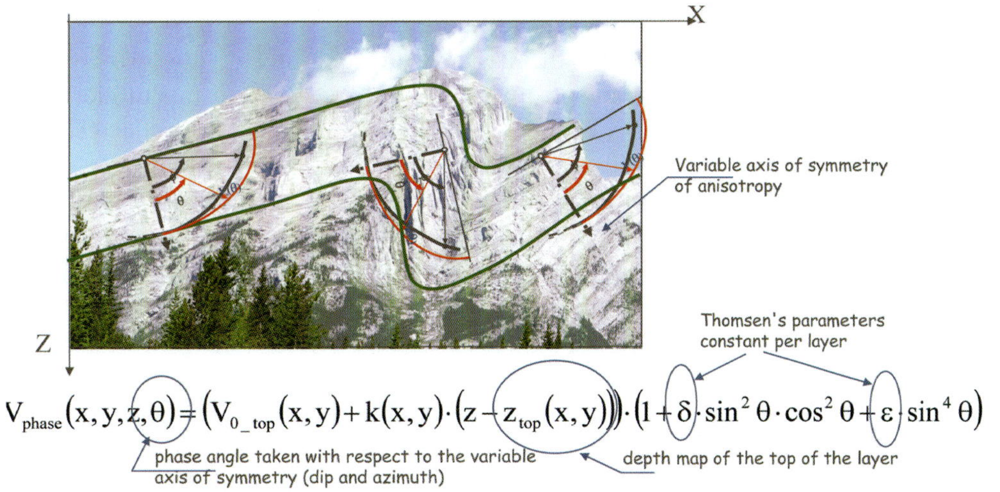

Figure 1.10: Tilted (TTI) or Structural (STI) anisotropy model. TTI is the generic term. In the STI model we assume that the axis of symmetry of anisotropy varies laterally, remaining perpendicular to the main dip and azimuth of the layer.

parameters in a reliable way and estimation of a single value per layer is already a challenge! The trend in the industry is thus to try and improve this situation.

One way to doing so is to allow the axis of symmetry to vary laterally. For this we have a 'geological guide'. If anisotropy is mainly a result of layering at all scales, it sounds reasonable to assume that the axis of symmetry of anisotropy remains more or less perpendicular to the dipping layers in tectonized areas. Several published anisotropic depth migration results confirm the validity of this assumption, which is illustrated in fig. 1.10. The 'generic' term is 'TTI' for Tilted Transverse Isotropy, but more and more frequently we use 'STI' for Structural or structurally conformable Transverse Isotropy.

Snell's law and rays in anisotropic media

The well-known Snell's law controls reflection and refraction angles at any heterogeneity interface in the Earth. Its application to anisotropic media is however somehow more complicated. The first point to note is that the usual *Snell's law applies to phase velocity* but we want to follow the propagating energy; hence, we need to compute group angles as well. One possible way to do this is illustrated in fig. 1.11. It shows a ray that hits an interface between two anisotropic layers and we want to know the angles of the refracted ray. Firstly, we need to compute the phase velocity that corresponds to the group velocity of the incident ray. This results from the solution of a first set of implicit equations that relate group and phase angles, group and phase velocity and the anisotropic parameters ε and δ above the interface. We can then apply Snell's law to phase velocity. Note however that contrary to the isotropic case, we again need to solve an implicit equation as phase velocity in the emergent medium depends on the angle too! When this is done, we compute the corresponding group angle, which gives the propagation direction and subsequently, the corresponding group velocity. This allows us to then propagate the ray further up to the next interface, where the whole exercise is done again. There are in fact more clever ways to jointly solve phase and group velocities but we develop this ray tracing exercise mainly to illustrate the fact that *accounting for anisotropy always makes imaging more complicated and time consuming*.

Anisotropy in itself, even without any heterogeneity, induces refraction and leads to unusual reflection angles, as shown in the following simple examples.

Fig. 1.12-a presents a two-layer model with interfaces dipping at 15°. The two layers have the same velocity (2000 m/s) and same VTI anisotropy ($\varepsilon = 0.3$). We 'shoot' in this model three rays at different angles. The figure shows that contrary to what we are used to, the reflected angle is not equal to the emergent one and a zero-offset ray is not a normal ray!

Fig. 1.12-b shows the same model, but now we have introduced an anisotropic contrast between the two layers to illustrate that refraction can be induced by purely anisotropic contrast, at least for rays not travelling vertically (VTI case). When we combine heterogeneity and anisotropy the effect may be even stronger or may also compensate: anisotropy behaviour is often counterintuitive and even more so for STI.

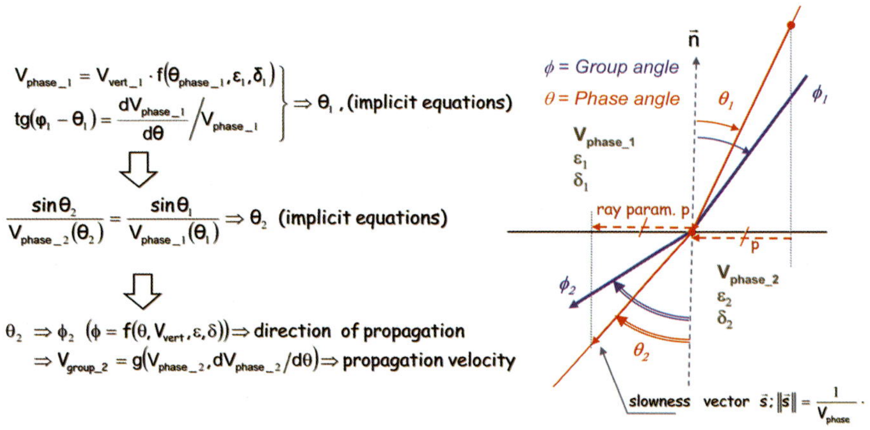

Figure 1.11: Refraction in anisotropic (VTI) media. Snell's law applies to phase angle, but equations must be solved jointly for both phase and group velocities, explicitly or not. The same applies for reflection when the normal to the interface does not coincide with the axis of symmetry of anisotropy.

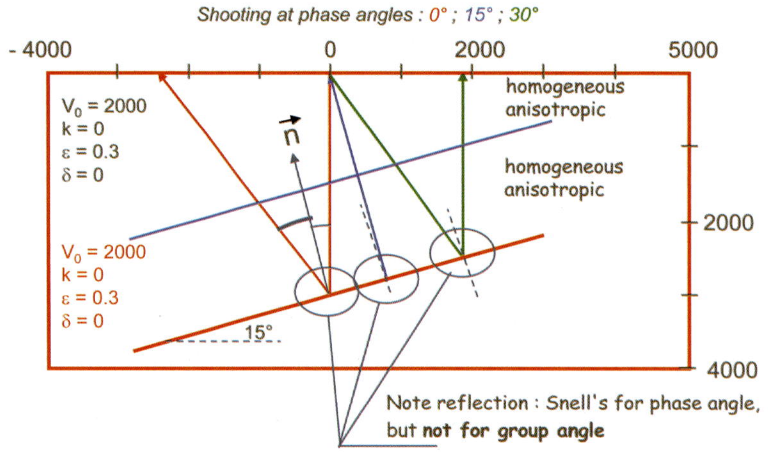

Figure 1.12-a: Example of ray tracing in homogeneous, but anisotropic (ε only) medium.

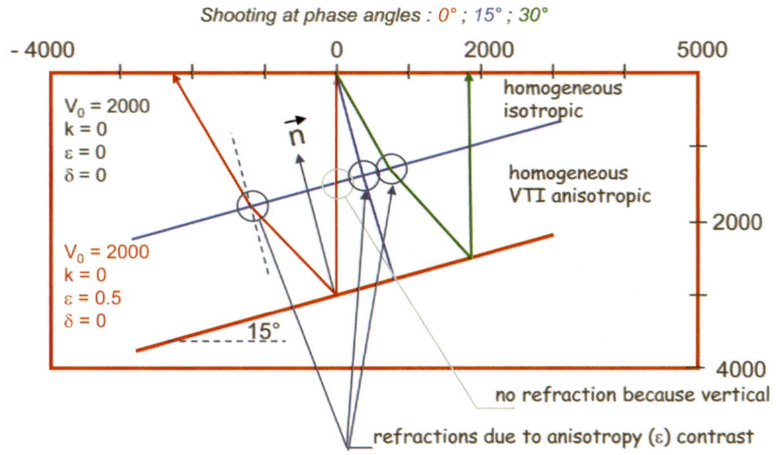

Figure 1.12-b: Example of ray tracing in homogeneous, but anisotropic medium with an ε contrast.

Rays, wavefront and seismic record relationships

Fig. 1.13 is another example of ray- and wavefront-tracing in an anisotropic heterogeneous model. The model is made of two layers, the parameters of which are displayed on the left of the figure. We assume a dipping (15°) reflector at a depth between 4 and 6 km. We show in red the downgoing wavefronts computed every 200 ms coming from a source marked as 'S'. Each of these wavefronts is a *snapshot*. The waves are reflected when they hit the reflector. The upgoing wavefronts are represented in blue, again at a 200 ms time interval. At the moment the upgoing wave hits the surface, geophones record the disturbance. The gather of these records makes up the shot record, which is displayed here with increasing time towards the top of the figure, contrary to the usual way as in fig 1.05.

The acquisition layout with receivers on both sides of the shot location is called *'split-spread'*. The image of the reflector is drawn as a red line on the shot record, schematically convolved with a wavelet at the receiver locations.

On top of the wavefronts, fig. 1.13 shows also two rays (in green) corresponding to reflections recorded by geophones placed at two different offsets.

We can make a few points, which will be useful in the following chapters, out of this modelling exercise:

- Firstly, note again the incident and reflection group angles, which are different due to the anisotropy of the layer.
- At the reflection location, the *downgoing and upgoing waves are coincidental,* by construction! These are marked by points A and B on the reflector.
- The reflection in B arrives at the receiver marked R, corresponding to an offset of about 4000 m after a time of 4.2 s, corresponding to the travel time t_s from the shot to the reflection point plus t_r from the reflection point to the receiver.
- More interestingly, one can measure the 'time-dip' of the reflection on the shot record, marked by a blue arrow. Time dip dt/dh is also called the *'ray parameter'*, denoted p. Its inverse dh/dt is known as the *'apparent velocity'* (it has the dimension of velocity).
- Using simple geometrical considerations, we can easily show that apparent velocity measured on the shot record and phase angle of the emerging ray measured at the surface are linked by the equation given in the figure.
- This means that if we know the velocity at the surface, one can infer from the apparent velocity of the event on the shot record the emergence angle of the ray or the wavefront of the corresponding reflection.

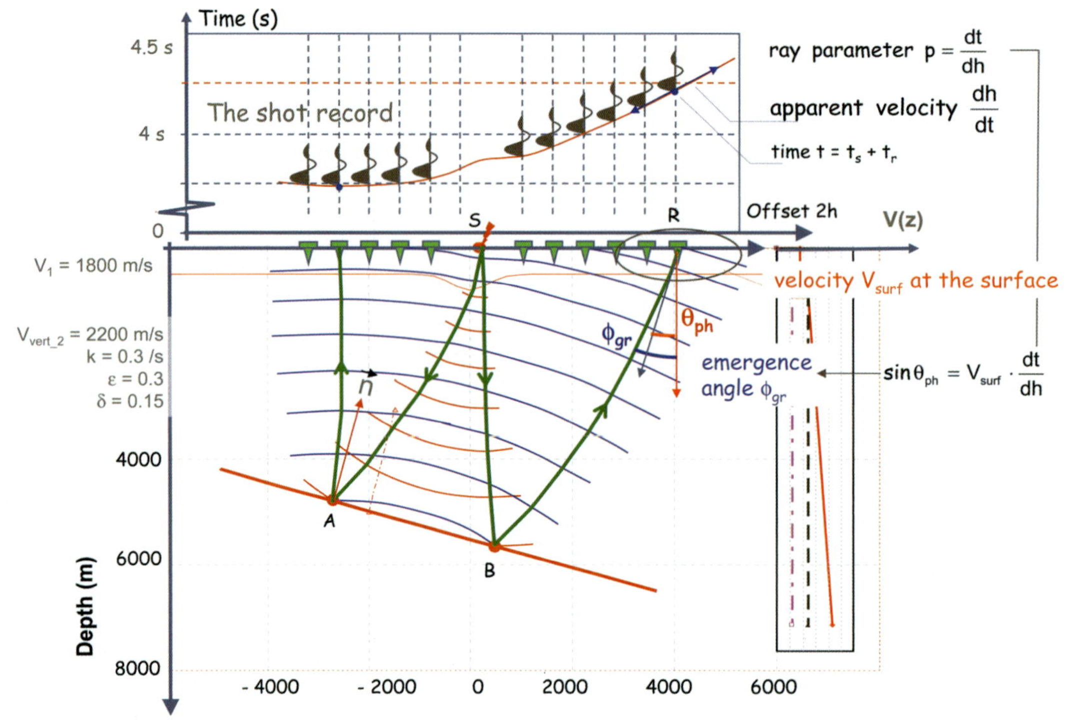

Figure 1.13: Wavefronts, rays and the seismic shot record. A synthetic example.

Some of these remarks and comments may appear trivial but will in fact be extensively used in the coming chapters and particularly those dedicated to 'ray-based' Beam Migrations (chapter 5) and Reverse Time Migrations (chapter 6).

2 Principles of Ray-based and Kirchhoff summation migrations

Before addressing the various approaches of Time- and Depth-Kirchhoff migrations, we will dedicate this short chapter to some basic concepts, particularly those concerned with reflectors and diffractors. This will lead us to some general considerations that we discuss as an introduction to Kirchhoff summation migrations and ray-based migrations.

Reflectors and diffractors in the Earth

Fig. 2.01 is a 2D migrated image that we will consider as an image of the 'geology' we will use to make some of our points. On this image we have highlighted in blue geological layers that have some lateral continuity, as we see them on any outcrop in the field. By analogy with optics, we call 'reflectors' the interface between two geological layers. In seismic terms, as discussed in the previous chapter, those interfaces correspond to changes in elastic properties (acoustic impedance) where the waves travelling from the surface will refract and partly reflect.
On the same 'geological image' however, we have pointed out red dots that correspond either to limits of reflectors or to features of a small size or small radius of curvature with respect to seismic wavelength. Remember the seismic wavelength is roughly between 10 and a few hundred metres.

Fig. 2.02 shows the '*zero-offset seismic response*' to this geology. Zero-offset means that we simulate through seismic processing, a thought experiment where each trace corresponds to the seismic record made at the surface on a receiver located very close to the shot that generates it. The crea-

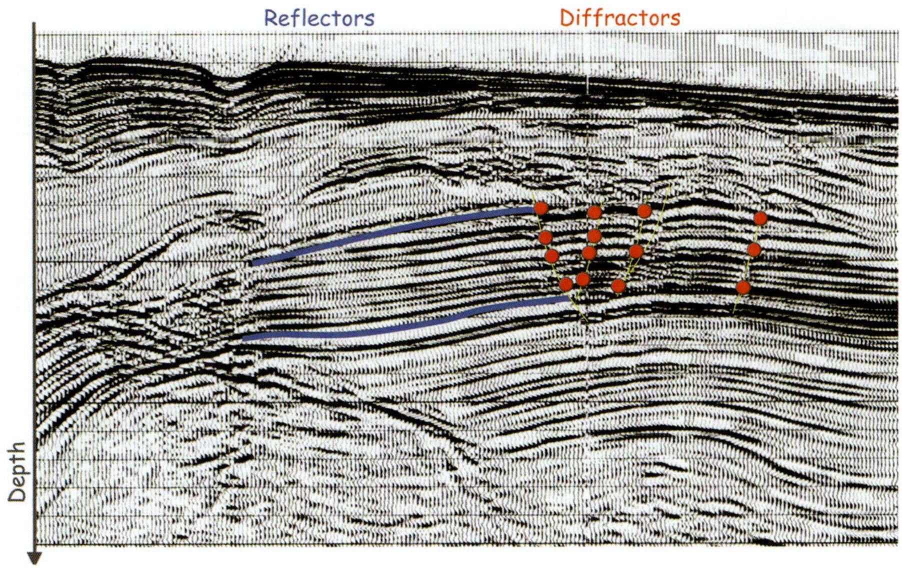

Figure 2.01: For the geophysicist, the Earth contains 'reflectors' that have some lateral continuity, but also localised 'diffractors' (for instance along faults).

tion of the zero-offset sections through *'DMO stack'* was a key step in seismic processing before the generalization of today's 'pre-stack' imaging methods.
Reflectors, unsurprisingly, appear as events with some lateral continuity. The limits of layers and the small features appear conversely on the seismic response as events with a typical hyperbolic-like shape: the so-called *'diffraction curves'*, again highlighted in red. We call *'diffractors'* the geological features that produce these diffractions.

The reflector model

Fig. 2.03 is a cartoon that shows on the left a very simple model with one reflector in a constant velocity isotropic medium. Zero-offset shot-receiver locations are marked by the symbol [SR].

The seismic image obtained with this recording layout is schematized on the right. It is a planar reflector, but now the vertical axis of the image is two-way zero-offset traveltime.
Let us consider an element D of the reflector in depth. Its image is an element of the reflector in time. Note that we have located at the surface the 'echo' of element D at a location noted 'X_{norm}' in the figure. This point is the emergence of the normal ray hitting the reflector in D: it is where an observer at the surface with a 'zero-offset recording system' would receive the 'echo' of the specific element D of the reflector. Because of the dip of the reflector, X_{norm} is not at the vertical of D. Consequently, if we want to locate the reflection coefficient in D at its correct location, which is the purpose of seismic imaging, we *need to 'migrate' the reflection* from the domain (X_{norm}, time) to the domain (X_{depth}, depth); hence the term 'migration' used for imaging.

Remember that we schematically trace 'normal rays' in fig. 2.03 because we assume an isotropic medium. In an isotropic medium, the zero-offset ray may not be 'normal' (see fig. 1.12).

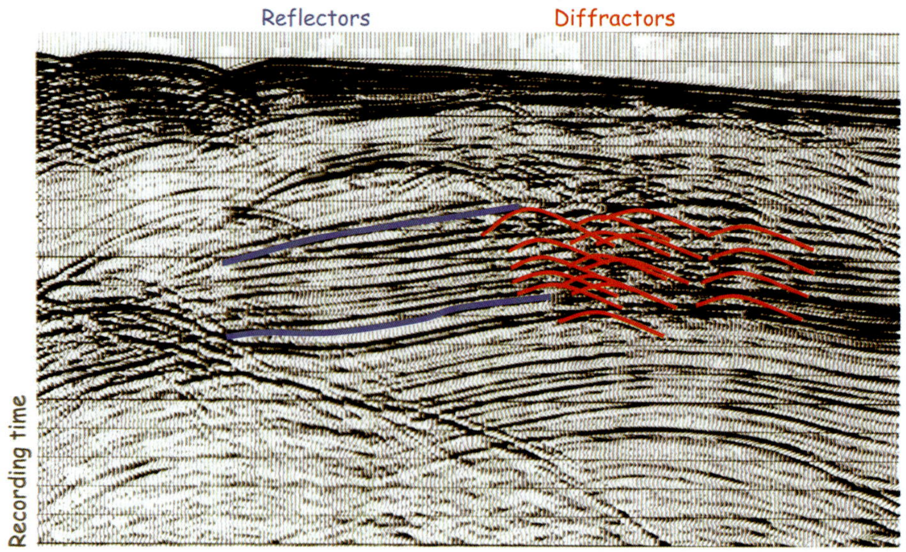

Figure 2.02: The 'zero-offset' seismic response of the 'geology' of fig. 2.01. 'Reflectors' appear as continuous events. Diffractors show up as 'diffraction pseudo-hyperbolas'.

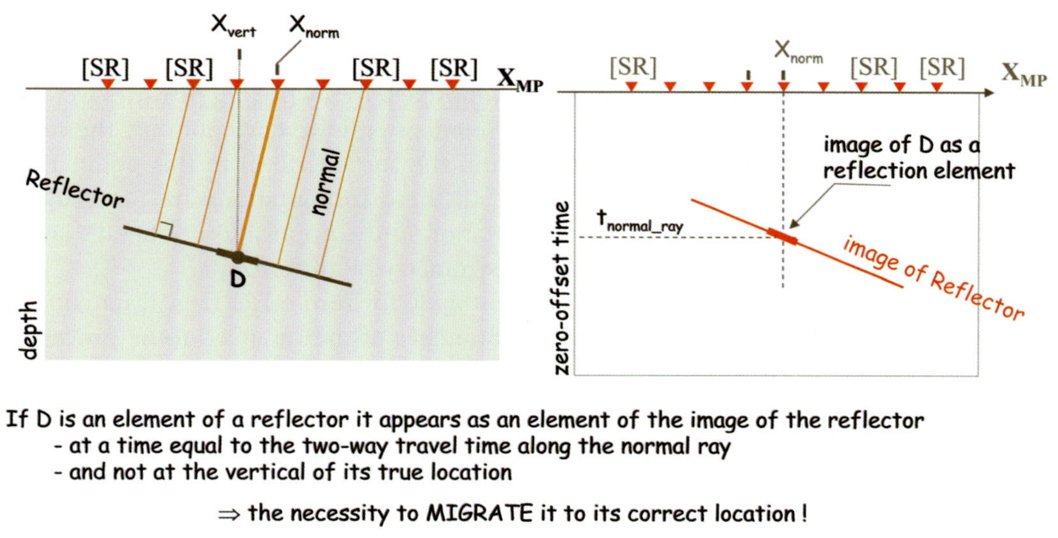

Figure 2.03: The reflector's model.

Ray migration

As a first hint at 'beam migrations' addressed in Chapter 5, we note that we can '*migrate*' element D in fig. 2.03. This necessitates a crucial prerequisite however: *the propagation velocity model*. How we obtain the velocity information will be discussed again and again in the following chapters. For the time being, we assume that we have it. We can then measure the 'time-dip' in the seismic response domain, as shown on the right of fig. 2.04-a, from which we can compute the ray emergence angle (see also fig. 5.01), then 'propagate' the ray through the velocity model during half of the zero-offset time picked on the seismic response. This halving factor takes account of the fact that energy propagates once down to the reflector and once again up to the receiver: the well known 'two-way traveltime'.

In fig. 2.04-a, we assume a homogeneous model. The migration ray is hence rectilinear. In real life, we deal always with variable velocity, possibly anisotropic, media. In this case, ray tracing is more tedious. An example of a possible scheme is schematized in fig. 2.04-b. The ray is computed piecewise by segments of constant traveltime δt. The ray is initiated at the surface, using the measured time-dip (or ray parameter) and the velocity at the surface that gives the emergence phase angle. Anisotropic parameters in the layer allow computing group angle and velocity, from which we can compute the first segment of the ray, assumed to be short enough to be rectilinear or an arc of a circle in a V_0+kz model. If velocity or anisotropy parameters have changed at the segment end, we apply Snell's law as described in fig. 1.11. The *reference direction in Snell's law is the normal to the phase velocity contrast*, marked as vector $\vec{n}$ in fig. 2.04-b. This vector is defined by the local velocity gradient $\vec{\nabla} V_{ph}$. Snell's law then provides the set of refracted phase and group angles and the respective phase and group velocities for the next propagation segment. Its length $l_2 = V_{gr\_2} \cdot \delta t$ is then computed.

The process can be repeated at the new segment end and iterated until we reach the reflection time t/2. When the ray has been traced for the duration of the propagation time, the end of the last segment gives the location of the 'ray-trace-based migrated element'. This is the ray version of the 'imaging principle' that we see again for wave extrapolation-based methods.

Note moreover, that the *dip of the reflection element is known*: in the isotropic case, it is simply normal to the ray. In the anisotropic case, it is normal to the phase angle and not the ray, which is defined by the group angle.

Another point worth noting is that this piecewise approach to ray tracing provides us with a 'calibration' of propagation time along the ray: at each segment end, we have not only the coordinates of the ray at that point, but also the travel time from the surface down to this point. Fig. 2.04-b is sketched in 2D but the reader must realize that all these computations are performed in 3D. Moreover if the model is TTI or STI, angle computation in the application of Snell's law at every segment end must take account of the local axis of symmetry of anisotropy.

Finally, this description of ray migration tracing is made for zero-offset rays. It remains valid for non-zero offset rays, with the difference that both branches of the ray need to be processed separately. This point will be addressed in Chapter 5.

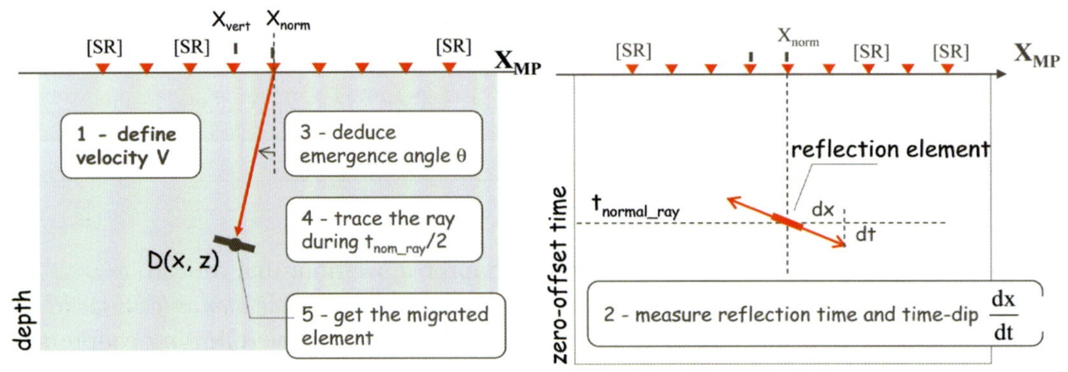

Figure 2.04-a: The concept of 'zero-offset ray migration'. The measure of 'time-dip' allows to compute the emergence angle and then to 'back propagate' the ray through the model.

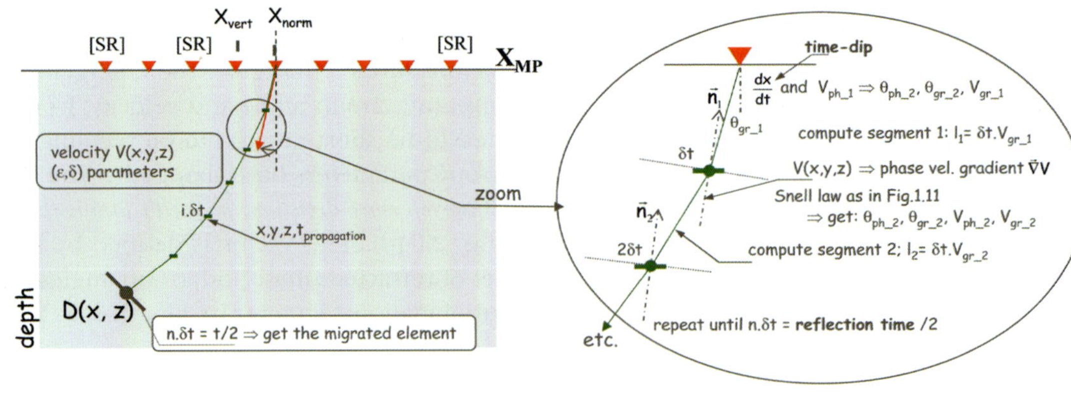

Figure 2.04-b: The concept of ray migration in heterogeneous models. The measure of 'time-dip' allows to compute the emergence angle and then to 'back propagate' the ray through the model.

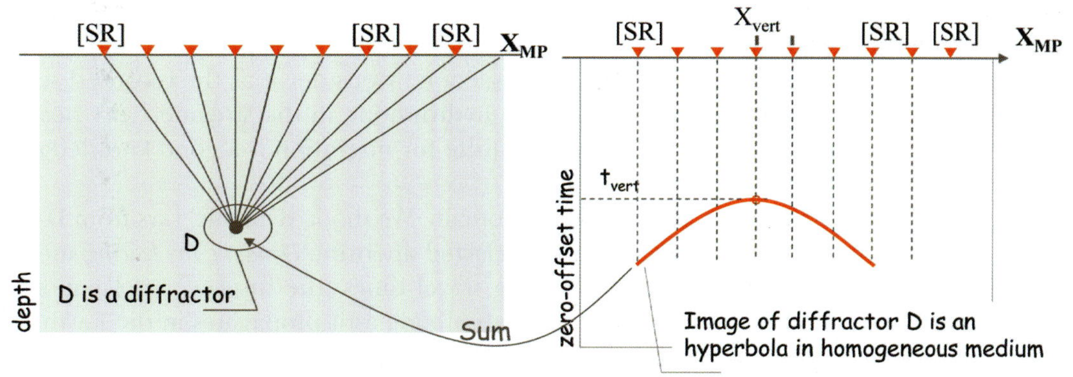

Figure 2.05: The diffraction model. If D is a diffractor, it appears on the zero offset seismic section as a 'diffraction curve'. Conversely, the reflection coefficient in D can be computed as a weighted sum of samples along the (hypothetical) diffraction hyperbola in the data domain. This is Kirchhoff migration, also called diffraction stack: we focus diffractions back to the diffraction point.

The diffraction model

Fig. 2.05 now investigates the case where D is no longer an element of a reflector but a diffractor, as defined earlier. The characteristic of such a diffractor is that it reflects an elastic wave in all reflection angles. The diffractor can be thought of as a small local sphere or more conceptually, as the superposition of an infinity of small reflectors, each with all possible dips and azimuths.
With this model in mind, an observer with a zero-offset recording system at any location at the surface will get an 'echo' from this diffractor. Consequently, its zero-offset seismic response lies on a hyperbola in the homogeneous isotropic simple 2D velocity model case that we have assumed in fig. 2.05 (simply the application of Pythagoras' theorem!). Note that the apex of the hyperbola is located at the vertical X_{vert} of the diffraction point D, where the echo time is the shortest!

In an homogeneous 3D model, the response of the diffraction point D would be a hyperboloid of revolution; in an heterogeneous anisotropic Earth, the response can become a complicated 'hyperboloid-like' complex, multivalued surface.

Now, if we want to reconstruct the diffractor in depth from its seismic image in time (again, this is the goal of imaging), we have to *'focus'* the diffracted energy back towards its location D in depth. This is why we often say that imaging consists in 'focusing diffractions'. This focusing exercise is known in the industry as 'Kirchhoff summation', which is addressed in the following paragraph.

Note in fig. 2.05 that the summation is along the diffraction curve. In theory, this diffraction curve is not limited in X, although the amplitude of the reflected wave decreases with travel paths according to spherical divergence. Hence, it is not necessary to sum from minus to plus infinity along the X axis (the whole surface in 3D). The maximum distance of summation actually used is called the *'migration aperture'*; this is an important parameter in imaging.

Kirchhoff summation concept

In the imaging techniques based on Kirchhoff summation, we actually assume that *each point in the subsurface is a diffractor* similar to the one in the cartoon in fig. 2.05. Imaging will then consist in 'focusing' each of these different points independently and the 'Kirchhoff - migrated image' will be the juxtaposition of each of these focused diffractors.

The principle is further detailed in fig. 2.06. The starting point is the pre-processed data on the bottom left of the figure: here again the 'zero-offset section'. Consider it as the recorded section of a seismic survey where the boat would tow only one hydrophone in the streamer very near to the source. As already said for ray migration, the prerequisite for migration is a good knowledge of the propagation velocity in the Earth.

Next, we consider a specific point D in the depth domain. We shoot 'dynamic' rays from D through the velocity model up to the surface and within the selected aperture. *Dynamic ray tracing* means that we compute not only travel paths and corresponding travel times (the *kinematic* information of the ray), but also parameters linked to evolution of wave amplitude as it propagates in the Earth, including particularly spherical divergence. Travel times together with amplitude form the so-called *'Green's function'* of point D to a point X at the surface through the velocity model V. Note that V can be any heterogeneous anisotropic 3D model.

The next step consists in taking the samples in the data domain (the zero-offset section) that lie along the calculated diffraction curve within the selected aperture, using the Green's function amplitude (and other pieces of information) to compute weights, to eventually compute the estimated reflection coefficient in D. This value is placed in the migrated image at the location X and depth Z of the current imaged point D. This exercise is repeated for all points D in the target zone of interest. The final image is the juxtaposition of the respective images of all points D.

In this final migrated image, reflectors simply appear automatically due to the fact that there exists effectively a non-zero reflection coefficient along them.
Note an interesting benefit of imaging by Kirchhoff summation: imaging can be performed for a limited area or selected sections only, which can be a serious time saver in the many iterations

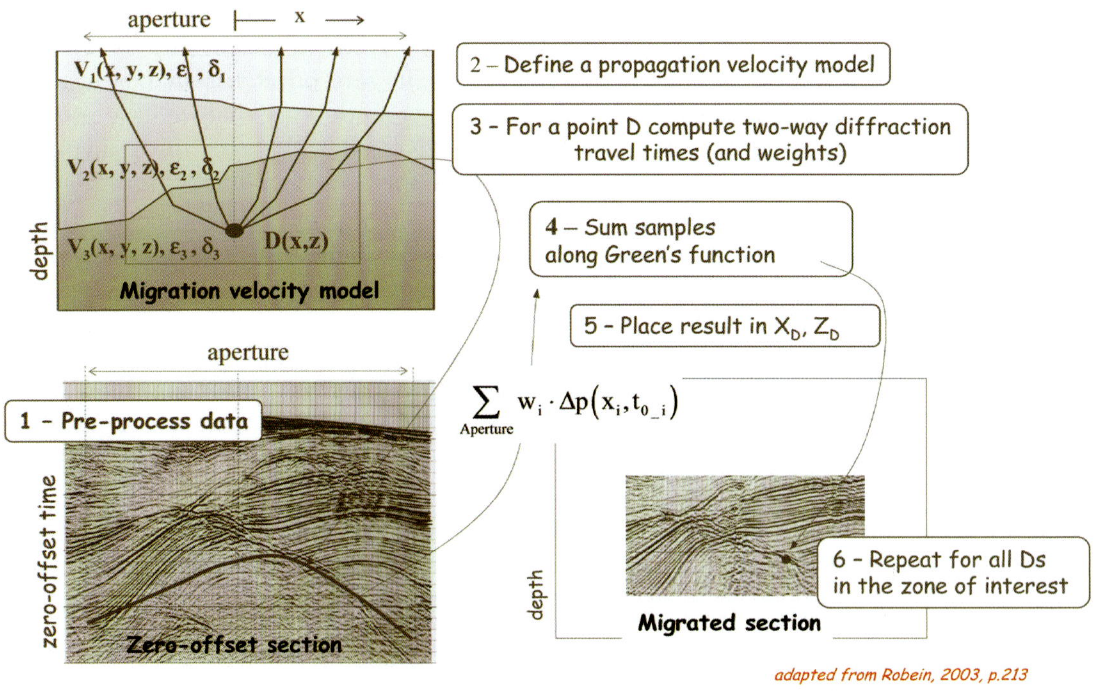

Figure 2.06: Conceptual zero-offset Kirchhoff depth migration.

required by velocity model building. The velocity model, however, needs to be defined everywhere from the surface.

We briefly mentioned that weights are used in the Kirchhoff summation. These weights not only depend on Green's functions, but also largely of the recording system lay-out. There is a great deal of work done in this domain to ensure a reliable estimate of the reflection coefficient. We discuss this point again in chapter 8 with 'Least Square Migations', but to discuss *'amplitude preserving migration'* issues goes beyond the scope of this book.

Some considerations on reflection and diffraction in Kirchhoff summation migration

Fig. 2.07 comes back to our simple initial model and superimposes the response of element D when considered as an element of a reflector (beam ray migration approach) or as an isolated diffractor (Kirchhoff summation-based migration).

It shows in particular that the image of the diffractor is tangential to the image of the reflector in X_{norm}. This could be proved by geometrical considerations but more interestingly for our purpose here, we look at it with a different perspective. Suppose that all points of reflector R in fig. 2.06 are considered as diffractors. Then the seismic response of each point D of R is a hyperbola; using a linear superposition principle, we claim that the response of R is the sum of all diffraction hyperbolas, which add constructively along the envelope of diffraction curves. This envelope is precisely the reflector; hence the tangency, according to the geometrical properties of the envelope (property already used in fig. 1.08).

Note that when we add diffractions to construct the image of the reflector, we perform a *'modelling exercise'*: we compute the response of the planar reflector; assuming that it is made of a series of close-enough diffraction points.

Coming back to Kirchhoff summation migrations, fig. 2.07 shows that for all points that are elements of reflectors, only samples belonging to the *'specular ray'*, i.e., the ray that corresponds to an actual reflection honouring Snell's law (specifically the normal ray for zero offset), will contribute to the summation. For band-limited data, with which we deal in real life, more traces contribute. They are traces with samples tangent to the reflector in a 'Fresnel zone' sense, i.e., such that they sum up constructively within a

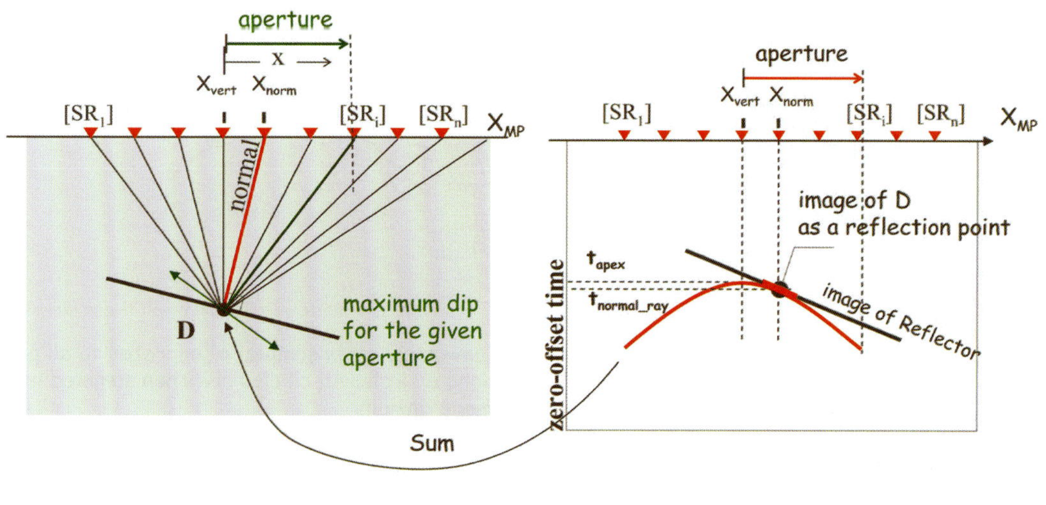

Figure 2.07: Reflection and diffraction in Kirchhoff migration. If D is an element of a reflector, only the specular part, tangency between diffraction and reflection (Fresnel zone), will contribute constructively to the summation. Aperture and maximum imaged dip are linked : steep dips require large apertures.

quarter of a wavelength. It is then hoped that all other samples outside this zone will sum destructively. In other words, because we do not know *a priori* the local dip of the reflector, we perform many 'useless' operations in summing lots of samples that do not constructively add to the image of the point under consideration. A second consequence is that since all these samples do not in practice perfectly sum up destructively, *Kirchhoff summation introduces background noise*. This noise may show up as smiles, particularly if there are events (noise; refraction; etc.) that do not follow the diffraction model.

A possible solution to sort this issue out would be to reduce the aperture. A shorter aperture means indeed less data manipulation and computation and hopefully, less migration noise. It is however important to realize the relationship between aperture and maximum dip to image. This is presented in fig. 2.07: steep dips correspond geometrically to large apertures. Migrating with limited apertures will filter out these dips, which is often not desirable. A less strong argument is that a limited aperture also reduces the focusing quality of diffractions induced by faults and consequently, their sharpness in the final image.

Aperture is hence an important parameter that is selected through series of tests to compromise between image quality and performance.

Another solution is to try and estimate the dip. We will show that we can do it, not in the migrated results but in the data domain directly. This consideration led to the development and to a large extent the success of *'beam migrations'*.

Implementation of Kirchhoff summation migrations

All the operations described above necessary to migrate data with a Kirchhoff summation technique request, as we see, a great amount of computation steps and data handling. There are lots of 'tricks' to optimize computing time and memory requirements. Particularly, Green's functions are not computed for each single point of the final image. They are instead tabulated on much coarser grids and interpolated on the fly as requested.

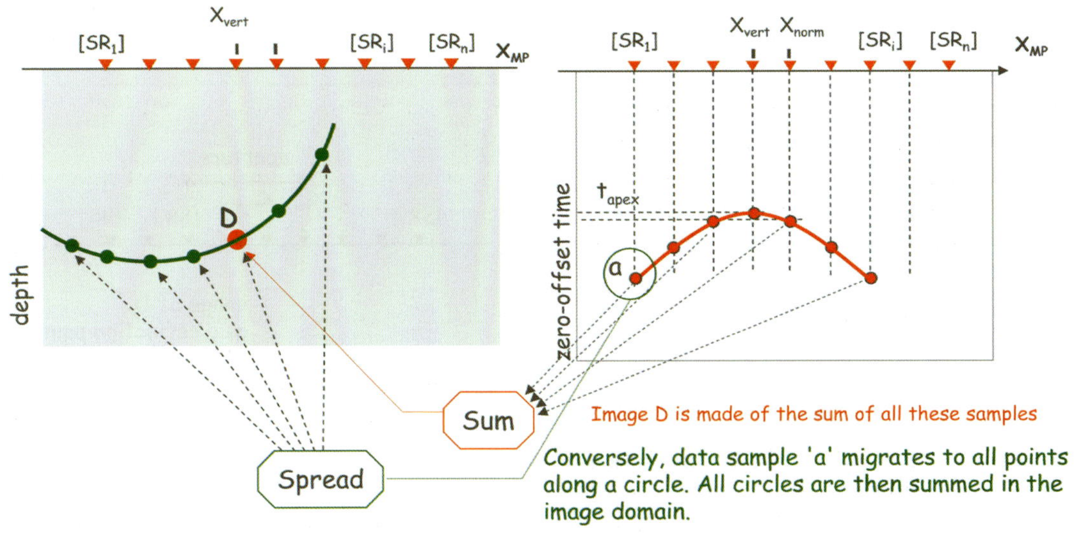

Figure 2.08-a: Two implementations of Kirchhoff migrations. In image domain-oriented approaches the image is created point per point; in data domain-oriented ones, data samples are 'sent' to their possible migrated positions one after the other.

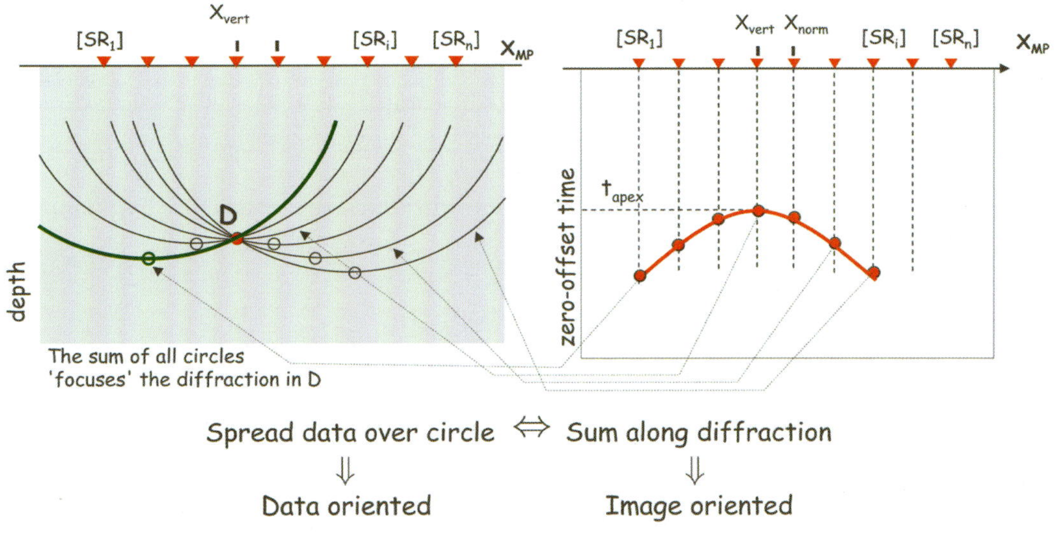

Figure 2.08-b: Summation of all circles gives a result similar to diffraction summation. The two implementations corresponds to different data handling and code organisations.

In order to optimize inputs/outputs, which are costly in terms of computing performance, another major implementation difference comes from the way data are handled in some practical implementation of the method. We describe above an implementation where the migrated image points (the pixels of the image) are computed one after the other. We call it an *image-domain oriented* approach. Many actual implementations take a different perspective, called *data-domain oriented* methods.

To understand it, let us consider sample 'a' in the data space on the right-hand side cartoon of fig. 2.08-a. Through Kirchhoff summation this very sample 'migrates' to point D. It can be shown however that pixels to which a given sample contributes in the Kirchhoff summation are not unique. In the specific 2D case of the homogeneous medium, the locus of all these pixels is a circle. It is in fact the classical 'smile' often referred to in migration text books, more properly called *'migration impulse response'* (see, for instance, Robein, 2003, pp.222-226).

An equivalent method to Kirchhoff migration that consists in the summation of samples picked along diffraction curves in the data domain is hence instead to 'spread' each sample along its corresponding circle, as schematized on the left-hand side of fig. 2.08-a. All circles are subsequently added to obtain the final migration (fig. 2.08-b).

In terms of data handling, in the diffraction summation approach, for each point 'D' to migrate, we need to search the right set of samples laying on the diffraction curves in the whole data set. In the second approach, each sample 'a' comes one after the other as recorded, is weighted and added to the image pixels along the circle to which it contributes. This makes the handling of huge modern data sets (typically hundreds of terabytes) much more efficient.

This idea is schematized in fig. 2.08-b, which shows the duality for the reconstruction of a given diffraction point between (a) summing samples picked in the data domain along hyperbola and (b) spreading each sample's amplitude along circles and summing the circles in the image domain.

These points are made here to emphasize and exemplify the importance of implementation in terms of both efficiency and to a lesser extent, quality of results for any type of migration. In Chapter 5, we will meet 'Gaussian Beam Migrations', which enter into the class of data domain-oriented migrations.

3 Pre-stack time migrations: principles and velocity analysis

With the considerations discussed in chapter 2, we are now equipped to address the most popular imaging techniques currently in use: pre-stack time migration or PreSTM in short. Pre-stack time migration principles include some assumptions, which limit the accuracy of the method. Since PresTM is still used and will remain so for some years to come, it is worth understanding what these limitations are. This chapter is hence dedicated to the presentation of PreSTM principles and its corresponding velocity analysis. Some pitfalls of the method for interpreters are addressed at the end of the chapter.

Definition of vertical time

When we mention 'time processing', it means that our objective is to build an image of the subsurface for which the vertical abscissa is not depth, as would be natural but *vertical time*. The concept of vertical time was introduced by Claerbout (1976, 1985) as a change of variable, which happened to be very useful until now.

Vertical time is defined as the (generally two-way) traveltime computed along a vertical trajectory without honouring Snell's law at velocity changes. This can be cast in more mathematical terms as a change of variable from depth to time and conversely, from time to depth, for a given propagation velocity field. The change of variable can be introduced in the wave equation. This change of variable, used in conjunction with some other approximations, led to simpler equations easier to integrate and used in the past for post-stack wave-extrapolation methods.

Fig. 3.01 gives an example of a model in depth and its equivalent version in vertical time. This new image would be the result of a perfect 'time imaging' process. We can already see on this cartoon some typical pitfalls of time imaging, such as the velocity-induced 'pull-up', or false structures under the overthrust fault.

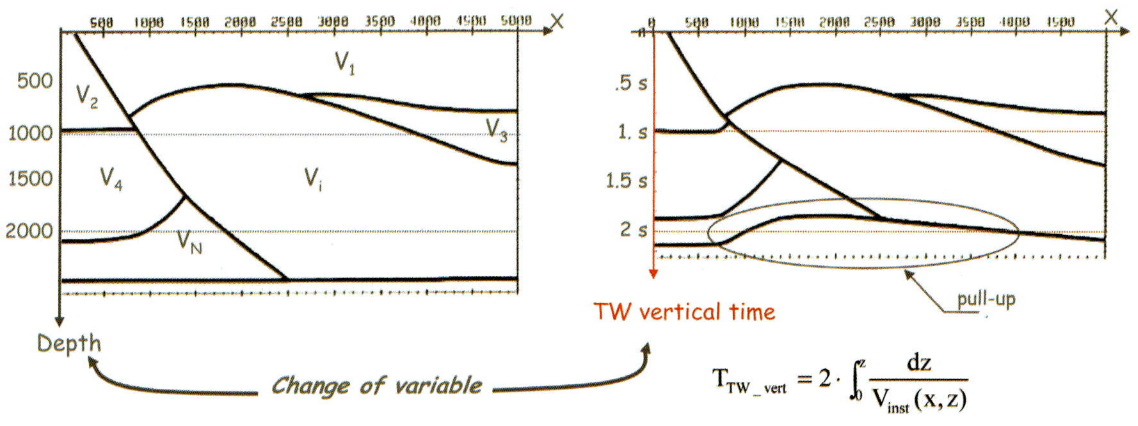

$$T_{TW\_vert} = 2 \cdot \int_0^z \frac{dz}{V_{inst}(x,z)}$$

Figure 3.01: Vertical time is the result of a change of variable from depth to time; this new image would be the result of a perfect 'time imaging' process.

Moreover, for wells being drilled in depth or rock volumes and reserves estimated in metric metres, it is clear that an image in vertical time will have to be transformed in some way to an image (often maps only) in depth. This is the objective of *time-to-depth conversion*, which is not addressed in this book (see for instance Robein 2003).

Despite of these obvious drawbacks, time imaging has for decades been the only imaging tool in practical use and remains very popular because of its relative efficiency, requiring limited human interaction.

Kirchhoff migration in the vertical time domain

Remember fig. 2.04, where we present the concept of Kirchhoff depth migration and note the differences between this figure and fig. 3.02, where we show the concept of 'zero-offset' Kirchhoff time migration.

The common point between the two approaches is that we keep the idea of Kirchhoff summation, i.e., that we obtain the reflectivity coefficient of a hypothetical diffraction point by a weighted sum of samples over the diffraction curve (surface in 3D).

The first difference is that we *place the result on the vertical axis at time* t_{vert} and no longer on the depth axis.

The second and main difference is that *we assume the Earth to be horizontally layered with no lateral velocity changes within the migration aperture*. Remember that the aperture is the area over which we perform the Kirchhoff summation. It is sketched as a triangle in the top left-hand panel in fig. 3.02.

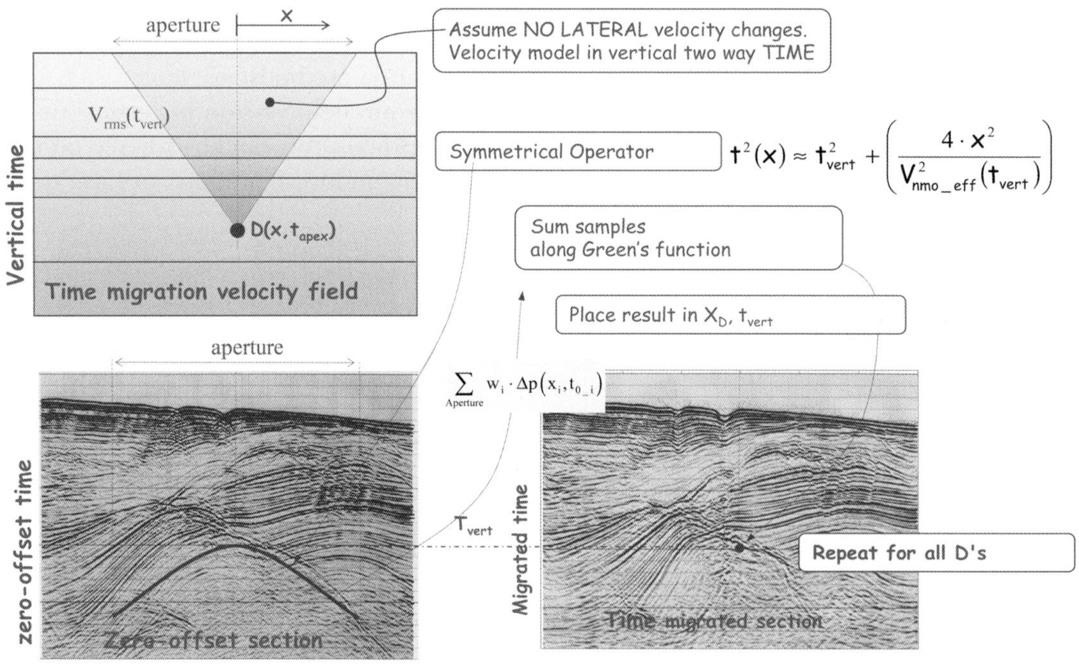

Figure 3.02: Principle of Kirchhoff Zero-offset (post-stack) time migration. The same concept as in Kirchhoff post-stack depth migration with two main differences: firstly the result of summation is placed at its vertical time location; secondly, one assumes no lateral velocity changes within the migration aperture.

A benefit of this '*multi 1D*' assumption is that the diffraction curve has an analytical expression, which is equivalent to Dix's formula for limited aperture. Under this assumption, the shape of the assumed diffraction curve is a hyperbola. For zero-offset data and with the no-lateral velocity changes assumption, t_{vert} is simply the *apex of the diffraction curve*. The shape of the hyperbola (for limited apertures) is controlled by the vertical root mean square or 'rms velocity' V_{rms} expressed as a function of t_{vert}.

This last point is the third main difference: contrary to what we said for Kirchhoff depth summation, in PreSTM we do not need an explicit (possibly anisotropic) velocity model but simply its smoothed, vertically 'rms-integrated' version as a function of vertical time and not depth.

The mathematical definition of the V_{rms} of a stack of n horizontal layers assumed to be homogeneous of velocity v_i and vertical travel time Δt_i (sometimes referred to as 'time thickness') is:

$$V_{rms}^2(t_{vert}) = \sum_{i=1}^{n} v_i^2 \cdot \Delta t_i \Big/ t_{vert}$$

If each layer is in turn vertically heterogeneous, it has its own vertical rms interval velocity $v_{rms,i}$, which should replace v_i in the equation above.

Note that V_{rms} has been historically defined and used when dealing with vertically heterogeneous but isotropic media. We define more precisely in fig. 3.10, another type of velocity denoted V_{nmo_eff}, which is an extension of V_{rms} for the anisotropic case. We will use this notation from now on.

Current implementations of time migrations relax on some of the points made above:

- Firstly, the $V_{nmo\_eff}(t_{vert})$ function is allowed to change laterally, though slowly. However, we still *assume a unique velocity and anisotropy function within the summation aperture*! This is a key characteristic of time migrations and particularly PreSTM that we discuss next;
- Secondly, we use ray tracing that yields a more accurate computation of Green's function for large offsets.
- Thirdly, again for more accuracy at large apertures, we apply more complex equations than the one in fig. 3.02, adding a second parameter called '*effective eta*' and denoted η_{eff}. This parameter takes more accurate account of heterogeneity and anisotropy for large apertures (fig. 3.10).

The common offset domain

So far we have presented 'zero-offset' migration. Pre-stack time migration is in fact implemented in the '*common-offset domain*', of which 'zero-offset' is a subset. Some reminders on data acquisition and sorting may be useful to specify what is 'common-offset domain' exactly. For the sake of simplicity, we assume a 2D acquisition although the vast majority of acquisition and PreSTM implementations are in 3D. Bear in mind however that dealing with 3D dramatically increases its degree of complexity!

Fig. 3.03 presents a regular 2D acquisition. A specific trace is characterized by two sets of parameters: the coordinates of its shot and those of its receiver. From these we deduce the offset equal to the shot to receiver distance and historically denoted '2h', and the coordinates of the mid-point between shot and receiver, X_{mp} in fig. 3.03.

The point here is that traces can be sorted by offset or more precisely in practice by offset range and all traces belonging to the same offset range are collected in the '*common-offset section* 2h'. Each trace is located in the section according to its mid-point coordinate X_{mp}.

For marine data, this collection of traces can be thought of as the record made by a boat with a single sensor placed at a distance 2h from the source, as sketched in fig. 3.04.

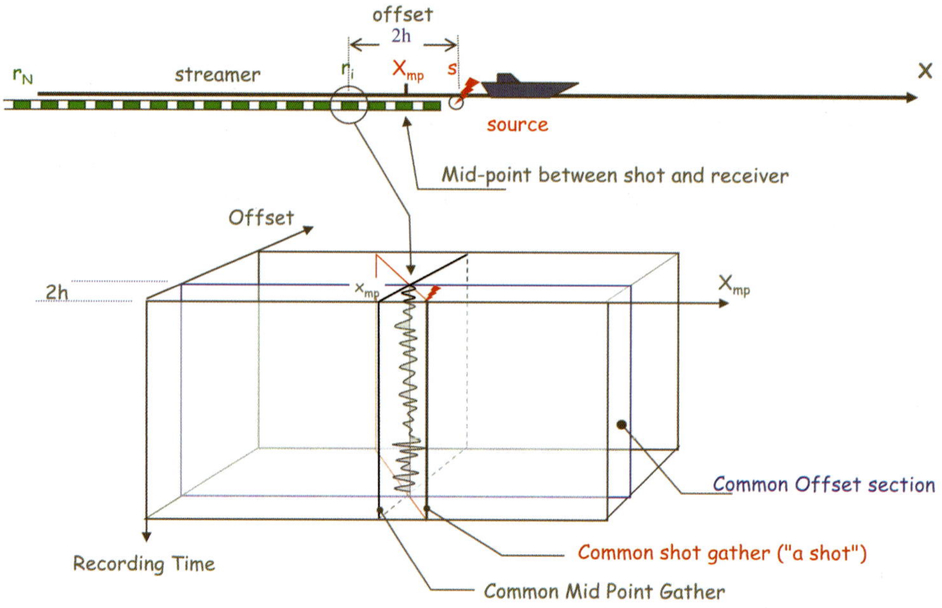

Figure 3.03: Acquisition and data sorting of a 2D seismic volume. All traces corresponding to the same mid-point between shot and receiver are collected in a Common Mid-Point gather. All traces recorded for the same shot appear in the data volume along a 'Shot gather'. All traces having the same offset appear in the Common Offset section.

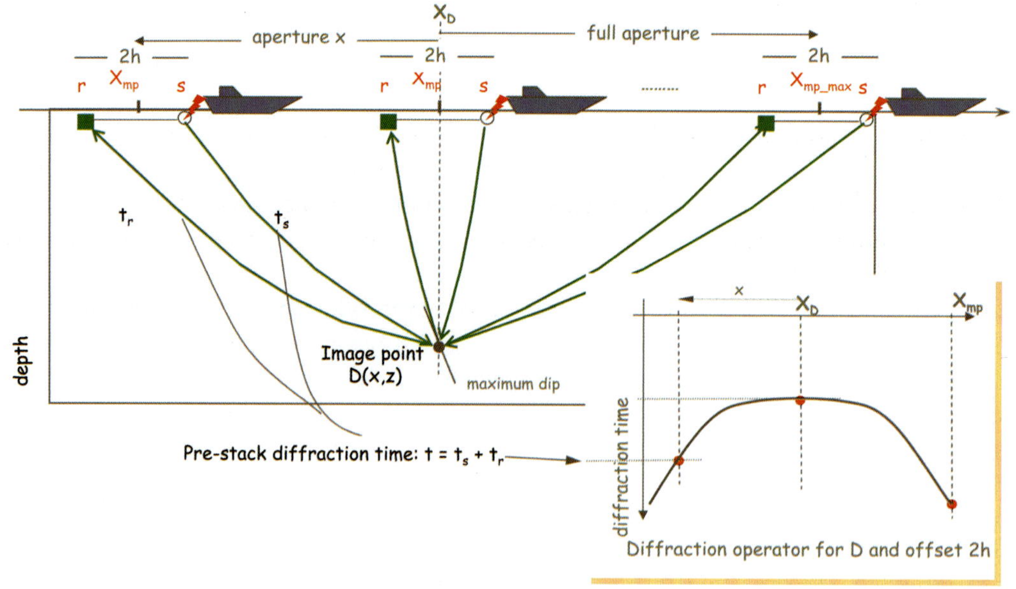

Figure 3.04: The diffraction in the common offset domain. At each image point location x_D, and for a given aperture x, the diffraction time is the sum of the travel time from the shot to the diffraction point D plus the travel time from D to the receiver at offset 2h. In PreSTM, the diffraction operator is symmetrical with respect to the image-point location.

Diffractions and Kirchhoff summation in the common offset domain

Again, we extend the principles of Kirchhoff summation, now from zero- to non zero-offset and we claim that we can obtain the reflectivity of the diffraction point by a weighted sum of samples over the diffraction curve. The weights are chosen such that if the diffraction point is part of a segment of a diffractor, one can 'reasonably hope' that we obtain the reflectivity for that specific offset. Recall that reflectivity is a function of angle and the reflection angle is in itself a function of offset. This is the basis of 'AVO' (Amplitude versus Offset) analysis.

The derivation of the weights in the Kirchhoff summation is a complex mathematical issue (Hanitzsch, 1997; Bleistein, 2001; ...). It goes beyond the scope of this book to discuss this issue of importance for 'amplitude preserved' Prestack time and depth Kirchhoff migrations.

We simply mention that the weights depend on and compensate for:

- The spherical divergence (chapter 1); hence data input to PreSTM should not be corrected for that phenomenon.
- Focusing and defocusing of waves due to velocity changes (such as caustics presented later in fig. 4.03 for depth imaging. In PreSTM, there are no caustics!).
- The fact that we consider D as a diffraction point or a reflector segment.
- The acquisition lay-out that impacts the illumination of the reflectors in depth: the Earth's reflectivity does not depend on the number of times the reflector has been 'hit' and recorded by our source-receiver system! We must consequently account for it in the summation.

The *'multi-1D' assumption* facilitates the development of practical solutions for the last two points in PreSTM. This is not the case for PreSDM in complex geological settings and explains why PreSTM is often considered as more 'amplitude preserving' than PreSDM, which would be true only where the above assumption is honoured.

Fig. 3.04 schematically presents a diffraction curve for non-zero offset. The diffraction time for a trace of offset 2h at location X_{mp} is the sum of the travel time t_s from the source to the diffractor and the travel time t_r of the diffracted energy back to the receiver at a distance 2h from the shot. Note the coordinates and definitions: the aperture x is the distance between the diffraction point (i.e., the point we are currently imaging) and the mid-point location of the trace that we consider. The shot is at a distance x + h (aperture plus half offset) and the receiver at a distance x – h. Note also the relationship between the maximum aperture and the dip of the reflector that would be imaged by that specific trace.

Having defined the diffraction curve in the constant offset domain, imaging is performed as for the zero offset case of fig. 3.02.

This is schematized in fig. 3.05. We keep the same key assumption, namely that the *velocity model does not include any lateral variations within the aperture*, using the velocity function $V_{nmo\_eff}(t_{vert})$ defined at the mid-point location.

The apex of the diffraction curve is no longer at the imaging vertical time t_{vert}. The difference is the familiar *'normal move out'* or *NMO* in short, which we define later in fig. 3.09.

Multi-offset imaging, common image gathers and stack

The first step in PreSTM is to sort the data into several common-offset sections, as presented in fig. 3.05. We do not discuss in the framework of this book the important, time consuming and sometimes difficult pre-processing steps that allow us to construct these common-offset sections in 2D and common-offset volumes in 3D.

Let us note however that in practice, acquisition is never very regular as sketched in fig. 3.03, especially in 3D land (obstacles, access) and marine (streamer feathering) acquisitions. See also

fig. 4.19 for the 3D case. Trace mid-points, offsets and azimuths may take any value. A common offset is thus in fact a *common offset class* that consists in all traces belonging to an offset range around a *'nominal offset'*. The objective is consequently to create offset cubes evenly and regularly populated in traces. This is important as holes or conversely, spikes of anomalously high amplitudes, create migration noise in the form of 'smiles' (the circles in fig. 2.08).
The various steps of this work include:

- The selection of the number and values of nominal offsets to process. This choice is dictated mainly by the acquisition parameters. A typical number of offsets lies between 50 and 100.
- Definition of the binning grid: bin intervals and grid direction. Again these are often decided as early as the acquisition design, according to the shooting direction and the expected seismic resolution.
- Binning: select all traces belonging to a given midpoint in 2D and bin in 3D (see for instance Robein, 2003, p. 134).
- Trace interpolation and regularization: create traces in empty bins;
- Trace depopulation: remove redundant traces in a bin;
- Bin centring: modify traces to correct for the fact that their mid-point is not exactly in the centre of the bin.
- Offset centring: time correct traces for the difference between their actual and the closest nominal offset it corresponds to.

The last two points are options that differ from one processor to the other. Some Kirchhoff codes assume a regular acquisition; others do not and take account of the exact shot and receiver coordinates. Common-offset sections and cubes need to be built accordingly.

Each section thus obtained is processed in the way described above (fig. 3.05), using operators of course that account for the offset value it corresponds to, but using for a given X_{mp} location the *same velocity function for all offsets*. This makes sense since the velocity function summarizes the propagation velocity in the Earth within the aperture and the Earth is unique!

In doing so, at a given X_{mp} location we have *as many migrated traces as selected offset ranges*. The collection of traces thus obtained is an object that plays a key role in pre-stack imaging. It is called *'Common Image Gather'* or 'CIG' or 'Image Gather' in short.

The CIG represents the images obtained for different offsets for the given velocity function. Note that this corresponds to adding a third dimension to the 2D data sets and a fourth one to the 3D data sets and even a fifth, when considering the shot-receiver azimuth.

The Earth being unique, if everything goes correctly, a given reflector should appear at the same 'true' vertical time, whatever the offset used. We say that we want the *'image gathers to be flat'*.

When gathers are flat, it becomes possible to sum the different traces in the gather to eventually obtain the migrated trace at location $X_{mp,}$ as shown in fig. 3.06. This stacking step increases both the signal-to-noise ratio and to a lesser extent, the primary-to-multiple ratio.

All events in the CIG should be flat; this nevertheless does not mean that all traces of a CIG should be exactly identical. Remember that the result of migration is supposed to be a 'log' of the elastic reflectivity of the Earth but reflectivity depends, as mentioned, on the angle of reflection at the interface. This angle in turn clearly depends on the offset. Consequently, if we compute and apply the proper weights in Kirchhoff summation, thus compensating for all perturbations in amplitude that the waves underwent during downward and upward propagation, the AVO analysis can be performed directly along the horizontal axis of CIGs. Consequently, stacking yields some average of the refelection coefficient for several reflection angles.

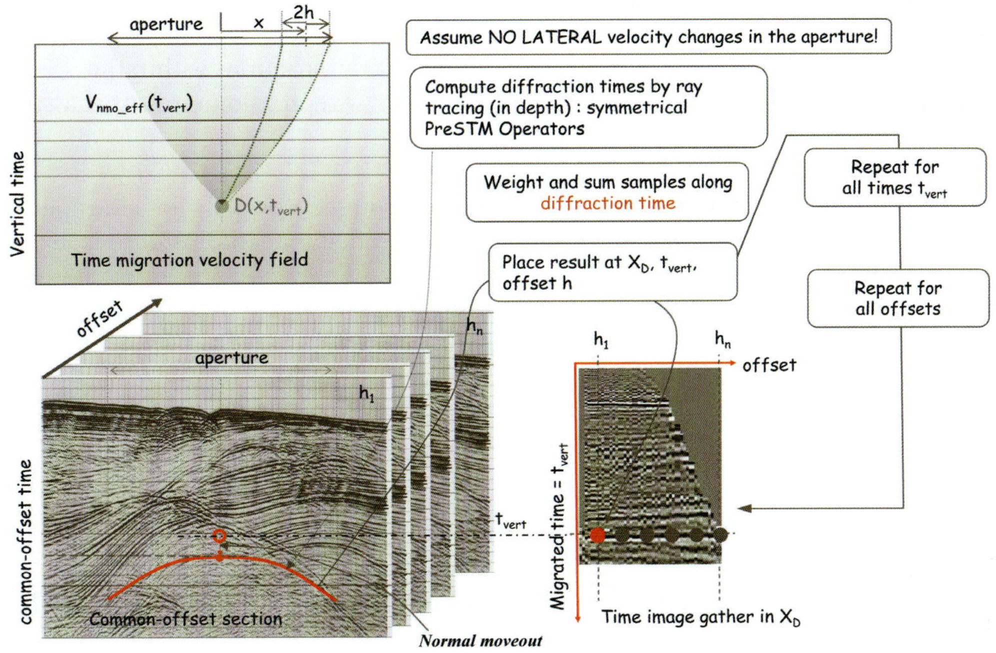

Figure 3.05: Principle of Kirchhoff PreSTM and construction of image gathers in time. Each common-offset section is processed independently, but with the same velocity field. The contribution of each offset section is kept in the image gathers.

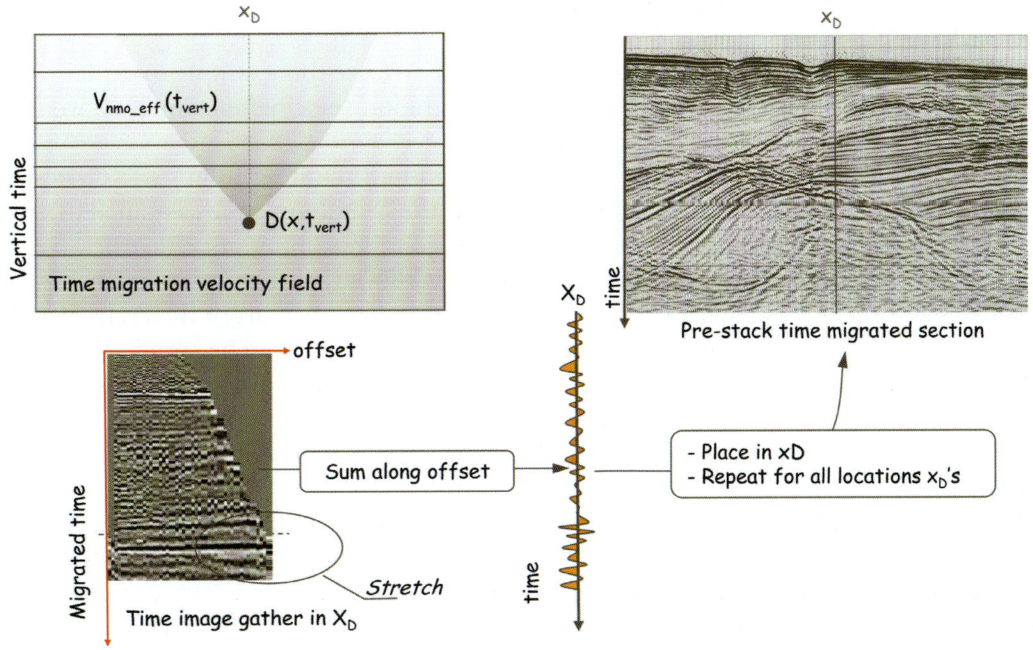

Figure 3.06: The stack process in PreSTM. Traces of the image gathers are stacked to get the 'pre-stack migrated section'. Stacking improves the image (better signal / noise) only if the gathers are 'flat'.

For the stacking process in fig. 3.06 to be efficient, it is crucial that image gathers are flat. Three factors can make it such that this is not the case in practice:

- A breach of the assumption that there are no lateral velocity variations within the aperture plus/minus half offset. Since the aperture may be as large as several kilometres, this assumption is very seldom met!
- A velocity function $V_{nmo\_eff}(t_{vert})$ that is not correct.
- The way we compute diffraction curves is not accurate enough, especially for wide apertures and/or large offsets in the presence of severe vertical heterogeneity, steep dips or anisotropy.

Leaving aside for the time being the first issue, we address the third one by adding a further parameter denoted η_{eff}, which describes more accurately propagation for large apertures. We then have to find the set of two functions $V_{nmo\_eff}(t_{vert})$ and $\eta_{eff}(t_{vert})$ that yields the 'flattest gather possible' and correlatively, the 'best time-migrated image possible'.

The goal of *PreSTM migration velocity analysis* (or MVA in short) is to optimize these two functions but before we develop MVA, we return for a moment to the third issue above, namely the derivation of diffraction curves in time imaging.

Derivation of the diffraction curve in PreSTM: the ray tracing approach

The diffraction curves are the '*migration operators*' that belong to one image point and that we apply to each appropriate data sample in order to migrate it and compute image gathers. In the following, we will use both terms equivalently. There are two classes of techniques to compute the diffraction curves:

- Ray-based methods, where rays are sent explicitly or implicitly in a modelled Earth.
- Parametric methods based on analytical solutions to ray propagation in 1D media, using a limited number of parameters.

Fig. 3.07 shows a PreSTM section taken from a 3D volume. We use this example to explain schematically how migration operators are computed, given a migration velocity function at a given location, referenced as 'VA' in the figure.

The section may come from an initial post-stack migration, resulting from early processing performed on board the acquisition vessel. The vertical axis is thus 'migrated time', assumed to be 'vertical time'. An interpreter has picked some horizons, which he considers important either because they are *key geological markers* or because they are horizons where he knows that there are possibly *strong velocity contrasts*. The selection of these horizons is an important interpretative part of the PreSTM process.

Velocity values are highlighted at the picked markers at location VA. The set of pairs of $\{V_{nmo\_eff}; t_{vert}\}$ defines the continuous velocity function $V_{nmo\_eff}(t_{vert})$, for which we want to compute the migration operators at each vertical time t_{vert}.

Fig. 3.08 schematically represents the ray-based technique used to compute the operator at that location.

In the left-hand part of the figure is a plot of the $V_{nmo\_eff}(t_{vert})$ function, with the markers picked on the section. On the cartoon to its right, we show the interval velocity computed using inverse Dix's formula (Dix, 1955; see also fig. 3.10).

Since we want to trace rays, we need to move to the depth domain. This is schematized on the next panel to the right in fig. 3.08. Transformation from time to depth is straightforward: thickness of a layer = (time thickness / 2) × interval velocity.

Again, there is an approximation here since inverse Dix's formula of V_{nmo_eff} yields interval V_{nmo} velocities and not the vertical average velocity that would be required. Note that interval V_{nmo} depends both on the vertical heterogeneity within the selected interval, controlled by its interval V_{rms} and the

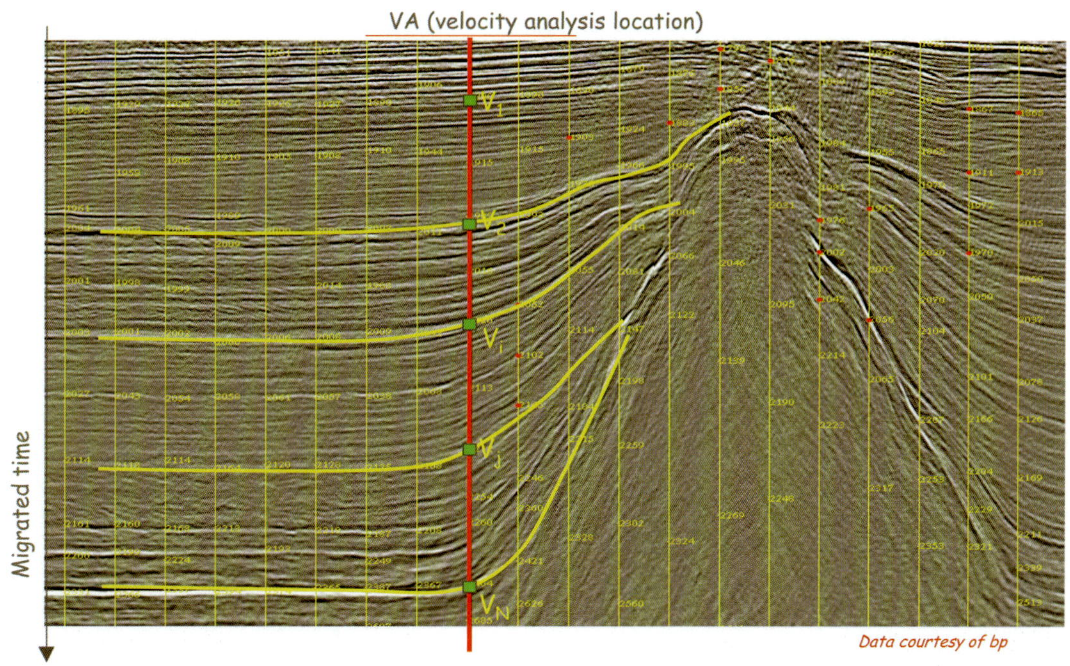

Figure 3.07: Derivation of the PreSTM operators by ray tracing. Step 1 : select horizons, which will drive the computation of the operator. Another today classical option is regularly and densely sampled horizons, typically every 50 to 100 ms.

δ anisotropic parameter and not ε (see equations in fig. 3.09 and 3.10). These 'details' are overlooked in practice.

Since the inverse Dix's formula is subject to instability, especially for thin layers, and PreSTM requires some degree of smoothness in the migration velocity field, smoothing is applied to the depth-velocity model. In practice, some editing may first be applied to remove aberrant values and then a 'heavy' smoothing is performed in a 3D sense in order to avoid sharp velocity changes from one migrated trace to its neighbours. How and how much to smooth the velocity field is part of the 'processor's skill'. Filtering and smoothing methods based on geostatistics are common practice (Le Meur *et al.*, 2003).

This being done, we are ready to 'shoot' rays from a series of image points D at various depths to a series of shot / receiver points at the surface. This shooting exercise provides us with tables of travel times for various points D in depth and various distances y of shots or receivers at the surface to the vertical of the image point (fig. 3.08).

Remember that we are looking for the diffraction curve belonging to a vertical time t_{vert} and an offset of 2h for various values of aperture x. Vertical time t_{vert} is simply twice the time for $y = 0$; which gives us the depth of D that corresponds to t_{vert}. As shown in fig. 3.04, the diffraction time is the sum of the time from the shot 's' to the diffraction point D plus the time from D to the receiver 'r'. Noting that in our scheme, the distance y for the shot is $y_s = x + h$, and for the receiver $y_r = x - h$, we deduce that the diffraction time is obtained by adding the tabulated times for t_{vert} at y_s and y_r : $t_{diff}(x, h) = t_s + t_r$.

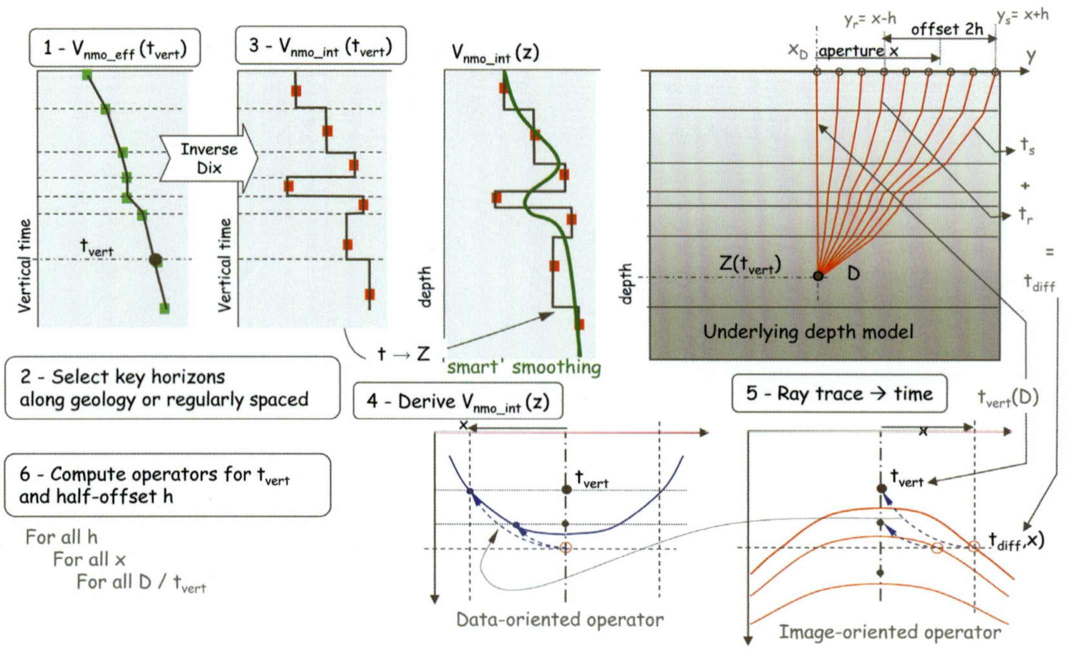

Figure 3.08: Derivation of the PreSTM operator by ray tracing. Firstly, build a 1D smooth velocity model. Secondly, compute one-way travel times by ray tracing. Thirdly, tabulate operator times for various t_{vert}, apertures x and offsets h. In 'data oriented' approach, we define the operator by looking at what distance x and vertical time t_{vert} a data sample at time t_{diff} must be sent to for various apertures x.

The whole diffraction curve is computed by repeating the operation above for all aperture x values within the maximum aperture and there is one such curve to compute for each vertical traveltime t_{vert}. Finally, all these operations need to be repeated for every offset to process and then to every mid-point or bin.

When moving to the next trace X_{mp} to migrate, since we allow the V_{mig} function to change laterally, the whole 1D model building, ray tracing and computation of the set of operators must be performed again. In practice, traveltimes are computed for a limited number of velocity functions, distances y and depth values z and tabulated. The requested values are interpolated as needed from the tabulated values computed on this coarse grid. As explained in chapter 2 (fig. 2.08-b) the summation may be carried out in two ways:

- We apply the summation along the diffraction curve. In this case the operator is computed as sketched in red at the bottom-right of fig. 3.08. We compute the reflectivity in D by searching in the data volume and summing together the samples corresponding to the migration operator. The image is built by repeating this operation for all points D.
- Alternatively, we can take the traces one after the other and 'spread' each sample at successive t_{vert} along its corresponding '*data-oriented operator*' in the image. The 'smiling operator' equivalent to the circles in fig. 2.08-b, is sketched in blue in fig. 3.08 for a sample at time t_{vert} of a trace in the common-offset section at the midpoint location corresponding to the image point abscissa x_D. Each data sample has its own operator and the migrated image is the sum of all these different 'smiling' operators.

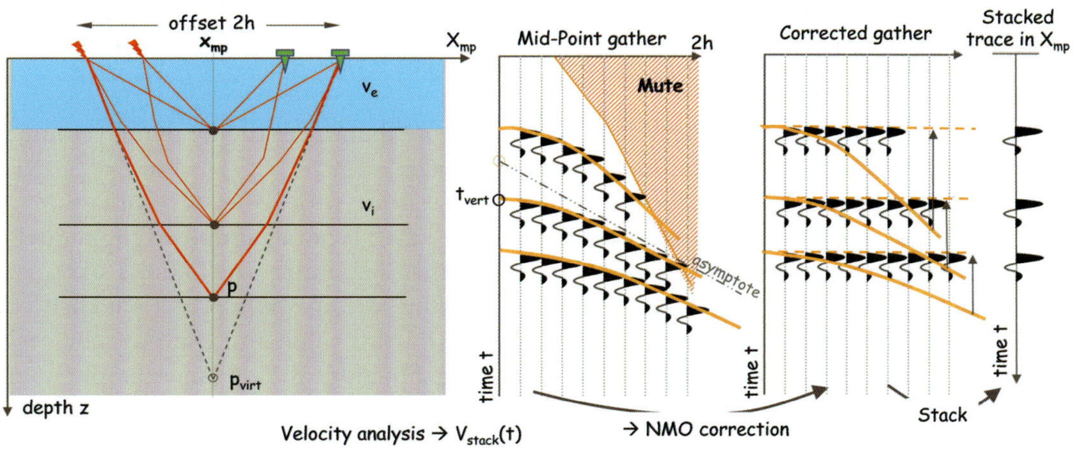

The NMO curve may be approximately a hyperbola.

Limited offsets: Dix with NMO controlled by $V_{nmo\_eff}(t_{vert})$ $\quad t(h) \approx \sqrt{t_{vert}^2 + \left(\frac{2h}{V_{nmo\_eff}(t_{vert})}\right)^2}$

Inverse Dix → nmo interval velocity $\quad V_{nmo\_int} = \sqrt{\frac{t_2 \cdot V_{nmo\_eff}^2(t_2) - t_1 \cdot V_{nmo\_eff}^2(t_1)}{t_2 - t_1}}$

Shifted hyperbola: 2 parameters V_e and $T_p(t_{vert})$: $\quad (t(h) + Tp - t_{vert})^2 = Tp^2 + \left(\frac{2h}{V_e}\right)^2 \quad Tp = \frac{2 \cdot z(P_{virt})}{V_e}$; V_e = slope of asymptote

Figure 3.09: Normal Move Out travel times and 'stacking velocity analysis'. 'One term' stacking velocity analysis aims at finding the velocity V_{stack} that flattens the gather best. 'Shifted hyperbolas' make an alternative to Dix valid for much larger offsets. The NMO curve is now defined by two parameters Tp and V_e. Tp is the 'time in the first medium of velocity V_e' of the virtual focal point (in the optics' sense) P_{virt}.

The latter approach explains why we sometimes see 'bowls' or 'smiles' in the migrated image. If a sample is anomalously high (streamer noise; burst; …) or low (missing or dead traces in the common offset volume) its corresponding smiling operator is not summing up destructively with its neighbouring ones and shows up on the section.

We have not introduced anisotropy so far in our derivation of PreSTM operators. This will come in the next paragraph. One can note, however, that there is no intrinsic difficulty in adding anisotropy parameters in the 1D model of fig. 3.08 or to shoot rays in these anisotropic models to compute diffraction curves or 'data-oriented' operators. The issue is to find which parameters we should use for this.

Derivation of the diffraction curve in PreSTM: the analytical approaches

An alternative to ray tracing is to use analytical solutions of traveltimes. Analytical approximations to traveltimes in 1D models are not new. Dix's formula (Dix, 1955) has been the most famous one and is still in use! The objective at the time was to address the issue of 'normal move-out corrections'. The idea, presented in fig. 3.09, is the following. If we assume the Earth to be horizontally layered, all traces recorded with the same mid-point between shot and receiver will illuminate the same points in depth, all at the vertical of the mid-point location. Traveltimes, however, will be different, increasing with offset.

Dix showed that if the offset is limited (for instance not greater than depth of the reflection point), traveltime as a function of offset can be approximated by a hyperbola that is controlled at a given vertical time t_{vert} by V_{rms}, the vertical root-mean-square velocity at time t_{vert} (now replaced by its extension to the anisotropic case V_{nmo_eff}, as mentioned earlier).

We met the notion of 'Mid-Point Gathers' in fig. 3.03 but note that traces there are not yet migrated, which makes a big difference with Common Image Gathers.

'*Stacking velocity analysis*' is the process that allowed seismic processors to estimate $V_{rms}(t_{vert})$, based on this simple model and approximation. Since the Earth is not as simple as the model used, in practice, the result of velocity analysis is the parameter that 'maximizes the flatness' of the gather. The process to find this optimal parameter consists in trying various values of V_{rms} at each vertical time, moveout-correct the traces of the mid-point gather for this value and compute the *semblance value of the corrected traces* (Taner and Koehler, 1969) as a function of V_{rms}. An example of such a '*velocity analysis panel*' can be seen in fig. 3.15. The parameter that maximizes the semblance is called '*stacking velocity*' V_{stack} (t_{vert}). Its value may depart grossly from V_{nmo_eff} in real life (Al Chalabi, 1994; Robein 2003; ...).

Once this velocity function has been estimated, traces from the mid-point gather can be corrected to zero-offset time, which is also vertical time for this simple model. Traces are then stacked together in order to improve the signal-to-noise and primary-to-multiple ratios. We have simply described here *the 'old' nmo-stack process*. This concept is still in use, although applied nowadays to migrated image gathers, as we will see below.

The version of the Dix equation presented in fig. 3.09 is an approximation valid for vertically heterogeneous media but limited offsets. Dix (1955) actually proposed a Taylor series expansion in even powers of offset 2h to approximate traveltimes at any offset. It is not used in practice because it does not converge fast enough.

Some authors proposed different schemes to address this problem. (De Bazelaire, 1988, 1994), and (Castle, 1994) popularized a concept that De Bazelaire called '*shifted hyperbola*'. It is a set of hyperbolas defined by two parameters for a given vertical time. These are in De Bazelaire's terminology:

- V_e: the velocity at the surface in which traveltimes differences are measured; also the slope of the asymptote of the hyperbola.
- Tp : the depth of the virtual focusing point in an optics sense, sketched in fig. 3.09, scaled in vertical time in a medium of velocity V_e.

The physical meaning of Tp and the corresponding moveout equation are given in the figure. In a 'classical stacking velocity' analysis based on this equation, the parameter scanned is Tp, for a fixed V_e. The reader may infer from the equations that a scan of Tp values corresponds to testing *parallel hyperbolas* for various t_{vert}, hence the term 'shifted hyperbolas'.

(Ursin, 2006) also showed that normal moveout traveltimes at large offsets are approximated with an accuracy at least as good as Dix's expansion to h^6 by simply *playing with both Tp and Ve parameters*. This led to the *'two-term' or 'bispectral'* scans.

Alkhalifah and Tsvankin (1995) analysed moveout in horizontally layered media for large offsets and for anisotropic VTI 1D models. They proposed an analytical solution, which has since become very popular as 'Alkhallifah's formula', also controlled by two parameters at a given vertical time:

- V_{nmo_eff} (t_{vert}): this effective normal move out velocity is an extension to the anisotropic case of $V_{rms}(t_{vert})$. The term 'effective' conveys the fact that it is integrated from the surface down to a depth expressed in vertical traveltime t_{vert}.

- $\eta_{eff}(t_{vert})$: an effective parameter that accounts for vertical heterogeneity and anisotropy in all layers.

Fig. 3.10 presents Alkhallifah and Tsvankin's 1995 model and equations. The η (eta) parameter is a new anisotropic parameter introduced by the authors as a combination of Thomsen's parameters ε and δ. They made the important point that η *is the anisotropic parameter measurable through time processing*. It is also sometimes called an 'anelliptical' parameter because it vanishes in the special case of 'elliptical', anisotropy that corresponds to $\varepsilon = \delta$.

In fig. 3.10, note also the critical difference between *'interval'* parameters V_{nmo} and η that apply to an interval or a specific layer and *'effective'* parameters V_{nmo_eff} and η_{eff} that integrate heterogeneity and anisotropy information from the surface down to the point in depth where they are measured.

Time processing yields estimates of these 'integrated' parameters. An extension of Dix's inverse formula is required to transform estimated effective parameters into estimated interval parameters.

Note that the interval V_{nmo} is a combination of the vertical heterogeneity of the layer (this is the interval V_{rms} in the equation) and the interval anisotropy parameter δ only. The effective V_{nmo_eff} being the rms average of the interval V_{nmo}, carries similar information. Consequently, measuring V_{nmo_eff} only (short offset analysis) does not give access to η or ε.

Fig. 3.11 and 3.12 illustrate in a very simple model the behaviour of V_{nmo_eff} and η_{eff} as a function of vertical heterogeneity and anisotropy. Fig. 3.11 represents a two-layer model. Each layer is isotropic. Normal moveout rays are traced for offsets up to 6 km. The corresponding traveltimes are plotted as a function of offset (red curve). The curve corresponding to Alkhalifah's analytical approximation is also displayed but superimposes almost perfectly with the ray tracing one: the approximation is good!

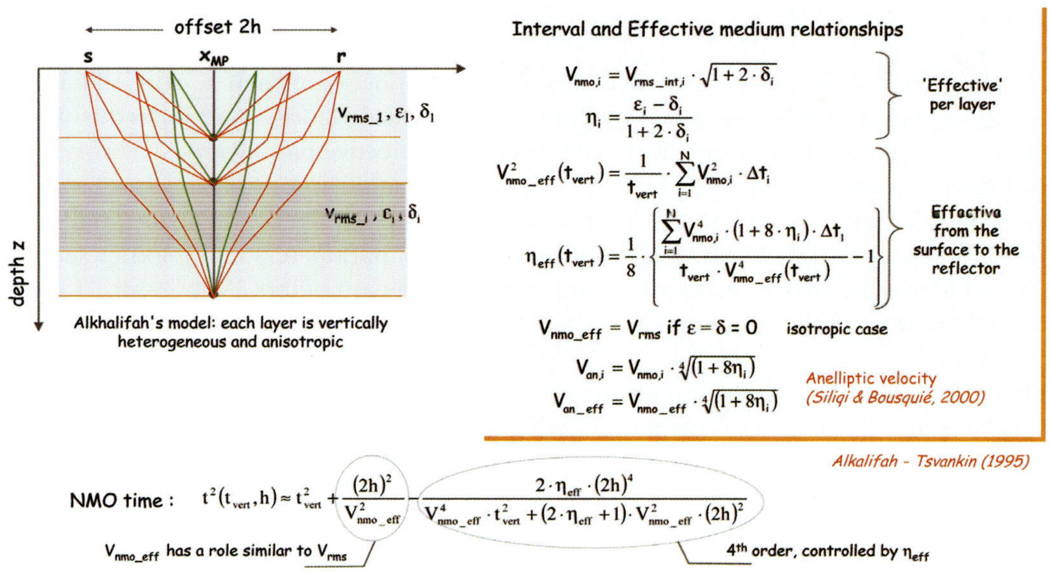

Figure 3.10: Alkalifah and Tsvankin's NMO anisotropic model. Vertical heterogeneity and anisotropy in each layer is characterised by V_{nmo} and η. η is a combination of Thomsen's parameters ε and δ. The overall 'effective' behaviour of the stack of layers is characterised by 'effective' parameters V_{nmo_eff} and η_{eff}. These two parameters control in turn the move out behaviour at all offsets.

The corresponding effective parameters are given on the bottom left of the figure; note that $V_{nmo\_eff} = V_{rms}$, but η_{eff} *is not nil because it bears information on the heterogeneous layering* of the model (check with the definition of η_{eff} in fig. 3.10). The time versus offset curve in blue is computed using the Dix's formula and the true V_{rms}; the plot at the bottom centre is a zoom on the difference between traveltimes computed with Dix and Alkhalifah, the latter being equal to ray tracing.

Fig. 3.12 is similar apart from the fact that we introduce anisotropy and vertical heterogeneity in the first layer. Note that Alkhalifah's approximation is still quasi perfect up to the maximum offset computed. V_{nmo_eff} however, is no longer equal to V_{rms} as an effect of anisotropy in the first layer; similarly, η_{eff} increases compared to the isotropic case.

The message is that *separating heterogeneity from anisotropy from estimates of* V_{nmo_eff} *and* η_{eff} *is not such an easy task*.

More recently, (Siliqi and Bousquié, 2000; Siliqi *et al.*, 2003) compared the shifted hyperbola and Alkhalifah's approaches and proposed a description of the moveout based on what they called 'anelliptic shifted hyperbola', now defined by the two following parameters at a given time t_{vert} (fig. 3.13):

- dtn = $t(h_{max}) - t_{vert}$, where $t(h_{max})$ is the moveout time for maximum offset;
- τ_0 that characterizes the summit of the shifted hyperbola, equivalent to Tp in fig. 3.09.

The claimed benefits of this set of parameters are:

- moveout corrections are independent of t_{vert}; shifted hyperbolas are computed by time shifts rather than dynamic corrections;
- no stretch is applied to the data, which means an apparent increase in resolution (the stretch is a physical phenomenon that cannot be removed from the recorded data);
- the sampling rate of τ_0 and dtn may be directly related to the seismic bandwidth;
- the parameters can be picked with less 'leakage' between them, allowing a more stable scan according to the authors.

Siliqi and Bousquié (2000) also established the relationship between their τ_0 and Alkhalifah's η_{eff}, introducing moreover another parameter, denoted 'anelliptical velocity' and related to V_{nmo} and η with the same equation, whether one deals with interval or effective parameters. The corresponding relationships are given in fig. 3.13.

These equivalency formulas link the two classical ways to parameterize analytical moveout functions and it is hence possible to express a result in one set of parameters into another set. The current practice in the industry is to refer to Alkhalifah's set of parameters $\{V_{nmo\_eff} ; \eta_{eff}\}$ even when shifted hyperbolas' parameterization is used in PreSTM.

Alkhalifah's formula and shifted hyperbolas were initially proposed to solve the normal moveout problem. They extend easily to the calculation of PreSTM operators, as shown in fig. 3.14. Note that in this case, the traveltime expression is the sum of two square roots. This is one version of the so-called 'Double Square Root' operator or DSR in short. We find this term again in WE-based PreSDM.

As a conclusion to the paragraph, we have introduced a set of two new functions: $V_{nmo\_eff}(t_{vert})$ and $\eta_{eff}(t_{vert})$. Based on these parameters, derived directly or via the anelliptical shifted hyperbola technique, we have at our disposal equations that define on the one hand moveout curves and on the other diffraction curves. The former one allows us to perform moveout analysis on CIG, as we will see in the next paragraph; a DSR version of the latter one is used in analytical derivation of PreSTM operators.

Using the DSR equations to compute the PreSTM operators makes migration and velocity analysis coherent, which might not be the case with explicit ray tracing.

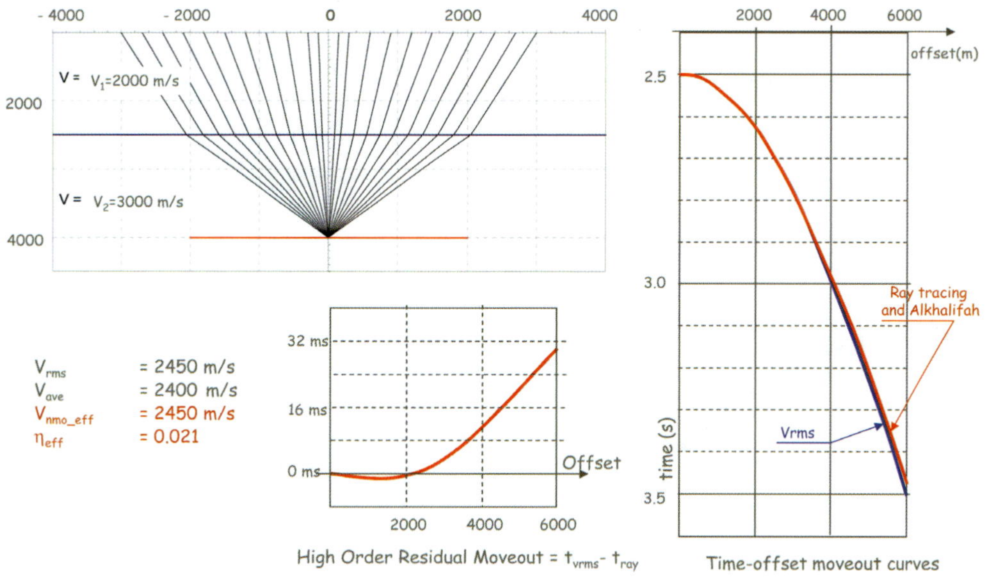

Figure 3.11: A model comparison of ray tracing, Dix and Alkhalifah moveout times. Synthetic example 1: vertical heterogeneity only. Note that η_{eff} is not nil because of heterogeneity. Alkhalifah's equation is valid at least up to 1.5 times the depth of the marker. Dix's formula fails for depth ≥ offset.

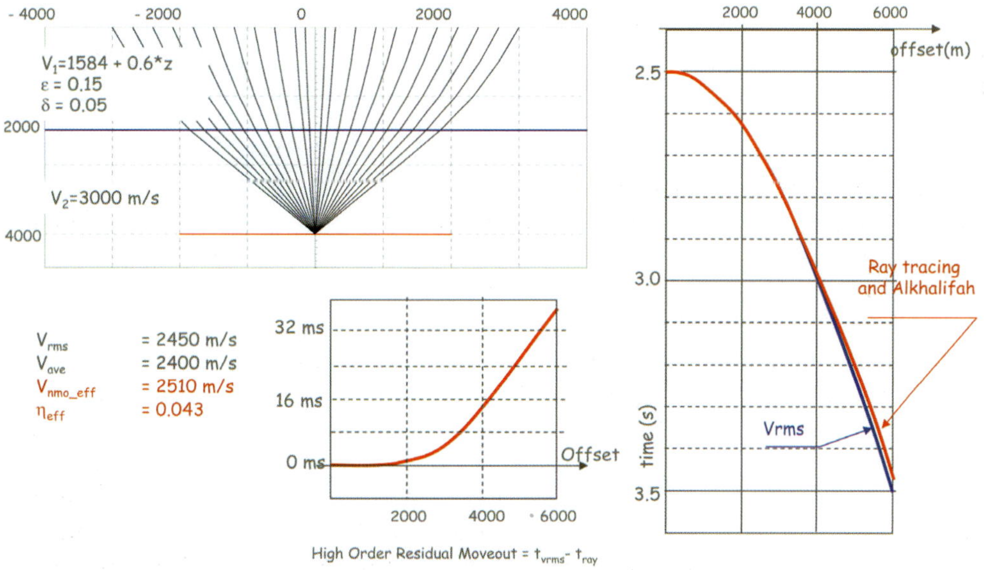

Figure 3.12: Comparison of ray tracing, Dix and Alkhalifah travel time. Synthetic example 2 : vertical heterogeneity plus anisotropy. The first layer is now heterogeneous (V_0 = kZ velocity function) and anisotropic. For the same vertical time of 2.5 s, η is twice as high as in fig. 3.11 and Dix's formula fails for even shorter offsets.

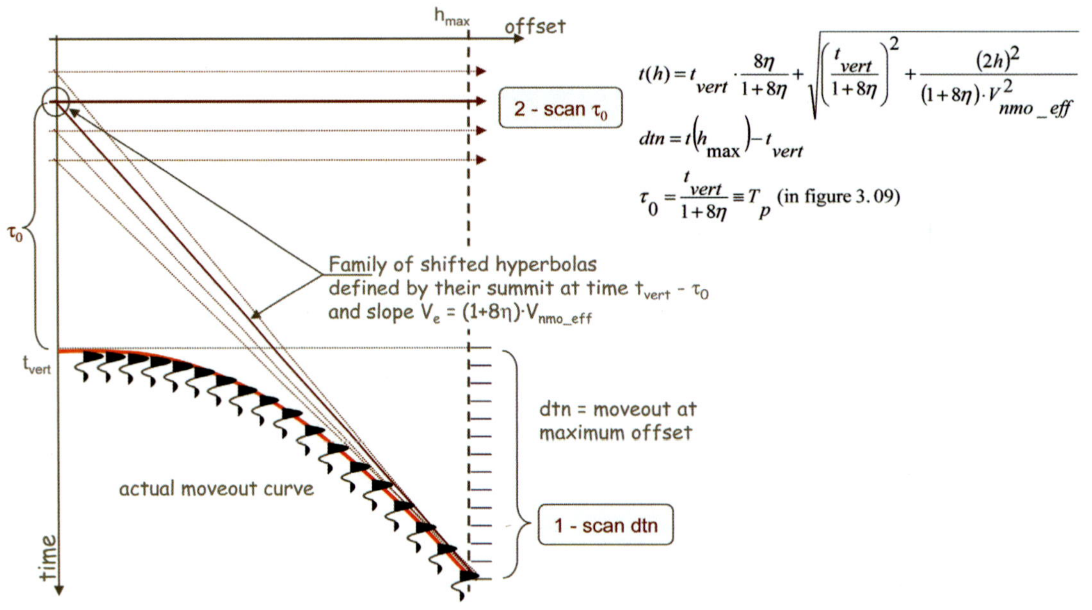

Figure 3.13: Siliqi's approach to shifted hyperbolas. The moveout curve is defined for a given t_{vert} by two parameters, dtn the moveout at maximum offsets and τ_0 that controls the family of shifted hyperbola. These parameters are related to Alkhalifah's V_{nmo_eff} and η_{eff} (Siliqi & Bousquié, 2000).

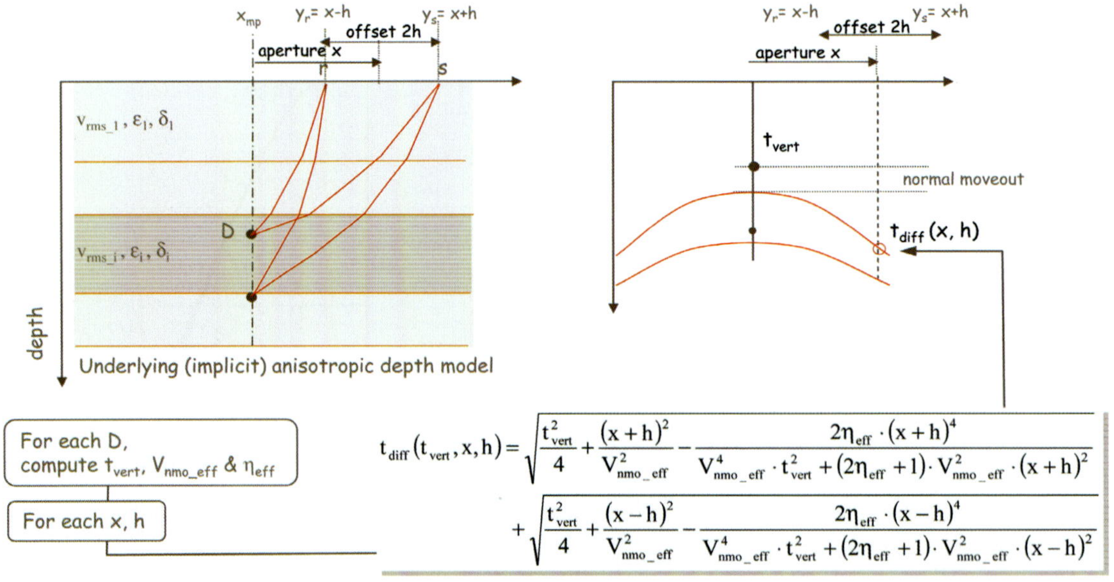

Figure 3.14: Application of Alkalifah's formulation to compute diffraction curves. This DSR (double square root) formula can replace ray tracing in the right hand panel side of fig. 3.08.

PreSTM migration velocity analysis

We address in this section the method to find the set of velocity V_{nmo_eff} and η_{eff} functions, which flatten image gathers best at any time and any migrated trace locations. We call it 'PreSTM velocity analysis and stacking'.

Fig. 3.15 gives an example of a typical workflow.

- The sequence of actions starts with the pre-processed data: data navigation merger, noise suppression, multiples attenuation, application of static corrections, source signature processing, etc., but no spherical divergence as mentioned earlier!
- Data are sorted by common-offset range cubes, with all the necessary work mentioned earlier to produce properly regular volumes.
- An initial velocity V_{nmo_eff} field is set up, resulting from a quick classical velocity analysis, elaborated in the pre-processing steps and sometimes done on board the acquisition vessel.
- With this initial migration velocity field, we can compute the corresponding migration operators, generally using the ray-based and data-oriented methods.
- These operators are applied to the data at each time sample to produce a selected series of common image gathers over a predefined grid. The sampling grid goes typically from 250 by 250 to 500 by 500 m. Some velocity lines are also output for QC purposes. In order to speed the procedure up, we benefit from the facility offered by Kirchhoff methods to output only the necessary subset of migrated data limited to the velocity lines.
- At this stage, image gathers are generally not flat and more steps are required. One possibility is a residual moveout analysis, the principle of which is described in fig. 3.22, but the industry's choice generally favours an *'inverse NMO route'* that goes as follows.

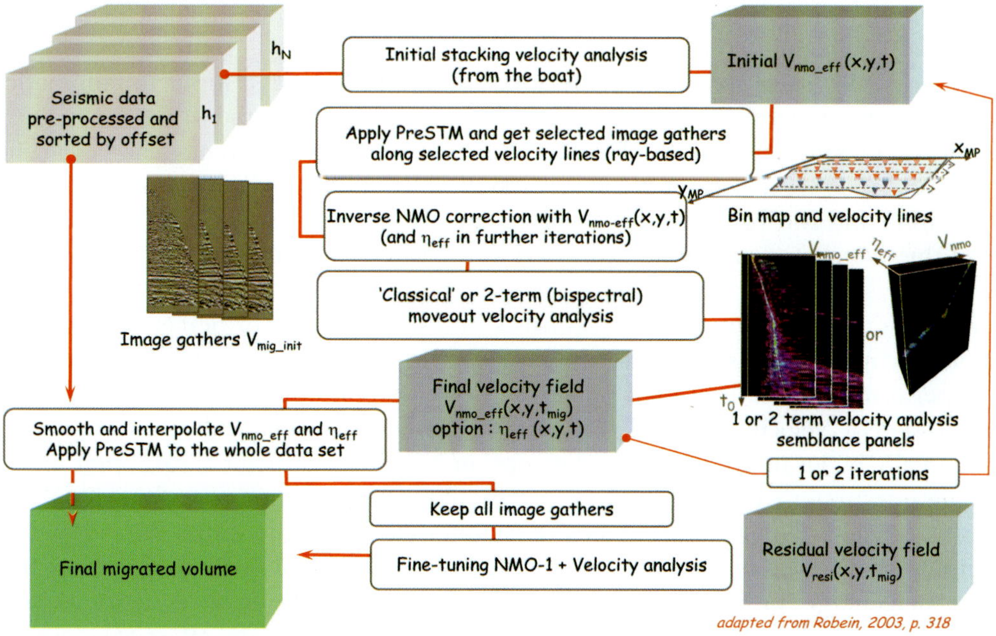

Figure 3.15: A schematic PreSTM Migration Velocity Analysis and stacking workflow. Many variations are possible and applied around this general scheme.

- The computed image gathers undergo *an inverse moveout correction* using the initial migration velocity at their respective locations and the original offset value. We then assume that this operation brings the gathers back to the situation sketched on the left of fig. 3.16, which is kinematically similar to the mid-point one sketched in fig. 3.09 or 3.10. Unfortunately, this assumption again has physical grounds only for a horizontally layered Earth with no lateral velocity variations. The rays sketched in fig. 3.16 on the left, would be the 'image-gather rays' (Robein, 2003, p. 325) only if time migration was 'perfect'. In real life, where dipping reflectors and lateral velocity variations occur, this may be *a very crude approximation that yields mispositioning and non-physical meaning of the derived velocity fields* but has the practical advantage of 'forcing' gathers to be as flat as possible.
- Next, and despite these approximations, a classical or preferably two-term velocity analysis is run on these gathers. Picking may be performed manually on gathers output on the initial grid. Nowadays however, they are performed *automatically on a very dense grid*, at each mid-point location if required. This high-density, high-resolution automatic picking ('HDHR velocity analysis' in short) nevertheless requires a close quality control of the result.
- The picks, whether performed by the interpreter or automatically, yield the new migration velocity values.

Some options – and actually combinations of them – are possible at this stage:

- The stacking velocity analysis can be classical, i.e., scanning the V_{nmo_eff} term only, and the process stopped or iterated: these are the one or two iterations featured by an arrow on the right of fig. 3.15.
- Another option is to 'freeze' the V_{nmo_eff} value either at the first run or after the first iteration and to search for the second term. This yields the second function $\eta_{eff}(t_{vert})$, in conjunction with the $V_{nmo\_eff}(t_{vert})$ function.

- A third option, called *'bispectral analysis'*, consists in a three-dimensional search whereby semblance values are computed for each t_{vert} as a function of both V_{nmo_eff} and η_{eff} or alternatively τ_0 and dtn. Fig. 3.17 is a display of a semblance volume of one CIG, input to automatic picking where the two parameters scanned are dtn(t_{vert}) and $\tau_0(t_{vert})$, presented in fig. 3.13. The result of an automatic picking on such an object will be a series of picked events, each event having its time, dtn and τ_0. Once all CIGs have been picked, we obtain a *volume of picked events* (often called the 'skeleton') and two volumes of the parameters dtn and $\tau_{0,}$ respectively. Remember that the result of this two-term scan can be expressed as the two other parameters V_{nmo_eff} and η_{eff}.

- A fourth option, the 'scanning route', consists in testing a series of percentages of the current V_{nmo_eff} and/or η_{eff} functions. It is described in fig. 3.18. The selection of the optimal percentage may be performed by the processor either on the corresponding image gathers or by interpreting a selection of 'velocity sections'. This may be achieved automatically or, as suggested in the figure, visually by an interpreter. This last option is used in difficult cases, often for land data where the interpreter's eye and brains are the only tools to compensate for poor data quality! This exercise yields 'volumes' of correcting factors to apply to the initial velocity field in order to compute the final one. Fig. 3.19 is an example of PreSTM sections resulting from three different percentages of a V_{nmo_eff} field derived after a first processing sequence as in fig. 3.15. Note how reflectors focus and defocus according to velocity: this is the effect of velocity on stacking. Reflectors also move or 'migrate' in our terminology, to different locations, illustrating the impact of velocities on migration itself.

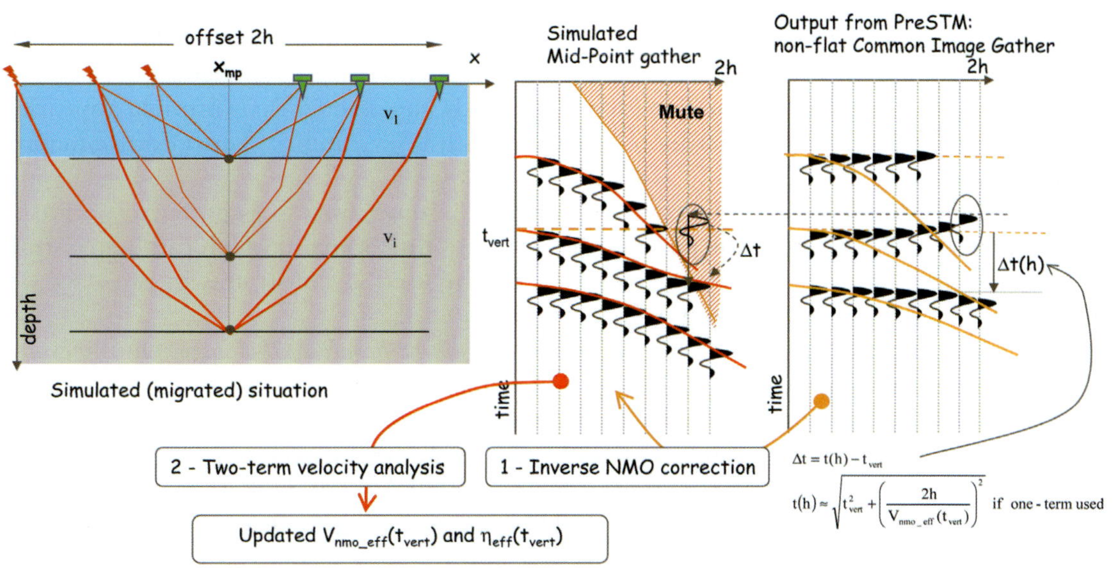

Figure 3.16: Normal Move Out Velocity Analysis after a first step of PreSTM. CIG are 'inverse NMO corrected' before undergoing a new two-term (or bispectral) velocity analysis.

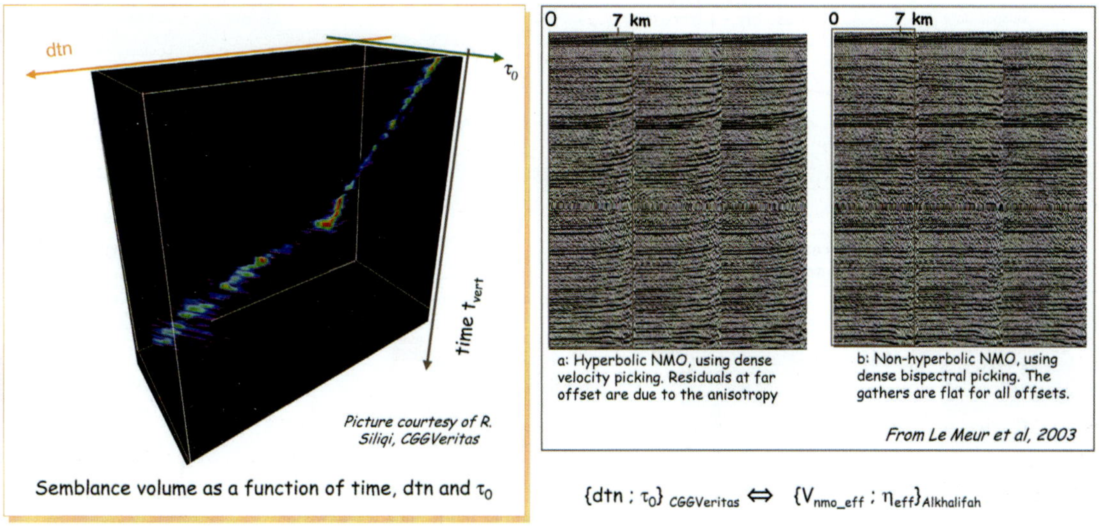

Figure 3.17: An example of bispectral analysis. Display of the semblance volume at a given velocity analysis location as a function of the two parameters τ_0 and dtn. Manual or automatic picking of the maximum semblance at each time sample t_{vert} yields the optimal dtn(t_{vert}) and $\tau_0(t_{vert})$. These parameters may be transformed in terms of $V_{nmo\_eff}(t_0)$ anf $\eta_{eff}(t_0)$. Note the limited 'leakage' between the two scanned parameters.

The processing sequence carries on as follows:

- The estimated V_{nmo_eff} and η_{eff} functions are interpolated at the final x, y and t sampling rate. PreSTM operators are computed using either ray-tracing, shifted hyperbolas' equation or Alkhalifah's formula and applied to the data to obtain the final PreSTM image gathers. Note that in the anisotropic case and for the ray-tracing approach, the inverse Dix's formula appearing in fig. 3.09 needs to be extended in order to transform both V_{nmo_eff} and η_{eff} from effective to interval values.
- This transformation has other objectives. For instance, some processors prefer to perform the editing and smoothing of the raw picks in the interval value domain, which can be constrained to follow geological layers. Interval values may also be used as constraints in pre-stack depth migration tomography, as discussed in chapters 5 and 8.
- In most cases, the stack along the offset dimension is *not performed at this stage*. On the contrary, all image gathers are preserved and undergo a final automated *Residual Move Out*. This final 'fine tuning' of the image gather flatness should be performed only if the corresponding correction is small, since data are not re-migrated. The RMO correction may be performed as described in the next paragraph and fig. 3.22 or as is more common practice, again by a reverse NMO followed by a one- or two-term final HDHR analysis. The corresponding fields of 'residual parameters' are part of the outputs of the processing, together with the V_{nmo_eff} and η_{eff} that were used for the final migration.

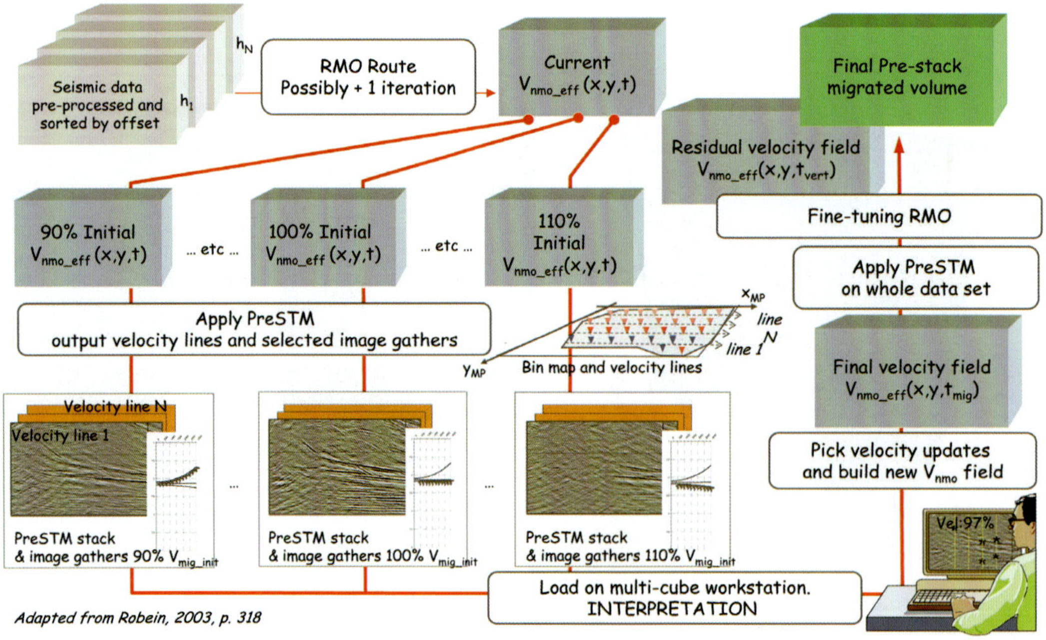

Figure 3.18: The 'scanning route'. Once a first guess of migration velocity and/or eta are estimated, their pick may be fine-tuned with a scanning technique in which the current velocity field is multiplied by a set of scalar values. PreSTM are re-run as many times and results are interpreted and the 'best' multiplying factors are picked manually.

The fields that result from automatic high-density picking include a number of artefacts, due to interferences, noise or multiples. As mentioned earlier (fig. 3.08), *smart filtering and smoothing of these raw picks is necessary* before we obtain the final ones. The V_{nmo_eff} section at the bottom right of fig. 3.19 illustrates the level of smoothness generally accepted for these fields. Remember that this field shows the juxtaposition of the velocity functions $V_{nmo\_eff}(t)$ that are used, each assumed to be laterally constant within the aperture of the migration operator. It should thus be seen as the 'multi-1D' velocity field. It is clear that this velocity should not vary too much laterally within the aperture. This 'rule' is a guide for the degree of smoothness imposed to the PreSTM migration velocity field.

Fig. 3.20, taken from Petroleum Geoservices Tech link publications, presents a more recent example. The top left-hand image shows a section of the current PreSTM stack overlaid by the 'skeleton' of all events picked for a maximum semblance and a time slice in the volume of the same parameter. A joint optimization of V_{nmo_eff} and η_{eff} is performed in this case, using a *simulated annealing* approach. A time slice through the 3D field of automatically picked η_{eff} values is shown at the bottom right of the figure. Finally, the results of editing, filtering, interpolation and smoothing of these raw picks are displayed on the bottom left. This is the field of values that will be used in the next iteration of PreSTM.

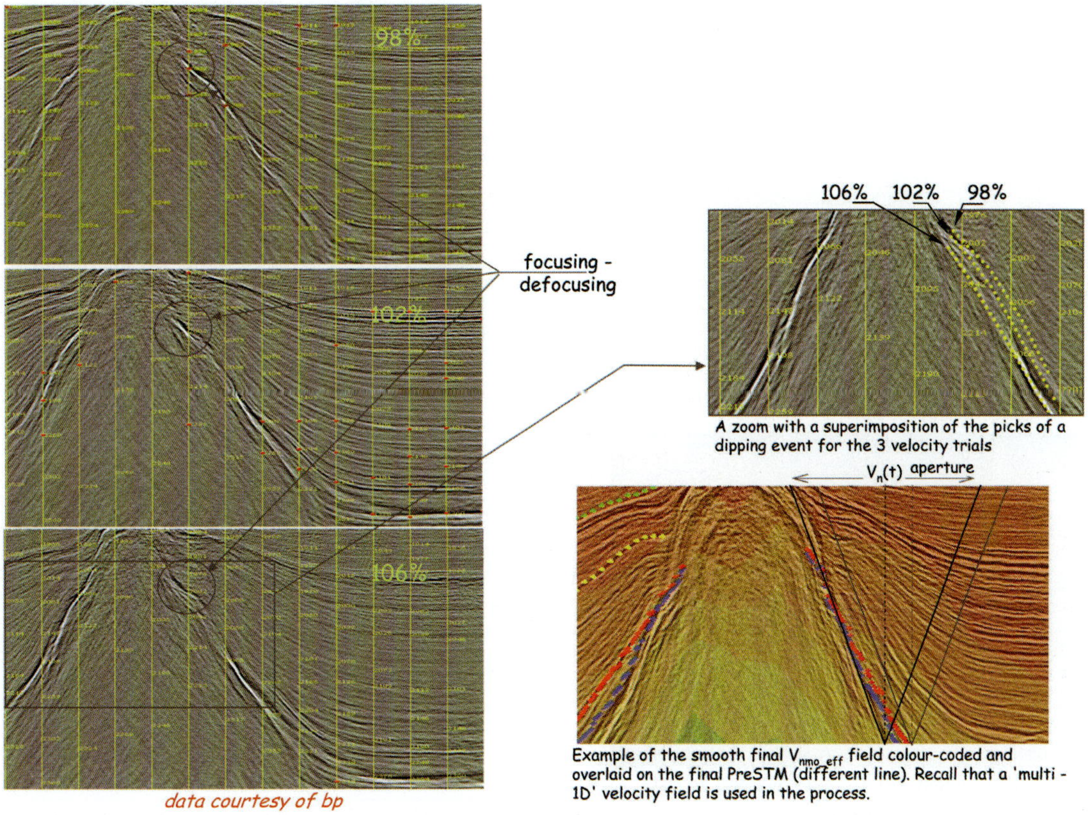

Figure 3.19: An example of the 'scanning route'. A series of percentages are applied to the current V_{nmo_eff} and 3D PreSTM is run with these migration velocity fields. One section is displayed for 3 percentages. Note both focusing/defocusing and 'migration' of dipping events with velocity.

Another example of both V_{nmo_eff} and η_{eff} fields resulting from high density automatic bispectral picking is shown as the left column of images in fig. 3.17. The figure also illustrates the possibility to transform the 'effective' fields into 'interval' fields of V_{nmo} and η parameters. The skeleton helps in defining what are the intervals of interest. As for Dix, this transformation may be very unstable for small intervals. Consequently, more stable ways to derive interval values are used, but these values need to be taken with care anyway.

These examples illustrate the high density at which events are picked nowadays and the degree of smoothness generally selected for such fields.
As a conclusion at this stage, the end-product of these different steps is multiple:

- A smooth version of the V_{nmo_eff} field, often referred to as 'migration velocity';
- a smooth version of the η_{eff} field or 'anisotropy parameter', also actually used to migrate the data;
- the two 'residual' V_{nmo_eff} and η_{eff} fields applied as the final fine tuning;
- a selection of corrected CIGs;
- and of course, the PreSTM migrated volume!

Principle of RMO

Instead of applying an inverse moveout correction followed by velocity analysis and new corrections, it is possible to run a residual correction directly. This is based on the assumption that

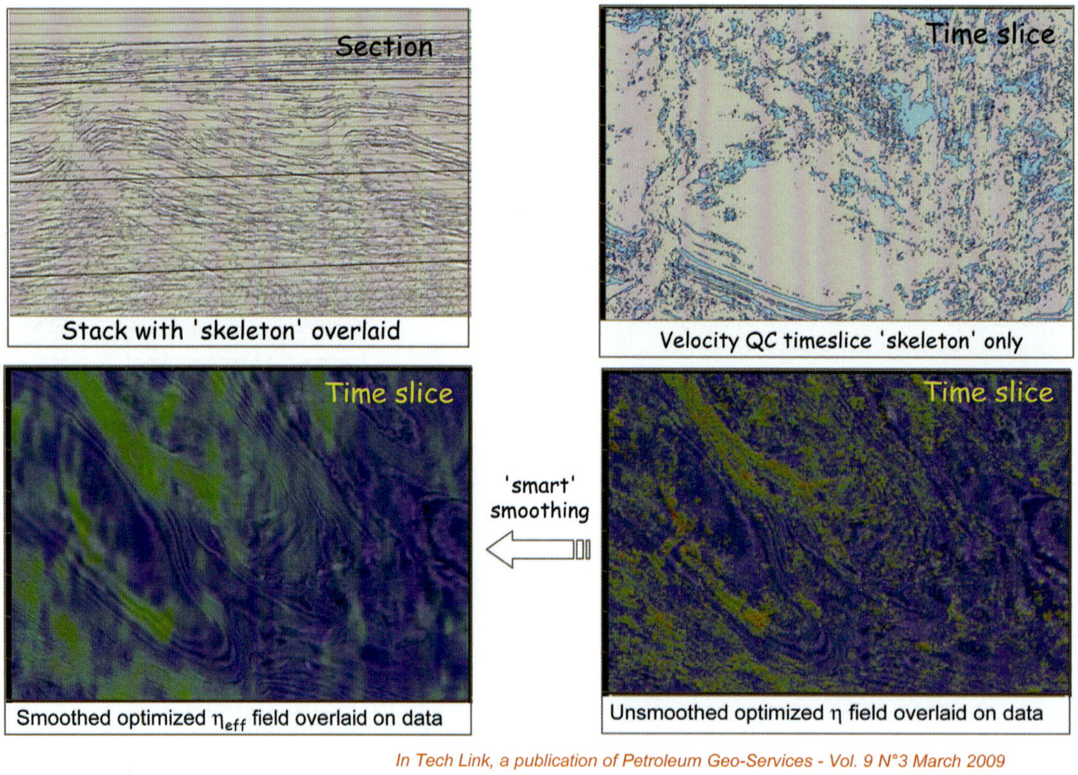

Figure 3.20: Automated Continuous PreSTM anisotropic velocity analysis process example from PGS; V_{nmo_eff} and η_{eff} parameters are estimated together, using simulated annealing.

the non-flatness of image gather can be parameterized by a function of one or two parameters. The 'classical' one parameter option is illustrated schematically in fig. 3.22. In this case, the non-flatness is assumed to be a parabola described by a unique parameter p. The RMO is very similar to the conventional stacking velocity analysis. Semblance is computed for each t_0 along hypothetical curves for different p values. The results are plotted on a 'semblance panel'. Maxima as a function of t_0 are then automatically picked to obtain the $p_{opt}(t_0)$ function, which optimizes the flatness of the gather. This correction is applied to the data, which can then be stacked.

In the 'higher-order RMO analysis, the residual moveout may be a 4th order polynomial of the type shown at the top of fig. 3.22, for instance.

When applied as the last step of fine tuning, the $p_{opt}(t_0)$ field can change rapidly from trace to trace, contrary to the V_{nmo_eff} and $\eta_{eff,}$ which should be smooth.

Benefits and limitations of PreSTM

It should be clear from the previous paragraphs that PreSTM can be performed with limited human interaction. This is thanks to the intrinsic assumption that velocity does not change laterally in the migration aperture. This assumption allows for the use of a simplified description of the propagation velocity field in the subsurface, which can be assessed quasi-automatically even though 'smart' and sometimes 'heavy' smoothing, controlled by processors, needs to be applied.

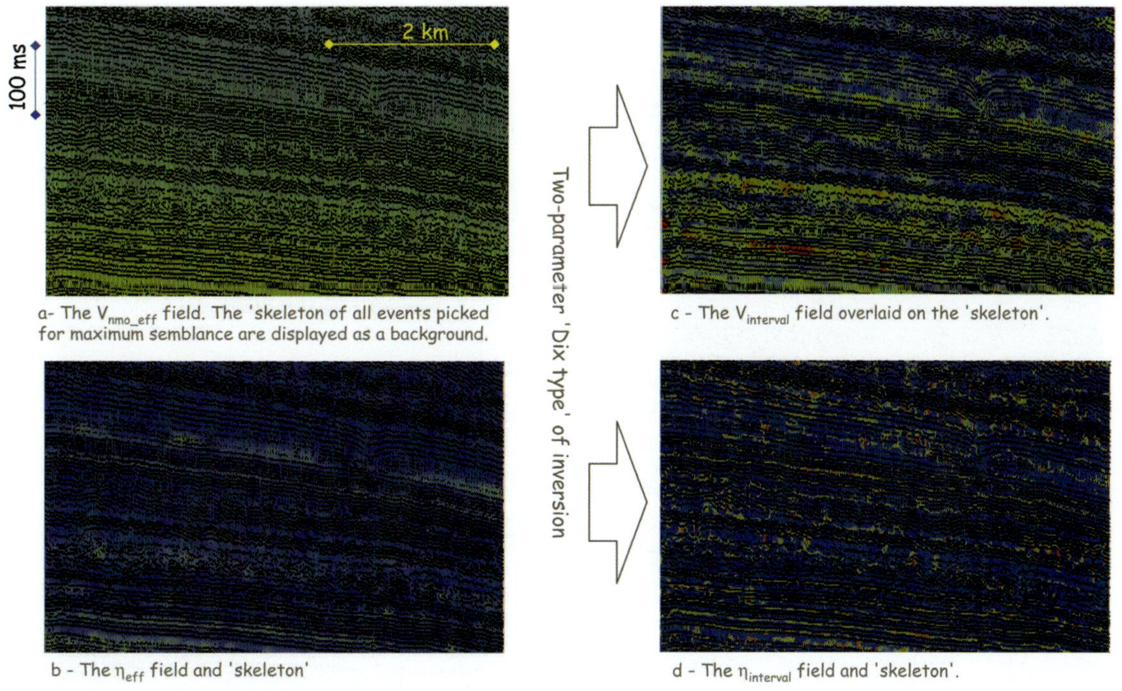

Figure 3.21: Anisotropic PreSTM yields two pieces of information: the V_{nmo_eff} and the η_{eff} fields. These fields can be transformed into interval V_{nmo_int} and η_{int} fields, using an inversion process. Note the high density of picked events.

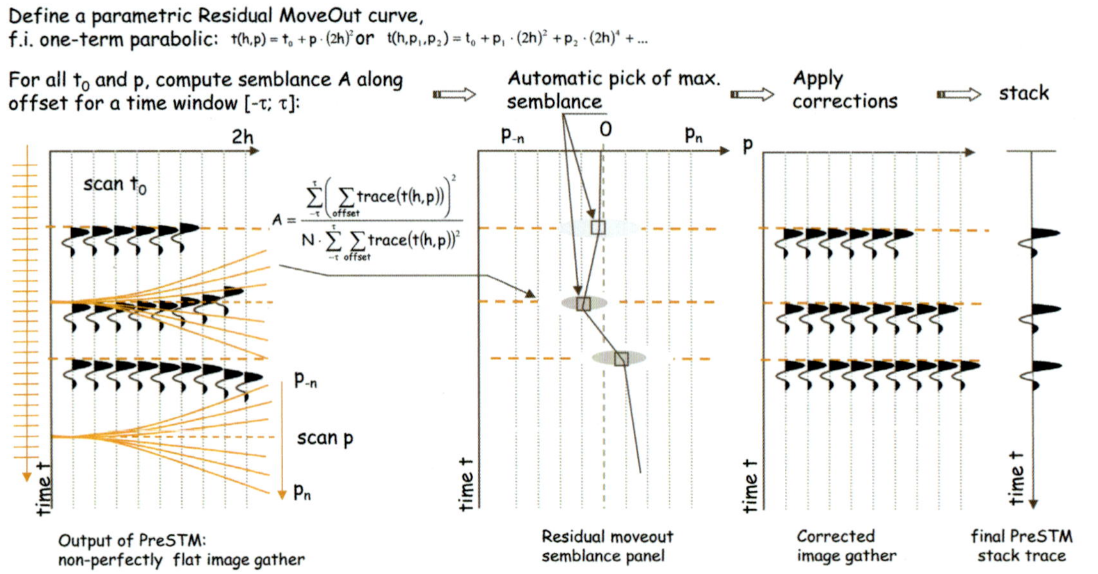

Figure 3.22: Principle of Residual Move Out Velocity Analysis for a one-term correction. Curves based on two-term equations are used more and more. The residual semblance becomes then a 3D object into which a 'bispectral' analysis is performed.

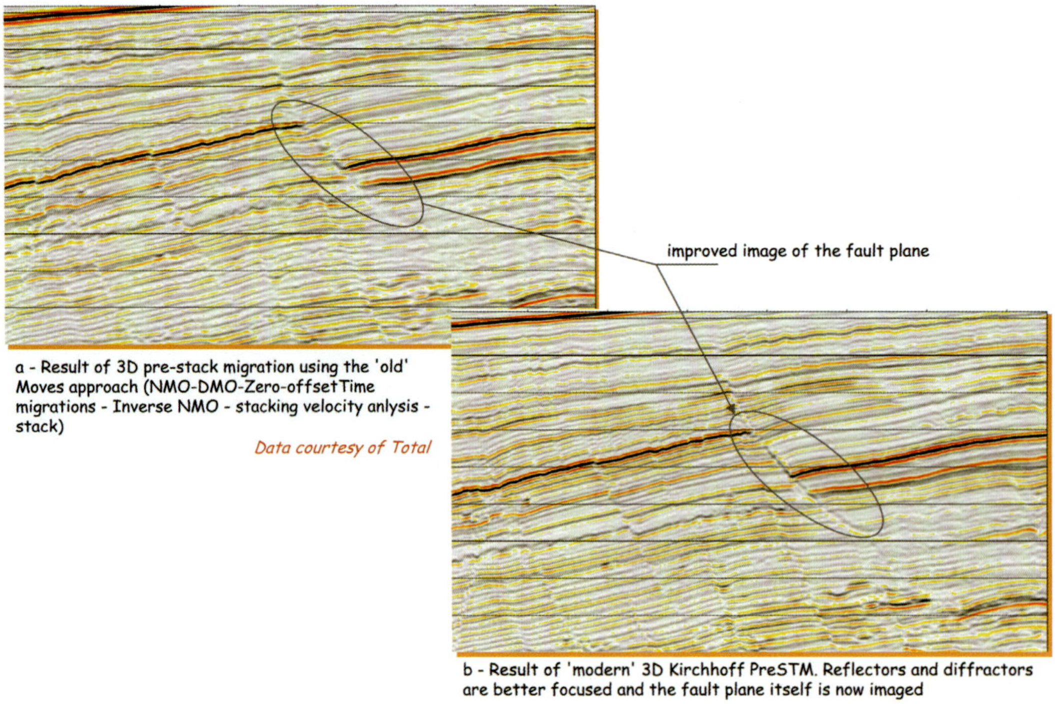

Figure 3.23: Benefit of Kirchhoff PreSTM, compared with the 'old' Moves technique.

Recent developments of the so-called 'anisotropic PreSTM' did not add a lot in terms of complexity to the way migration velocity fields are established, while leading at the same time to improved images for large offset surveys, strong *vertical* heterogeneity and the presence of steep dips. Fig. 3.23 illustrates improvements in terms of image quality due to continuous technology developments. Here we compare the 'old' pre-stack migration (panel a) with modern PreSTM (panel b) with high-order operators. Note the increase in vertical and lateral resolution; reflectors and diffractors along faults are better 'focused'; the steeply dipping fault plane itself is now imaged as a reflector.

This is because PreSTM solves to some extent the issue of conflicting dips. This point is schematized in fig. 3.24. The cartoon shows two events arriving at about the same time for short offsets but corresponding to reflectors at different dips (here a salt flank, but it can be a fault plane, as in fig. 3.23). They exhibit different move-outs because the wave imaging reflections at A and B travel through different parts of the Earth, where propagation velocities may be very different (discarding for the sake of argument the dependence of moveouts on dip). Rays reflecting on point 'A' for instance, travel through deeper and faster layers.

PreSTM reduces ambiguity, as the reflections are now displayed in a migrated location. They appear on gathers located at approximately the vertical of the reflection point. Move-out analysis and RMO flattens these gathers as much as possible, hence improving the final images of reflectors in A and B. This explains why the fault plane is well imaged in fig. 3.23 and why PreSTM is still so popular.

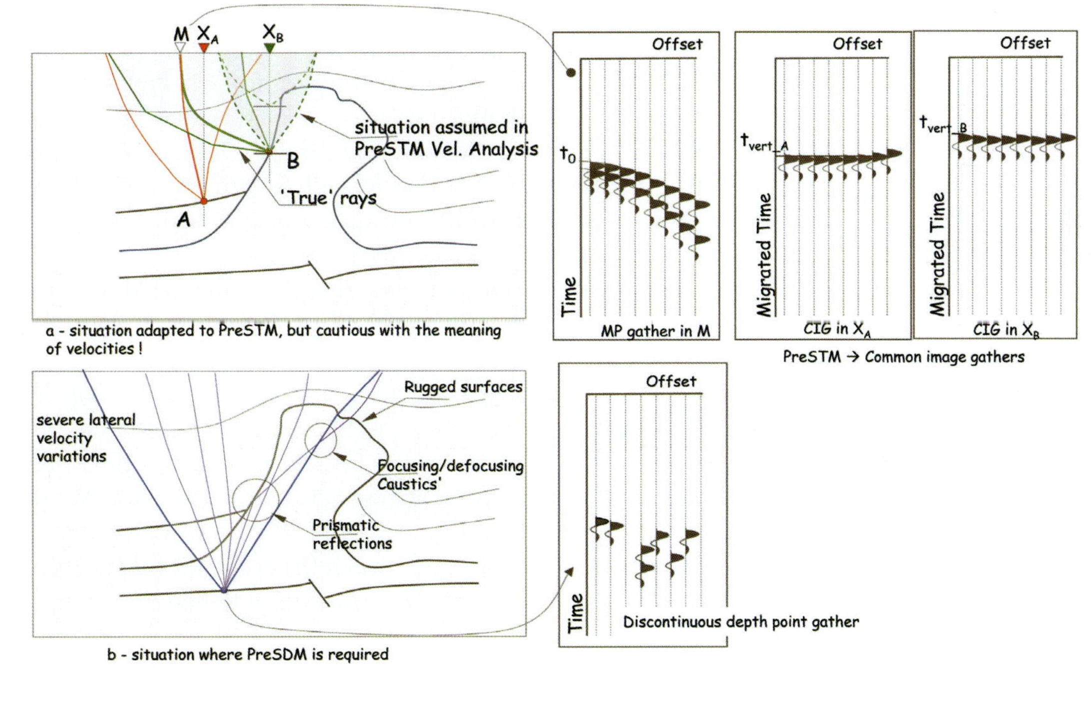

Figure 3.24: PreSTM will remove ambiguity when several arrivals are coincident on mid-point gathers. The multi- 1D assumption however does not allow to measure 'true' geological velocities. PreSTM fails altogether in complex geological settings.

These intrinsic benefits however, hide serious drawbacks, often overlooked. They are of three types:

- *Mispositioning* of markers in the migrated image: as soon as there is some lateral velocity variation in the subsurface, i.e., almost all the time, seismic reflectors are not in their targeted vertical time location.
- *Migration velocities remain 'processing parameters'* and must be taken as such when we need to interpret them in a 'geological sense'.
- *Poor imaging below complex structures*

The causes of these limitations are to be found in the basic assumption of PreSTM, illustrated again in fig. 3.24. Looking at the reflection at point B, we note that PreSTM operators are computed in the aperture zone sketched by rays in green dotted lines in the figure, assuming moreover no lateral velocity changes within this aperture. The true propagation took place however along the rays sketched in green continuous lines: the actual waves travel through a different part of the Earth (possibly with lateral velocity changes) and it is this part of the Earth that the move-out times that control our process carry information from.

The mispositioning issue is not new but there are probably still wells missing their target because of it. We illustrate the phenomenon in fig. 3.25 with a North Sea case published by Gancarski and Kaculini in 1998. Here, the well was aimed at testing a down-thrown compartment of a field and was designed on the time migrated image, as was current practice at the time. Results from the well were unexpected. In fact, it most probably hit the up-thrown compartment, as analysis of the projection of the well in the time-migrated domain showed afterwards. This was possible thanks to the development of a Total's in-house time-migration simulation code, unfortunately completed after the well was drilled (Kaculini and Arnaud, 1997)!

In fig. 3.26 we use a more caricatured, though real, example to illustrate points one and three above. The image at the top left comes from a 3D PreSTM and on the bottom right a 3D PreSDM in a salt tectonics geological setting. On the right and left of both images, lateral velocity variations are limited; images are of equivalent quality. The main difference is in the poor rendering of the relative thicknesses of layers in the time image. The pull-up (see also fig. 3.01), due to high velocity in the salt body, is particularly visible on the right. Although these artefacts will be compensated for in the time-to-depth conversion step that will come afterwards, they may still mislead interpreters.

In the middle of the section, lateral mispositioning and poor imaging due to PreSTM artefacts are much clearer. The image of the salt base in the PreSTM is visible but firstly poorly focused and secondly, totally mispositioned. We have drawn hypothetical vertical well trajectories on the figure to emphasize difficulties in well-to-seismic ties. Markers in depth marked as coloured dots are (visually) mapped back onto the PreSTM image, showing clearly that the projection of the well trajectory in migrated time is a long way from the vertical stretch generally used! In fact, the well trajectory is broken. This is because the mispositioning depends on the complexity of the true propagation velocity, on the PreSTM optimal migration velocities and as importantly, on the dip of the reflectors. This last point is often ignored.

As a consequence, vertical time-to-depth conversion does not work in these circumstances and more sophisticated techniques need to be used. The reader may refer, for instance, to Robein (2003, pp. 392-394) for a discussion on this topic.

Fig. 3.26 is a case of strong lateral velocity variation, but less severe ones may still have serious consequences, as mentioned earlier. The message is then not to be surprised sometimes by difficulties in tying wells to seismics; a dipping water bottom, such as the one noted again in fig. 3.26, may suffice to introduce serious mispositioning, especially for dipping or deep reflectors, just to name one classical pitfall.

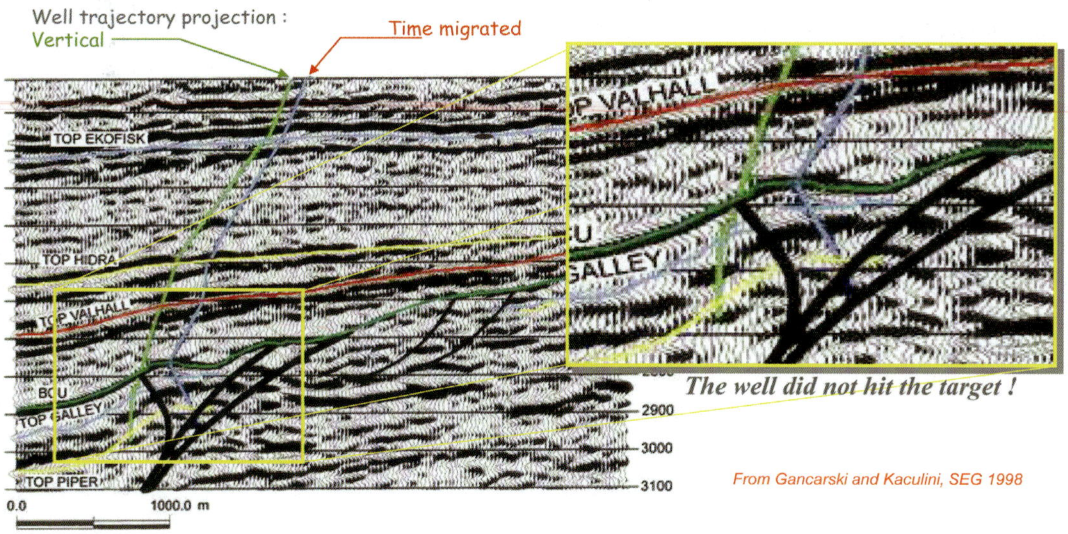

Figure 3.25: The vertical axis of time migration is not 'vertical time' in many cases. This needs to be taken into account in well-to-seismic tie and well design based on time-migrated images.

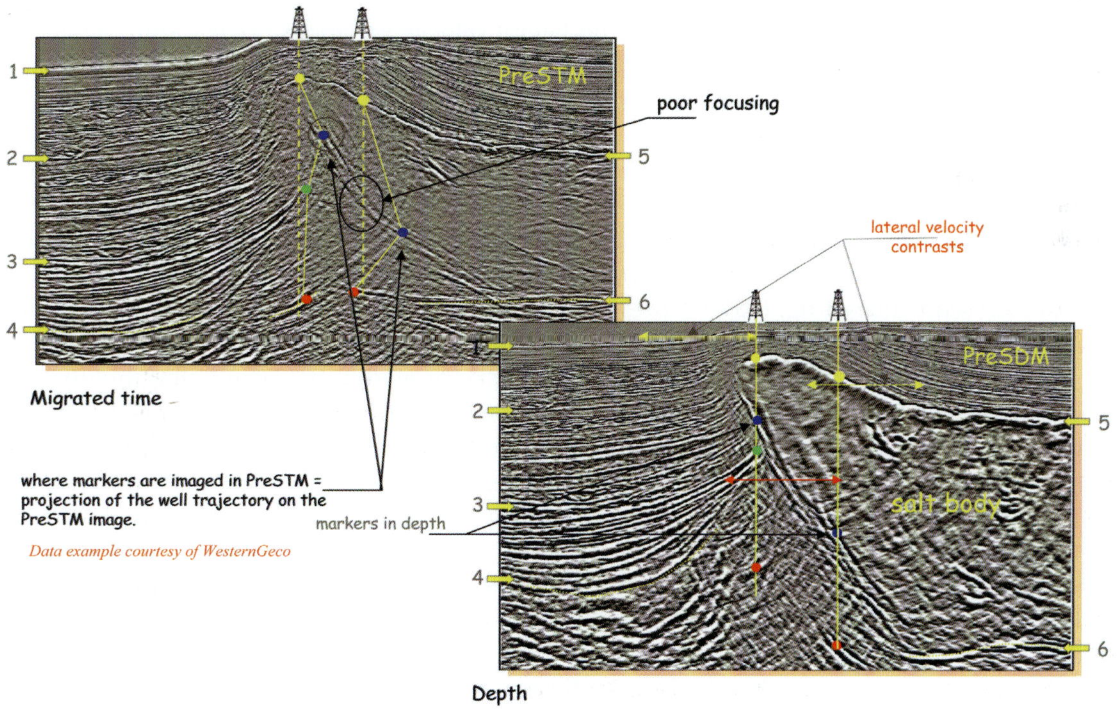

Figure 3.26: Artefacts in PreSTM. On the left and far right of the PreSTM section, assumptions are met: note however the relative thicknesses of layers in PreSTM versus PreSDM (salt body pull-up). A mapping of events picked on the PreSDM image back to the PreSTM one gives an idea of where we should draw the well trajectories on the latter and of the amount of mispositioning.

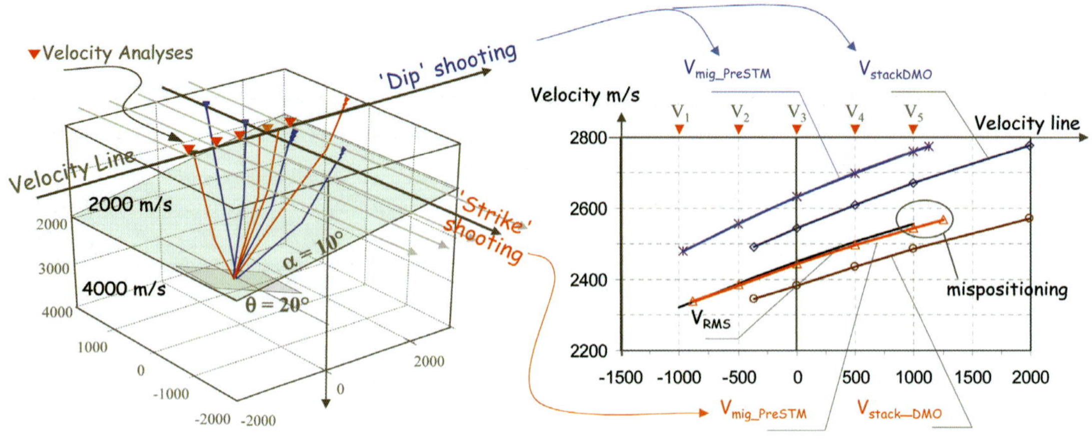

Figure 3.27: In PreSTM, migration velocities depend on the shooting direction. Here, PreSTM migration velocities for a simple two layer isotropic model are computed assuming either 'dip' or 'strike' shooting. V_{rms} and the 'old' stacking velocities after DMO are displayed as references. In this case, dip velocities are as much as 200 m/s higher than V_{rms}. Strike shooting yields velocities close to V_{rms}, but slightly mispositioned.

Fig. 3.27 addresses the issue of the meaning of PreSTM velocities. Migration velocity fields are optimized to flatten image gathers, assuming that no lateral velocity variations occur within the aperture. In theory, such a perfect flatness could hardly ever be achieved with this assumption. In real life, however, we deal with limited bandwidth data, which means that flatness accuracy may be limited to a few milliseconds or more. Introduction of the η_{eff} parameter leaves another degree of freedom to force flattening. Finally, remember that high-order residual moveout correction is applied systematically to compensate for non-perfect flatness. As a consequence, although PreSTM velocities are computed in a quasi-migrated position, they are still 'provelocities' (for processing velocities, Al Chalabi, 1994), as we used to call stacking velocities in the 'DMO' time of the 90s.

On the left-hand side of fig. 3.27 we present a two-layer model with an interface and a reflector dipping in opposite directions. We test two shooting directions: dip- and strike-shooting. Maximum offset is close to reflector depth. Migration velocities are computed using an in-house tool that simulates in this case the kinematic behaviour of isotropic (one term) PreSTM. Optimal migration velocities are computed at five locations along a velocity line placed in the in-line direction. Results are displayed on the plot on the right of the figure. Note the large difference between PreSTM velocities measured when shooting in the dip direction and velocities when shooting in the strike direction.

In the former case, velocities depart a great deal from rms velocities that are also plotted in the figure. In the latter case, velocities are close to the vertical rms velocities. This is because the corresponding rays (in blue) do not 'see' any lateral velocity variations, contrary to the dip shooting case (red rays). The velocities are nevertheless not exactly equal to the rms velocities because we are dealing with 3D migration operators; a close look at the figure would show that image-gather rays do not hit the marker exactly at the same location on the reflector.

In real life and complex geological settings, pure dip- or strike-shooting never exists and the results of PreSTM Migration Velocity Analysis are a mixture of both: velocities depend on the acquisition lay-out; they cannot therefore be considered as 'geological' velocities. The message to be obtained from this simple example is that PreSTM velocities are to be handled with care when using them, for instance, in time-to-depth conversion or pore-pressure prediction projects.

A last point is that since operators are computed in 1D models, PreSTM cannot account for multi-pathing and caustics, terms and concepts that we will develop in chapters 4 and 5.

Finally, despite of all the benefits that make PreSTM still a widely used tool, in complex geological settings and in the presence of lateral velocity variations and/or anisotropy, it may be impossible to obtain a good stack at all: PreSTM reaches its limits. This is sketched at the bottom of fig. 3.24 and illustrated by the real case of fig. 3.26. The alternative for obtaining a good image from the subsurface is then pre-stack depth migration, starting with the ray-based techniques addressed in the next two chapters and following with the Wavefield Extrapolation methods discussed in chapters 6 and 7.

4 Ray-based Kirchhoff Migrations and Tomography[1]

As we have seen in the previous chapters, migration deals a lot with 'propagating' events back to the location where the reflection happened. For this, we need two things. Firstly, techniques to simulate propagation and 'back-propagation'; secondly, an explicit velocity model in which to perform the propagation.

The velocity model may be very complex, as on the real example in fig. 4.01. This is a necessity since it is precisely when the geological setting is complex that we need PreSDM.

The techniques used to propagate waves are of two classes: ray-tracing on the one hand and direct integration of the wave equation on the other. This fourth chapter deals with Kirchhoff migration, a method that uses the first class of techniques. In the second part of the chapter, we present different ways to build propagation velocity models, again using ray-based methods. These are known under the generic name of 'tomographic inversion' or simply 'tomography'. We conclude the chapter with some important considerations related to diffraction imaging and the three-dimensional character of seismic imaging.

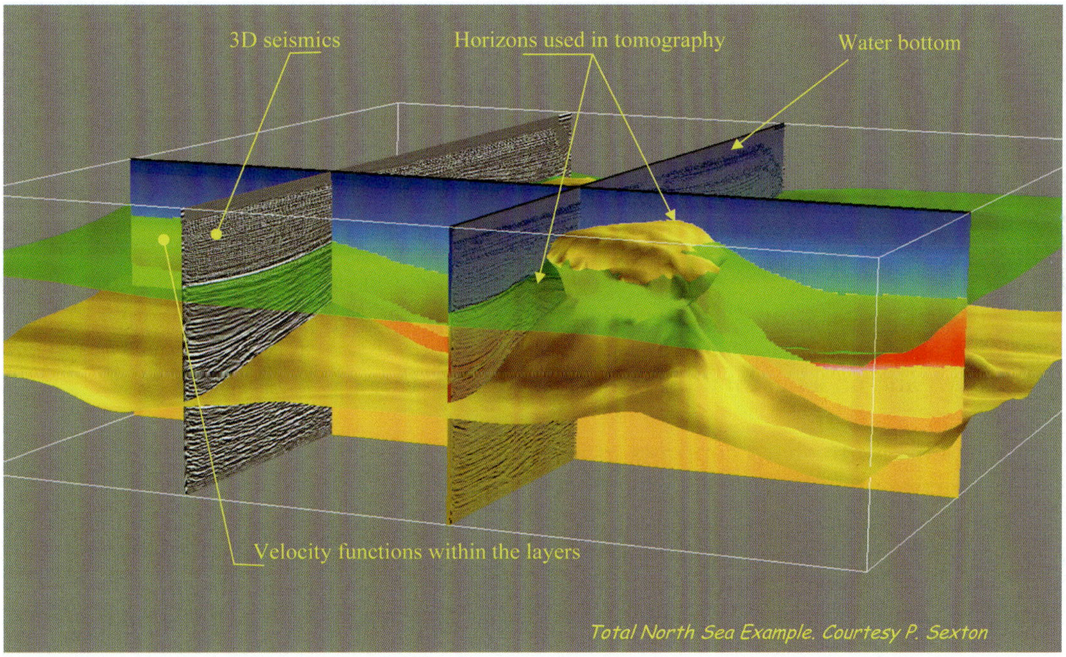

Figure 4.01: PreSDM requires an accurate velocity model and the right algorithms to migrate the data. This image is the result of layer-based tomography and Kirchhoff PreSDM. Some of the horizons used in tomography are displayed here as surfaces; velocity values are colour-coded and superimposed to the seismic image.

1 The paragraphs on tomography have been kindly co-authored by **Pierre Hardy**.

Principles of Kirchhoff PreSDM in the common-offset domain

Principles of Kirchhoff migration based on the summation along diffraction curves have been extensively discussed in chapters 2 and 3. Fig. 3.05 in particular presents the general PreSTM scheme. Fig. 4.02 shows a similar scheme but now for Pre-Stack Depth Migration. The similarities are the following:

- Data are sorted into several common-offset 3D volumes. Each of these volumes is 'regularized' to contain traces of comparable fold and possibly azimuth range, placed on a regular grid in x and y.
- Each point 'D' is processed as if it was a diffraction point. Reflectivity in any point 'D' is computed by a weighted sum along its corresponding diffraction surface; the limits of the diffraction surface are governed by the aperture of the migration operator (fig. 3.04).
- Each common-offset volume is processed independently but with the same velocity model; we keep the offset dimension and compute common image gathers (CIG) at predefined locations. They will be used for quality control and tomographic velocity model building.

However, the key differences are:

- The need to *build an explicit propagation velocity model in depth*. It can be as complicated as necessary and should include anisotropy.
- The computation of traveltimes and weights (the 'Green's functions') for each imaged point considered as a diffraction point 'D' by explicitly shooting dynamic rays through this 3D velocity model in depth.
- The result of the Kirchhoff summation along diffraction curves is *placed in the common image gathers directly at its corresponding depth* Z_D *and offset of origin 2h*.
- The vertical axis of image gathers is therefore depth, as is the axis of the migrated trace that is the sum along the offset of the image gather.

The diffraction surface for a given point D at depth z, aperture x and offset 2h is computed from the sum of the traveltimes t_s along the ray from the shot to the imaged point plus the traveltime t_r from the imaged point to the receiver, in a way similar to the one shown on panel 5 of fig. 3.08. The key difference however is that traveltimes are computed by ray tracing performed *through complex 3D models* and no longer multi-1D ones, as for PreSTM.

Ray tracing is sensitive to sharp contrasts in velocity and high incident angles. It is consequently necessary in practice to smooth the velocity model. This is illustrated with a synthetic 2D example in fig. 4.03. The desired model might be complex, as the one shown in panel (a) of the figure but rays will be computed through its smoothed version presented in panel (b). Some rays and wavefronts shot from a point Xs at the surface are shown, which illustrate the following points:

- The complexity of rays and wavefronts in complex models. Wavefronts computation is a classical alternative to ray tracing. Note wavefronts that are re-interpolated as they develop in depth on the figure.
- More than one ray may link two points, for instance Xs and D in the figure. The consequence is that *traveltimes may be multivalued* in such complex models. For multi-values to occur, rays must cross at some locations in the subsurface. Such points, like D in fig. 4.03, panel (b), are called *caustics*. They are commonplace in complex models.
- Smoothing is necessary but has an impact on the PreSDM result, since it modifies the time-to-depth conversion part of migration. To limit these effects, firstly, smoothing is performed in the slowness (1/velociy) domain and secondly, although ray paths are computed in the smoothed

velocity field, traveltimes along these rays are computed using the un-smoothed one. The degree of smoothness applied and the method remain important parameters to control.

The consequence of the presence of caustics on Kirchhoff migration is that the diffraction curve can be multi-valued, as schematized in the bottom left panel of fig. 4.02. Multi-valued diffraction curves are not straightforward to handle in Kirchhoff summation. Three main options exist in commercial codes for this:

- use the shortest time only, well adapted to traveltime computation based on integration of the *eikonal* equation; the eikonal equation is the version of the wave equation that accounts for traveltimes only (kinematic part);
- use the shortest travel distance; this trick avoids possibly complex ray paths, especially in salt tectonics, which may have a weak energy;
- select and use the branch that corresponds to the highest energy;
- effectively handle multivalues.

The last two options require a special effort in ray tracing and data storing in the computer. The second option is the most popular one. The third one is costly in terms of code performance and not often applied.

Poor handling of multivalued diffractions is *one weakness of Kirchhoff PreSDM*.
Fig. 4.02 and 4.03 are in 2D, but we come back at the end of the chapter (fig. 4.19) to some specificity of the computation of Kirchhoff migration operators when dealing with 3D data, as in most real-life cases.

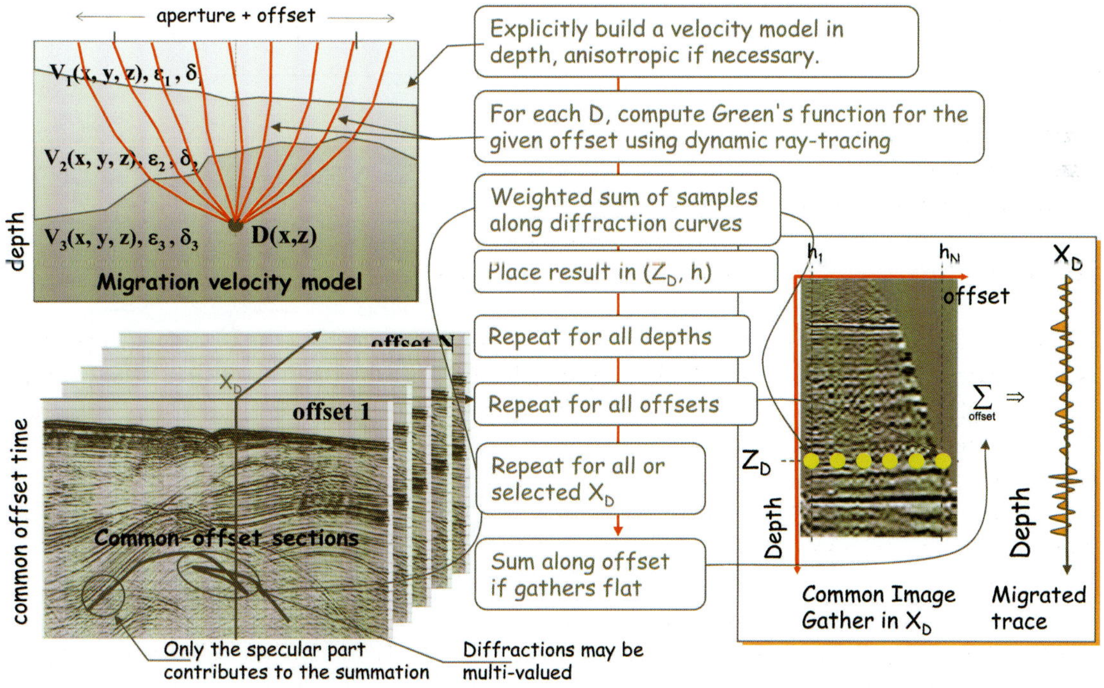

Figure 4.02: Principles of Kirchhoff Pre-stack Depth migration. The first step consists in building image gathers using ray tracing through the current velocity model. Traces are stacked if the gather is flat.

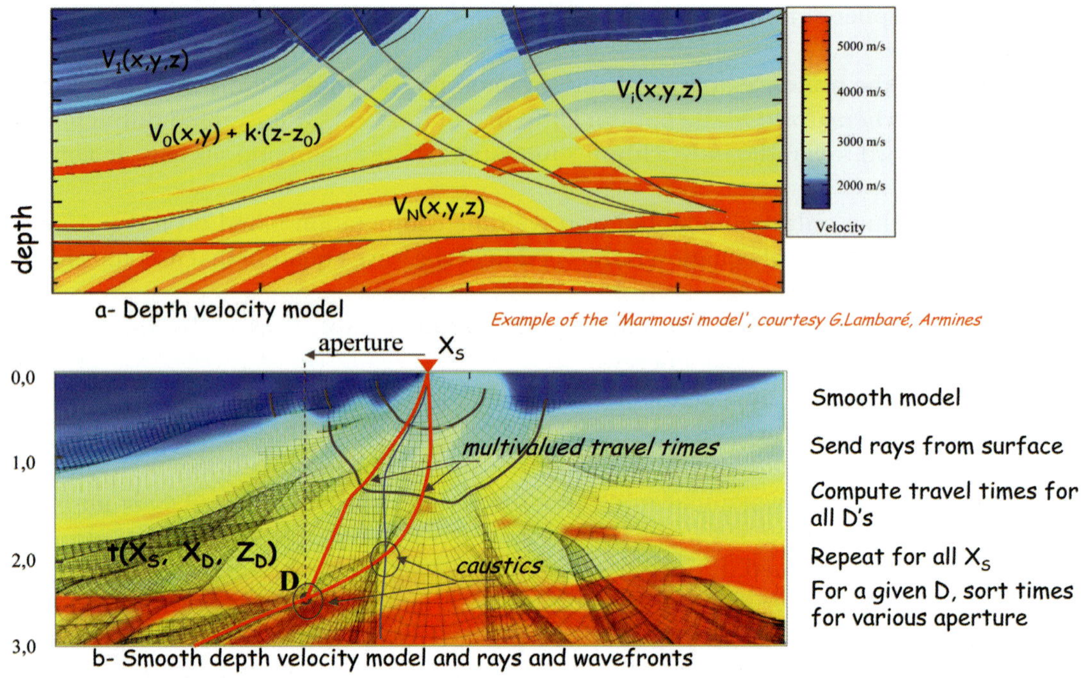

Figure 4.03: Computation of travel-time in depth imaging. In complex models, several rays may link a point in depth and a point at the surface (caustics), leading to multivalued diffraction curves.

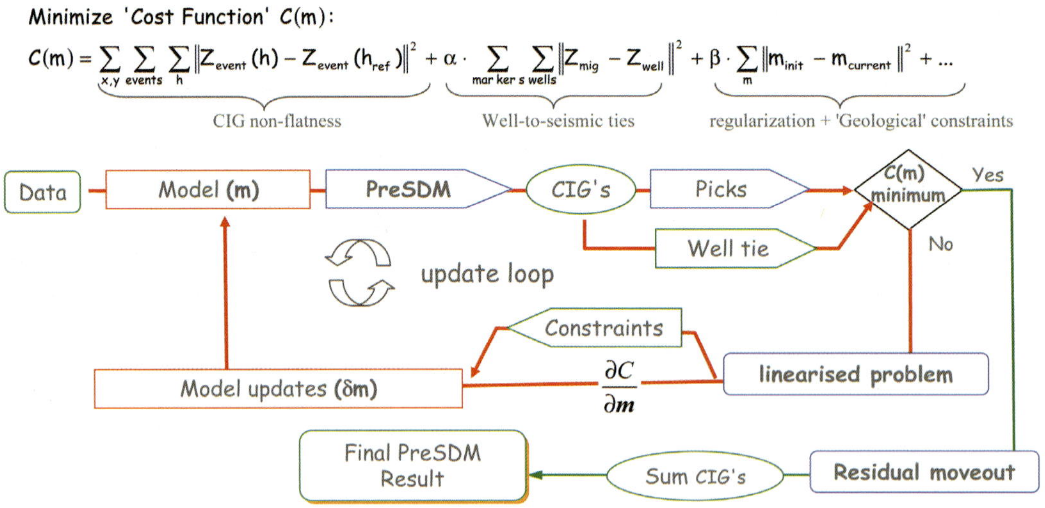

Figure 4.04: Principles of linearized tomography. A first model is set up. PreSDM is run with this model. Common Image Gathers are analysed in terms of event flatness. The model is updated in such a way as to reduce the cost function C(**m**). The process is iterated until C(**m**) reaches a minimum value. A PreSDM is needed at each iteration.

Image domain-oriented versus data domain-oriented implementations

Reverting to fig. 2.08, we observe that the sequence of actions for Kirchhoff Pre-Stack Depth Migration illustrated in fig. 4.02 is of the *'image-domain oriented'* type. This means that the image is built point by point. The location of all data samples contributing to this point needs first to be computed; it is the explicit expression of the diffraction surface. Next, samples must be searched one after the other in the whole data set, which is generally huge (possibly hundreds of terabytes).

In the less intuitive data-oriented approach, we take conceptually one sample after the other as it is sorted in the data volume. Each sample has its acquisition 'properties' (shot and receiver coordinates or alternatively mid-point coordinates, offset and azimuth) and traveltime. The problem to solve is to find to which locus this specific data point 'migrates'. In the simple case of chapter 2, this locus is a circle in 2D or a sphere in 3D. In real complex models, it is similar to a *'wavefront'* that may be complex and also multi-valued. Fig. 4.03, panel b presents examples of such wavefronts for the 2D zero-offset case.

The implementation issue in this case is no longer to search the data sample in the huge data volume, but to build migration wavefronts on the regular output grid, using large tables of pre-computed traveltimes of rays with their own 'properties' (shot and receiver locations, traveltimes, amplitudes). These tables are computed on reasonably coarse grids along the offset-azimuth-time dimensions. Fast search techniques and smart interpolations are necessary and this means serious code development challenges, especially when traveltimes are locally multivalued! Remember moreover, that we deal with dynamic ray tracing in order to compute not only Green's function traveltimes but also *weights*, which act as multiplying factors to the sample, varying along the wavefront.

Tomographic inversion of image gathers

As mentioned in fig. 4.02, the stack along offset of each Common Image Gather (CIG) produces the PreSDM trace, directly in depth. This summation yields a valid result only if events in the gather are horizontally aligned. We say that the 'CIG is flat'. As in PreSTM, the game is then to *find the velocity field that best flattens the gathers*, but now this field is an *explicit propagation velocity model defined in depth*. This goal is achieved through a 'depth migration velocity analysis' or 'tomographic inversion' or simply 'tomography'. Tomography is posed as an optimization problem: we want to find the anisotropic velocity model that minimizes a 'cost function' (also called 'objective function') that mathematically quantifies the non-flatness of the gathers while honouring specific soft, hard and geological constraints.

Constraints are important to address the non-uniqueness of the minimization problem (there can be several models that minimize the non-flatness of gathers, given the limited seismic resolution) as well as to ensure a good seismic-to-well tie with the migrated result and to input geological 'knowledge' into the process.

Fig. 4.04 presents schematically the overall tomographic process. A version of the cost function C(**m**) for a given model **m** is given at the top of the figure. **m** is written in bold letter as the *vector of all parameters* $\{m_i\}$ defining the velocity model. C(**m**) includes three terms:

- a measurement of the non-flatness of the CIGs;
- 'hard' constraints for well-to-seismic ties; we want the seismic image to conform as much as possible to well data when available;
- 'soft' geological inputs. The example chosen in fig. 4.04 is a 'regularization term' that controls the departure of the new model from either the initial one or the current one at each iteration. There are many other possible geological inputs, as we will see later.

In most cases, it is not possible to reach a minimum of C(**m**) in one single step but this is instead performed through an iterative process in which *the model is updated at each iteration*, hopefully converging to an optimum at the end. We run a PreSDM with the current model at every iteration of the model update loop and common image gathers are produced at predefined locations. The computation of CIGs is necessary to produce the residual moveouts, which are input into the linearized inversion scheme (see below). The process is iterated until one estimates that C(**m**) has reached an acceptable minimum. Generally, less than ten iterations are sufficient to reach such an acceptable minimum of the cost-function but this number is highly dependent on the complexity of the imaging problem to be solved.
The scheme in fig. 4.04 looks simple but there are many practical issues here, including the following ones:

- The definition of the *velocity model* **m.** It is the object that is updated as iterations progress. Its description in the computer, i.e., the set of parameters that fully defines it and are modified at each iteration, drives the whole process.
- The optimization procedure and the inversion strategy are related to some extent to the velocity model parameterization. It includes the model update: which parameters to change at each iteration and how much in order for the cost function to decrease as quickly as possible (efficiency) but regularly (stability).
- How to measure and quantify the 'non-flatness' of gathers: residual move-outs are key inputs to the process.
- What constraints and regularization and how to implement them in the tomography.
- And finally, the need for 'cosmetic' final residual moveout before the final stack of gathers.

We discuss these points in the next paragraphs.

Different models to represent the anisotropic velocity field

There are three classes of velocity model parameterization: layer-based, gridded models and a mixture of both, often called hybrid models.

Layer-based models are defined by a set of interfaces, generally represented as depth maps digitized on a regular (x, y) grid, plus velocity functions within the layers between the interfaces. The '$V_0+k{\cdot}Z$' parameterization presented in chapter 1 (fig. 1.09) is an example of such a velocity parameterization. Factorized VTI or STI anisotropy are easy to add to each layer, as described also in chapter 1 (fig. 1.10). In this case, current practice is to use a single set of Thomsen's parameters per layer and for STI, to assume that the symmetry axis is normal either to the dip of the top or of the bottom horizon or even to some average of both. Fig. 4.01 is a real example of such a layer-based anisotropic model. The model parameters **m** that are inverted include therefore the V_{0_top}(x,y) and k(x,y) values at the grid point of these maps.

Anisotropic parameters ε and δ, generally set constant in each layer, are either given or inverted too. In these models, anisotropic ray tracing takes account of refraction both within layers and at interfaces. It is then necessary to check at each point of the ray whether it crosses an interface or not. Application of Snell's law requires accurate dip estimation at the crossing point. This is a *source of ray instability*, cause of one of the limitations of ray-based imaging methods.

In gridded models, velocities and anisotropy parameters are defined at points of regular volumic grids. It can be a set of spline nodes located on a coarse grid (typically with a few hundred metres interval). In this case, the velocity value is computed at any point as requested using *spline interpolation* functions.

The parameterization may also be simply a finer grid, though generally still coarser than the final image grid. This is schematically represented in fig. 4.05. We find three levels of gridding:

- A dense velocity grid, with a grid interval typically of 50 x 50 x 10 metres. Velocity values at any point can be linearly interpolated using the values of the neighbouring grid points. This grid bears as many details as possible from previous studies (often coming from a first PreSTM project and its geological interpretation). This is the grid in which rays are being sent for Green's function computation and tomography.
- A sparser one that bears the *velocity updates* $\{dm_i\}$ only; this trick allows one to limit the number of unknown parameters to invert for the tomography process, as discussed in the next paragraph.
- A third, even sparser one may be the support for anisotropy parameters updates ($d\varepsilon$, $d\delta$). The anisotropy axis of symmetry is driven by the dip field (see the paragraph on high-density rmo analysis). This assumes that anisotropy varies more slowly than velocity. There is no geological or physical evidence to support this assumption but it is simply an acceptance of the poor resolving power of seismics to estimate these parameters!

Tri-linear interpolation is a classical way to compute velocity at any location from the values at the neighbouring grid nodes. In such gridded models, ray tracing does not 'see' layers. Refraction results from velocity changes controlled by the values at the neighbouring grid points. Gridded models avoid ray instability at interfaces, at the cost however of using a smoother velocity field that may not be fully representative of wave propagation at sharp velocity contrasts. They are also consistent with the velocity parameterization used ultimately in all migration techniques.

Hybrid grids are a combination of regularly gridded velocity and anisotropy values. Layers defined again as z(x, y) digitized maps are included in the gridded velocity volume. Conversely, the velocities within a layer are given by the grid points lying between the interfaces delineating it. Interfaces may or may not be considered as refractors in ray tracing, depending on the options chosen. They are however, 'seen' by rays as reflectors for tomography.

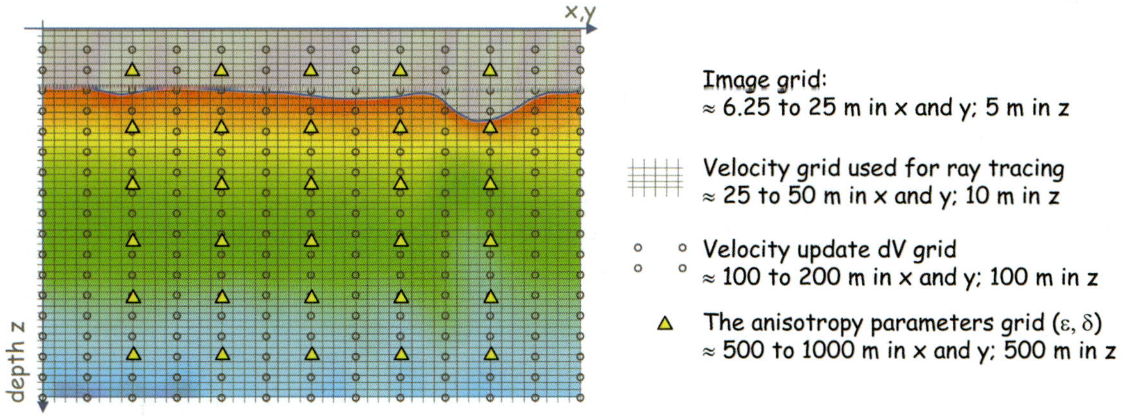

Figure 4.05: Gridded model description. The velocity model used for ray-tracing is defined on a grid generally coarser than the final image grid. The velocity update resulting from grid-tomography is digitized on a even coarser grid, reducing the number of parameters to invert for. Anisotropy is again defined and possibly inverted on an even coarser one. Velocity at any point may be computed through interpolation.

Linearized tomography

In this paragraph we look a bit further into the scheme of fig. 4.04; we will need some (basic) maths for this!

Recall that the goal is to flatten CIG. Let us call $Z_{event}(x, y, h, \mathbf{m}_{cur})$ the migrated depth along the offset 2h for a given event picked on the CIG at location (x,y), resulting from a PreSDM iteration with the current anisotropic velocity model $\mathbf{m}_{cur}$. We characterize the 'non-flatness' of the event by comparing its depth Z_{event} for all offsets 2h with a reference Z_{event_ref} at a given offset h_{ref}. The zero offset is chosen most of the time, despite the fact that this offset is not recorded. This set of values as a function of offset forms the 'residual moveouts' Δz:

$$\Delta Z_{event}(x, y, h, \mathbf{m}_{cur}) = Z_{event}(x, y, h, \mathbf{m}_{cur}) - Z_{event\_ref}(x, y, h = 0, \mathbf{m}_{cur})$$

Ideally, we would like to find directly a new $\mathbf{m}_{final}$ such that ΔZ_{event} are nil everywhere after re-migration. If one assumes to be close to the solution, i.e., if the current ΔZ_{event} are small, we can linearize the problem by saying that small linear corrections to the current velocity parameters $\mathbf{m}_{cur}=\{m_i\}$, denoted $\{dm_i\}$, will lead directly to the solution $\mathbf{m}_{final}=\{m_i + dm_i\}$. The $\{dm_i\}$ are often called *velocity perturbations*.
In mathematical terms, this is written as a Taylor series limited to the first order (linearity):

$$\Delta Z_{event}(x, y, h, \mathbf{m}_{final}) = \Delta Z_{event}(x, y, h, \mathbf{m}_{cur}) + \sum_{m_i} \frac{\partial \Delta Z}{\partial m_i} \cdot dm_i \equiv 0 \quad (4.01)$$

In this equation, the 'gradients' $\partial\Delta Z/\partial m_i$ tell us how much a given residual moveout ΔZ somewhere in the gathers changes if one perturbs one single parameter m_i of a small quantity dm_i somewhere in the current model. And again, we would ideally like this equation to hold for all offsets of all events of all CIGs.

The set of model updates $\{dm_i\}$ that would lead to the solution $\mathbf{m}_{final}$ are the unknowns of the problem. Their number N may be much smaller that the number of parameters defining the whole model, as shown in fig. 4.05.

On the other hand, we have as many equations (4.01) to estimate $\{dm_i\}$ as the number M of measurements of moveouts on CIGs. These measurements are denoted $\mathbf{\Delta Z} = \{\Delta Z_j\}$.

Finally, there are N times M gradients $\partial\Delta Z_j/\partial m_i$, since each of the N model parameters may in principle 'influence' any of the M residual moveouts in the CIGs. The matrix of gradients is called the Jacobian of the problem.

Assuming that all the necessary gradients have been estimated, in order to find the right $\{dm_i\}$ values, we need to solve the set of M linear equations with N unknowns. This set of equations is written in a matrix notation:

$$\left[{}^{\partial \Delta Z_j}\!/\!{}_{\partial m_i} \right]_{\substack{i=1,N \\ j=1,M}} \bullet \begin{bmatrix} \vdots \\ dm_i \\ \vdots \end{bmatrix}_{i=1,N} = - \begin{bmatrix} \vdots \\ \Delta Z_j \\ \vdots \end{bmatrix}_{j=1,M} \quad (\bullet \text{ denotes matrix product}) \quad (4.02)$$

In practice, this set of M equations must be completed with a further set of (non-linear) equations reflecting the constraints and regularization terms in C(**m**).

Sticking to equation (4.02), we may already note some issues here:

- the system is huge: the number N of unknowns may be a few million for large surveys, as can the number of moveout measurements M for dense moveout analysis;

- the gradient matrix may have many zeroes though, reflecting the fact that a given area of the model in practice influences only a limited number of CIGs;
- the system may be either undetermined (N> M: more unknowns than measurements or equations) or over-determined (N<M), depending on the model parameterization. It has in any case no unique solution that would perfectly flatten all CIGs.

In order to address these issues, one solves the system at each iteration *in a least-squares sense*, i.e., we look for the set of $\{dm_i\}$ values such that the sum of the squares of all ΔZ_j reaches a minimum instead of being exactly zero, which is not achievable. Doing so corresponds precisely to the first term of the cost function C(**m**) in fig. 4.04.

This new problem is cast into the classical *'normal equations'*, which are written in matrix form:

$$\left[\frac{\partial \Delta Z_j}{\partial m_i}\right]^T \bullet \left[\frac{\partial \Delta Z_j}{\partial m_i}\right] \bullet \begin{bmatrix} \vdots \\ dm_i \\ \vdots \end{bmatrix} = -\left[\frac{\partial \Delta Z_j}{\partial m_i}\right]^T \bullet \begin{bmatrix} \vdots \\ \Delta Z_j \\ \vdots \end{bmatrix} \tag{4.03}$$

(The superscript T denotes matrix transpose).

We now have a set of N equations for N unknowns. It can be solved directly by numerical methods, but one has to realize that any such classical method reaches its limitations when dealing with matrices with $N > 10^5$. Note that the extra constraint and regularization equations are included in this system, acting as a preconditioning of the $\{dm_i\}$, which tends to stabilize the resolution of the set of equations (4.03).

As an order of magnitude, a survey of 30 × 30 km (a typical size in today's standards), processed to a depth of 10 km, with a model defined on a grid of interval $\Delta x = \Delta y = \Delta z = 100$ metres, corresponds to $N = 9 \cdot 10^6$; M would be 64 times greater than N (close to a billion!) for dense grid-tomography (see next paragraph). Note that the (possibly non-linear) constraints increase the number M of equations (4.01) but not the size N×N of the system (4.03).

Iterative solvers such as 'conjugate gradients' are alternative approaches often used. It goes beyond the scope of the book to discuss numerical issues but it should be clear at this stage that it is *numerically beneficial* to limit as much as possible the number N of model parameters. This does not mean that it is beneficial for the accuracy of the final PreSDM result too, as we see later on. It is common practice to test and set up the parameters on a properly selected limited area of the project before going to full-scale tomographic inversion of the whole data set.

CIG's flatness and residual move-out analysis

Physical intuition leads to the conclusion that it is preferable to have an over-determined system, i.e., more measurements than parameters, to update the velocity model. At the same time, the greater M, the more gradients we need to estimate. The impact on the overall efficiency is nevertheless small, as the gradient computation and the system set up are fast and easily parallelizable steps in the whole chain.

The industry trend is therefore to augment the number of residual moveout measurements **ΔZ**. This is what we discuss in this paragraph. Note, however, that only independent parameters will bring valuable information to the inversion procedure.

Residual moveouts **ΔZ** are picked on the gathers after each iteration. Fig. 4.06 is an example of results of an iteration of tomographic inversion. It shows a *'velocity line'* (i.e., a section through the 3D volume used to visualize the current result of PreSDM) with two Common Image Gathers displayed as a 'fourth dimension'. Note that image gathers are reasonably flat for the upper part of the section but flatness is not yet reached in the bottom part. There is a good reason for this: the upper layers must be well resolved before inverting the velocities of the deeper ones! These are the CIGs on which the $\Delta Z_{event}(h)$ are measured. The process is called equivalently 'residual moveout' (rmo) analysis or 'non-flatness' analysis. We currently find two approaches: *single-offset* and *multi-offset residual moveout* analysis.

The principle of *'single offset' moveout analysis* in PreSDM is very similar to the case of PreSTM discussed in chapter 3, particularly with fig. 3.22. The two differences nevertheless are that firstly, the vertical axis is now depth instead of time; and secondly, one assumes possibly a hyperbolic as well as a parabolic shape of the moveout curve.

We call it 'single-offset' moveout analysis simply because only one parameter is used to approximate the depth of an event as a function of offset in a common image gather for the whole offset range. On the right hand side of fig. 4.07, we present a cartoon of a non-flat event. Residual moveout analysis consists in the search for the parameter C_2 such that samples in a depth window around the investigated depth have a maximum semblance along a trajectory defined by C_2. As mentioned, the trajectory may be hyperbolic or parabolic. As early as 1989, Al Yahya introduced the concept of non-flatness, showing that in the case of a horizontally layered Earth or small velocity errors and any dip, an erroneous migration velocity leads to a curved moveout of reflectors in a CIG in the shape of a hyperbola. He calibrated the hyperbola by a parameter he denoted γ, equal to the ratio between the migration velocity and the true harmonic average velocity (i.e., the average of slowness). Processing companies still use the parameter γ to quantify and quality-control CIGs. Events on a CIG are flat when $\gamma = 1$. As a rule of thumb, a γ value kept in the range [0.95 – 1.05] within the whole model volume indicates a reasonably good overall flatness.

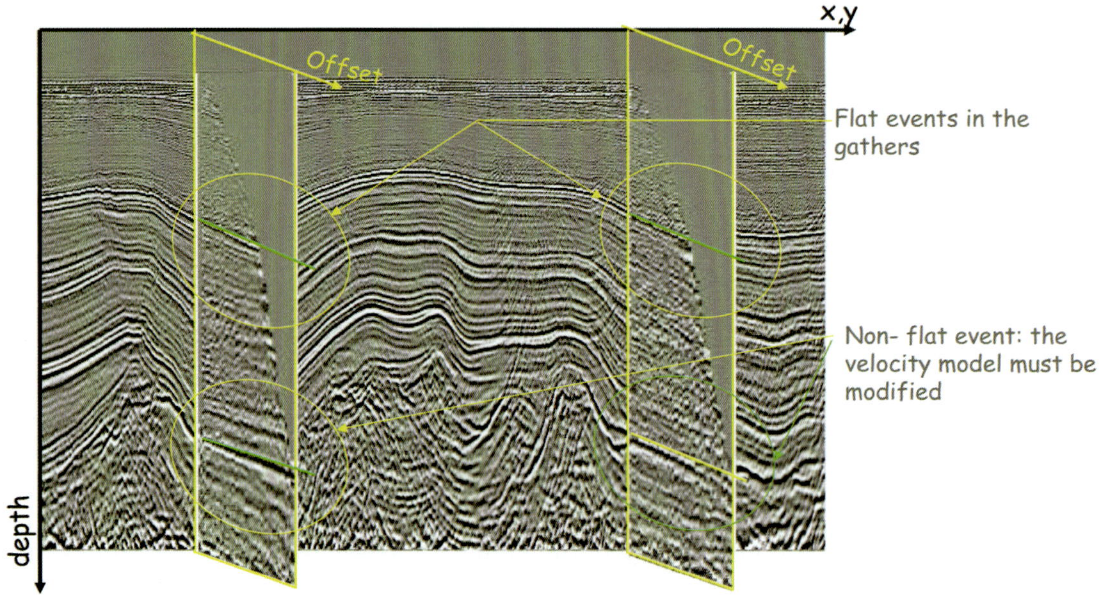

Figure 4.06: An example of PreSDM with two Common Image Gathers displayed as a 4th dimension object (x, y, z, h). Temporarily keeping the offset dimension allows us to verify the quality of the current model. Values of the residual moveouts, a measurement of the 'non-flatness' of the gathers are inputs to tomography.

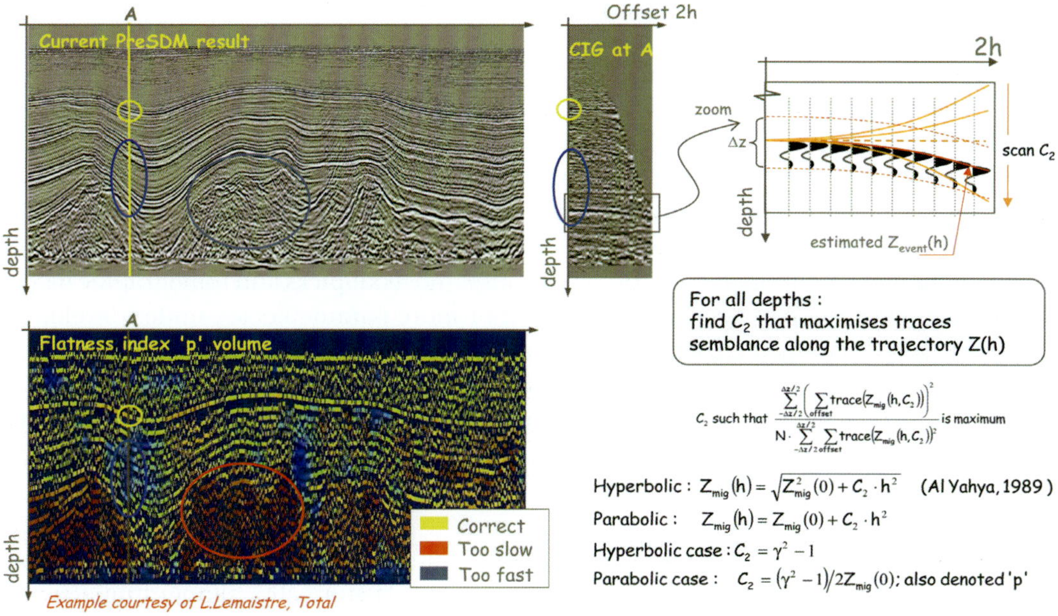

Figure 4.07: 'Single offset' residual move out (rmo) analysis : the flatness index 'C2' is computed at each depth of each location where a CIG has been computed. Two options are used to describe rmo curves: hyperbolic or parabolic. C_2 fields are displayed for QC purposes, but more importantly, are inputs to tomography inversion.

Others use a parameter denoted 'C_2', the coefficient of the h^2 term in the hyperbola's equation. Others again assume that the hyperbola can be reasonably well approximated by a parabola that is also calibrated by a single parameter denoted sometimes 'p' , or 'C_2' again!
The various equations and relationships are given in fig. 4.07.
The choice of the shape of the rmo curve is not crucial but two things are important to note however:

- These simple equations allow a very fast automatic search for C_2 everywhere in the migrated data volume. The result is a 'C_2 volume' (or 'γ volume' or 'flatness index volume', according to the equation and terminology used). An example of a cross-section through such a 3D 'flatness index volume', courtesy of Total, is presented at the bottom left of fig. 4.07. This volume can be quickly checked to detect areas of major velocity problems in the iterative tomography process and to concentrate efforts on these areas.
- The ΔZ(h) measurement is 'synthesized' into a single value per event for every CIG. The full ΔZ_{event} (h) can be reconstructed for any offset as needed, in order to feed the system of equations (6.03) in the inversion process; for this we apply the equation used for flatness index estimation. Despite this 'synthesis', the size M of the vector of measurements **ΔZ** is still huge. It is reduced in the case of layer-based tomography by taking the flatness indices only along horizons. In the case of grid-tomography, we select the most 'relevant' events. We discuss these two approaches in the next paragraph.

To assume that the residual moveout curve is close to a hyperbola is a crude approximation in many complex geological settings. The current trend in the industry is to move towards more systematic *multi-offset RMO analysis*. Fig. 4.08, modified from Brittan J. and Yuan J., (2005), illustrates this point with a synthetic 2D test with buried velocity anomalies. A zoom on a CIG at a given iteration of Kirchhoff PreSDM shows the difference between the actual pick of residual moveout at each offset (in red) and the estimated picks from the parabolic assumption (in green). Both sets of picks are fed into tomography inversion and the results are compared. Fig. 4.08-c shows that gathers are now much flatter with the multi-offset approach, hence resulting in an improved summation. Comparison of the inverted velocity models (4.08-d) indicates that the model coming from multi-offset picks is closer to the true model. However, the buried anomalies are not yet fully resolved. They are 'spread' over depth or more precisely, along the rays used for the gradient computation, as a result of limited resolution of tomography along rays (see also fig. 4.10).

Despite this limitation and its increased cost, multi-offset moveout picks and tomographic inversion are developing rapidly. RMO curves characterized by 2 or more parameters are under development. One difficulty is to select parameters as 'orthogonal' as possible to make multi-parameter search codes efficient and stable. Remember that we are looking for sets of parameters in a volumetric sense.

Furthermore, high-density volumetric picking in conjunction with grid tomography increases the vertical resolving power of tomography.

Layer-based versus grid tomography

In this paragraph, we enter the domain of inversion strategy. Historically, tomography developed as a layer-based method but nowadays, grid tomography has become the standard. We discuss both methods below.

Layer-based tomography may use either a layer-based model (no surprise!) or more often nowadays, a hybrid one. Fig. 4.09, a slightly more pictorial version of the schematic workflow of fig. 4.04, illustrates one iteration of such a layer-based tomography in a *hybrid velocity model* made of horizons and velocities defined as regular spline nodes in the layers between horizons.

The strategy here is layer-by-layer inversion, starting classically from the top layer and progressing downwards. Other strategies are possible; for instance, invert layer 1 then layer 2, then layer 1 and 2 simultaneously, then layer 3, etc.

The cartoon in fig. 4.09 assumes that we are currently inverting the third layer. The parameters $\{dm_i\}$ to invert are the updates of the parameters at the nodes of the spline functions, which allow the computation of velocity values everywhere. A Kirchhoff PreSDM with the current model is run and produces the migrated section and CIGs at selected mid-point locations (in fact every mid-point location in modern practice). These are loaded on an interpretation workstation. The latter are visually checked for QC purpose; the former is interpreted and the depth of the horizon is picked by an interpreter. High-density single- or multi-offset depth residual moveouts are computed, yielding a flatness index volume. The residual moveout along the picked depth horizon is extracted from the full volume: the $\Delta Z(x,h)$ in the figure. Some editing or smoothing may be useful at this stage to eventually provide us with the M measurements $\{\Delta Z_j\}$, making up the right-hand side matrix in equation (4.02).

Next, we need to build the 'Jacobian' matrix of 'gradients' $\partial \Delta Z_j / \partial m_i$. These are computed using ray theory, as schematically explained in fig. 4.10. Conceptually, we would perturb the node (i) in the model. We look for the ray(s) crossing the neighbouring cell and define the segment l that is influenced by the perturbation. The travel time t of the ray is the invariant. We also know all the parameters of the ray, particularly the reflection angle and the dip at the reflectors. Using ray perturbation theory (the dm_i is small), we can derive the new reflecting element and particularly its depth at the vertical of the initial reflection point (note that traveltime t is invariant in first approximation along the reflector's segment in application of Fermat's principle). In this way, we obtain the ΔZ_j for this specific CIG and offset corresponding to the given dm_i, which is what we are looking for! Some trigonometric considerations yield the necessary equations in isotropic media but again,

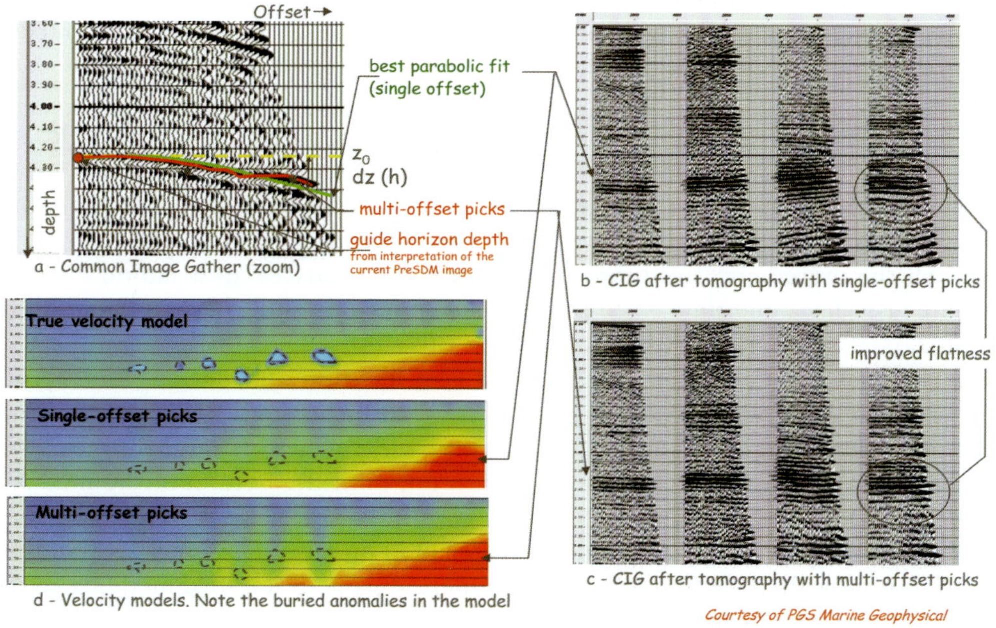

Figure 4.08: A 2D synthetic example illustrating the impact of multi-offset picks in inversion tomography (from Brittan *et al.*, 2006).

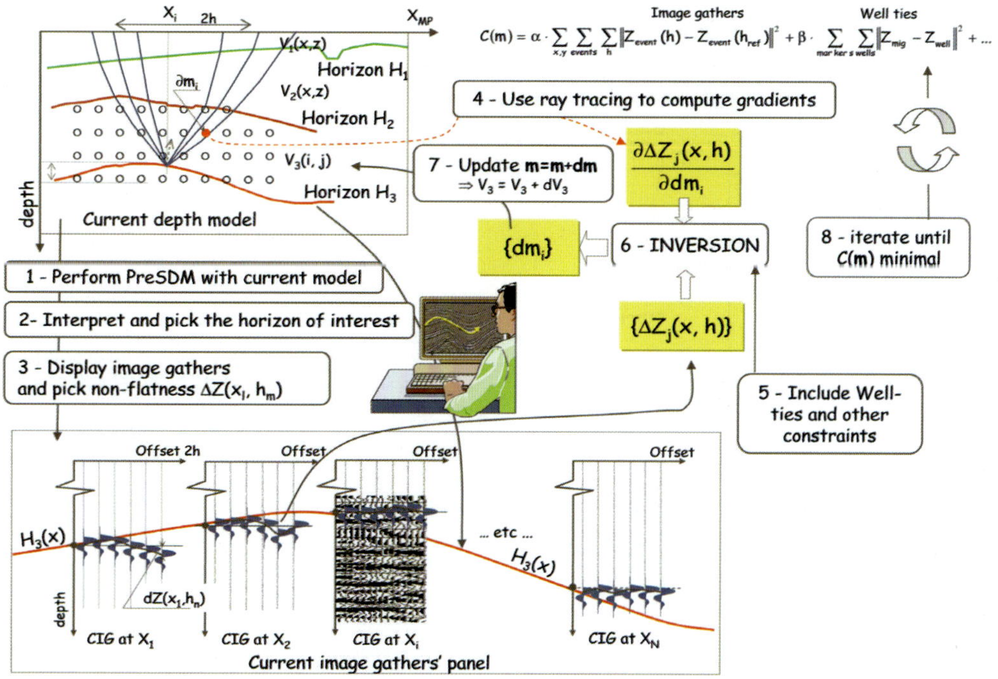

Figure 4.09: A scheme of 'hybrid' CIG Inversion (tomography). The model parameters **m** are defined on a grid of spline nodes, but inversion is driven by non-flatness indexes along selected reflectors only. The process is run top-down, layer per layer.

things are a bit trickier when dealing with anisotropic models. Similar computations are repeated to compute all the necessary gradients. In practice, the computation of gradients is executed in a more efficient series of actions.

It is interesting to note in fig. 4.10 that any similar disturbance dm_i along a given ray yields some ΔZ residual moveout; this is one of the facts that limit the resolving power 'along rays' of tomography, as mentioned in a comment to fig. 4.08. This situation naturally improves when using volumetric rmo analysis jointly with grid-tomography, topics that we discuss below.

We now have all the ingredients necessary to build the 'big matrices' in equation (4.03) and to let the computer run for some time … to provide the expected result, i.e., the 'best' updates $\{dm_i\}$, in a least-squares sense, to the velocity model at this step. We run the Kirchhoff PreSDM again, compute and pick the CIGs, compute the new cost function and decide whether it has sufficiently decreased to stop the inversion or whether a further iteration is needed. If this is the case, the whole procedure is repeated again; if not, we pick the new horizon 3 on the PreSDM and move on to the inversion of the fourth layer and so on, until the bottom of the model and the end of the project.

Note in passing the three levels of iteration: the outer loop on layers, an inner loop of linearized inversion to minimize the cost function and an innermost one if iterative methods are used to solve the large linear system of equations (4.03).

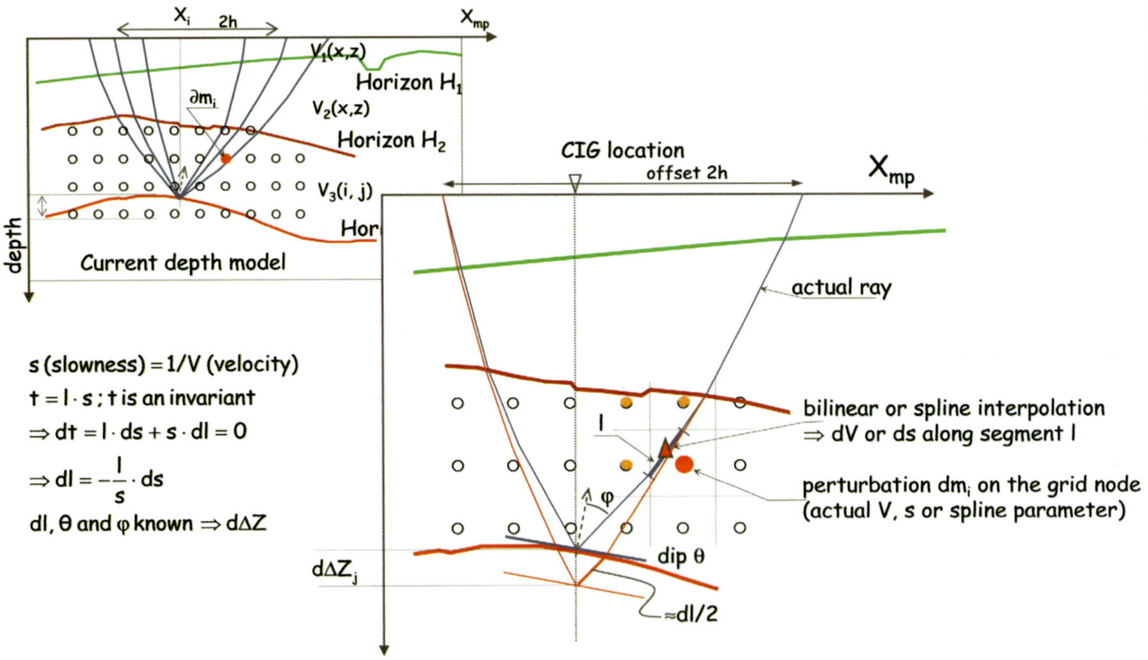

Figure 4.10: Ray tracing computation of the gradients. A perturbation dm_i on the ith parameter of **m** induces a small velocity change dV (or a change of slowness ds) along a segment l of neighbouring rays, which in turn induces a change dl in the overall ray length. We know the ray direction and dip at its end, hence the new reflection element, its modified depth at the CIG location and hence ΔZ_j for the input perturbation dm_i.

As for PreSTM, fig. 4.04 indicates that a *final fine tuning 'residual moveout'* is run on all image gathers to improve the final summation. We should add that to apply an adapted mute in the CIG domain is an important step to avoid to 'adding noise to the signal', especially for far offsets.

The reader will have noted that when inverting layer 3, we have 'frozen' grid values above it (i.e., they have not been updated in the process). This process is often called 'hybrid tomography' since it includes the inversion of velocity parameters on a regular grid, but is constrained by rmo values picked along horizons. The latter is useful and necessary for the following practical and historical reasons:

- geophysicists are used to picking and working with horizons; in sub-salt imaging for instance, picking the top and base of salt bodies is an integral part of velocity model building;
- as presented in fig. 4.10, we need reflectors on which to compute the gradients that will be input in the matrix inversion.

Working with continuous reflectors is no longer a constraint with *'grid-tomography'*. Indeed, in this case the *dip information* is picked automatically, together with non-flatness. This may be done by looking for semblance along an hyperbola as indicated in fig. 4.07, but now for a 'macro-CIG' composed of traces within a 3D 'probe' of size [Δx, Δy, Δz]. Once this 'average' rmo is computed, traces might be summed along offset after application of the picked residual moveout. Next, these optimized PreSDM traces are used to estimate the dip and azimuth of the corresponding seismic event every Δz. We call *'facet'* (Hardy, 2007) this event assumed to be locally planar. An efficient manner to estimate dip and azimuth is Claerbout's (1992) now classical plane wave destructor technique.

The non-flatness volume of fig. 4.07, is now replaced by a *volume of vectors*: at each event location in the volume, we have the non-flatness index (for instance C_2), the local dip θ and azimuth α. It is current practice to also add parameters assessing the 'quality' of the event, such as the amplitude and coherency along the picked facet.

The result of the process is schematized in fig. 4.11. It shows the 'RMO-optimized PreSDM stack section' extracted from the 3D volume at a given iteration. The central CIG is represented as before as a fourth-dimension object, but this is only part of the column of data of section [Δx, Δy] in which facets are picked. A probe, such as the one shown in the figure, is a subset of that column in which facets are picked.

Amplitude and quality are used to select the events that are considered as the most relevant and reliable ones (best signal / noise). These selected events form, in the same way as in the layer-based case, the M terms of the right-hand side matrix in equation (4.02). The difference however, is in the size of M, which can now be more then several hundred million as mentioned!

Fig. 4.12 is a real 3D data example of facet picking. The panel (a) in the figure presents a volume extracted for inspection and analysis from the full 'C2 volume'. The latter is the volume of the single-offset residual moveout measurements of all picked facets, represented on a regular dense grid, possibly after some interpolation, editing and smoothing.

The colour scale in panels (a) and (b) in fig. 4.12 corresponds in fact to $1/\gamma$ value, γ being defined in fig. 4.07. Note the high density and high resolution of picks made possible with the facet approach. Panel (b) is a section through the C2 volume with the 'optimized RMO stack' in the background. This is the stack obtained by applying the C2 correction to the CIG and summing the traces together. Panel (c) illustrates the 'quality' parameter that quantifies the coherency of data along the facet. Finally, panel (d) is the same optimized stack section as in panel (b), but now with the dip of the facet superimposed as colour-coded information. To be more precise, it is the apparent dip dx/dz along the section, since a facet includes a plane with its dip

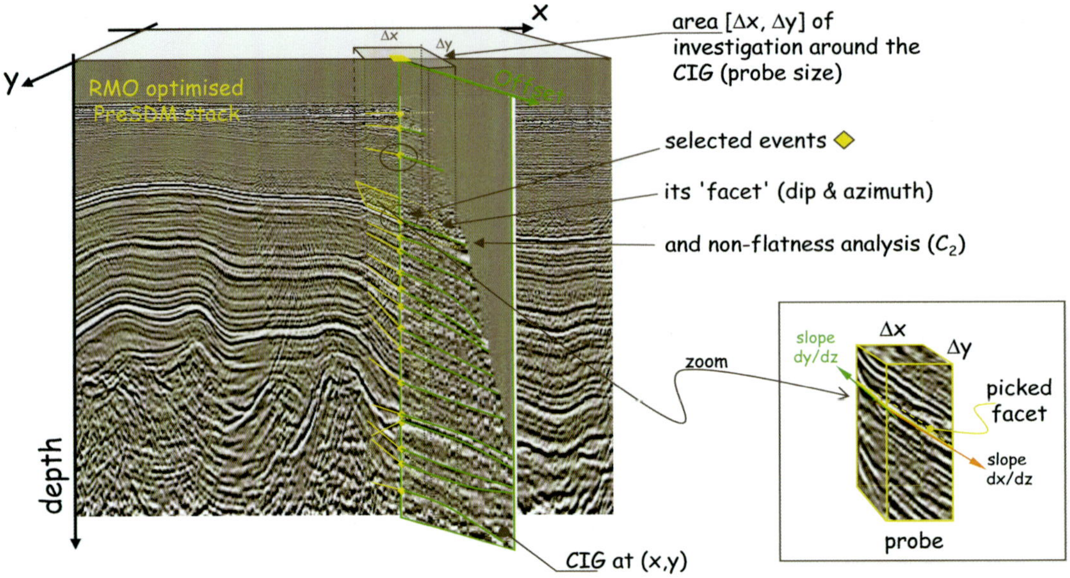

Figure 4.11: 'Facets' picking. As an input to grid tomography, we pick not only de non-flatness of events, but also the dip, azimuth and other 'quality' parameters of the current PreSDM image in a window around the CIG. This is repeated for all CIGs, yielding a volume of vector properties (C2, dip, azimuth, amplitude, coherency, ...).

and azimuth in 3D. All these elements are used in high-resolution tomography:

- C2 allows to compute the 'measurements' ΔZ_j in equation 4.03;
- Dip and azimuth replace the dip and azimuth of picked markers in fig. 4.09 and 4.10, thus allowing the computation of gradients $\partial \Delta Z_j / \partial m_i$ for lots of reflecting elements densely positioned in the 3D model. This is a key benefit of high-density facet picking.
- Quality may be introduced as a weighting factor to favour measurements that have a greater reliability than others.

Since the number M of measurements is large, it is possible to reliably invert a much larger number N of model parameters $\{dm_i\}$ than for the layer-based approach. The model can then be more finely sampled and hopefully, more details in the model will be revealed by tomography. This is at the computational cost nonetheless of solving much larger linear systems, at the limit of current computing power (and numerical accuracy).

We should emphasize that grid tomography works on data picked at all depths. This allows a more stable estimate of the evolution of velocity with depth. The 'k' parameter of the $V_0 + k{\cdot}Z$ model is intrinsically taken care of and estimated.

We describe above a 'global inversion' strategy, whereby one estimates the set of model parameters for the whole model in a single run, but tomography strategy analogous to the top-down layer-based approach is possible too. The sequence can be, for instance, the following:

- Invert globally all measurements and model parameters down to a depth Z_1.

- Load the PreSDM result on a workstation and pick a first key horizon, say H_1.
- Next, all model parameters above this horizon H_1 are 'frozen'.
- Run a second 'global' inversion of all measurements and parameters between H_1 and a second depth Z_2.
- Load the corresponding PreSDM image and interpret and pick a second horizon H2.
- And so on, down to the bottom of the model and the deepest depth of the PreSDM project.

This technique is also called 'hybrid tomography'.

Fig. 4.13 shows a synthetic example of grid tomography on a 2D model simulating seismic acquisition in a complex geological compressive environment, including severe topography. The complex velocity model is presented at the top right of the figure (Gray and Marfurt, 1995). The tomographic inversion is initiated with a simple velocity model with a single vertical velocity gradient and no lateral velocity variation. The velocity model obtained after five iterations is shown on the bottom right of the figure. Most of the complex features of the true model have been retrieved. The decrease of resolving power in terms of velocity estimation from top to bottom, is well illustrated here. However, the overall vertical resolving power is increased as compared with layer tomography. This is because measurements at all depths are used in the inversion process.

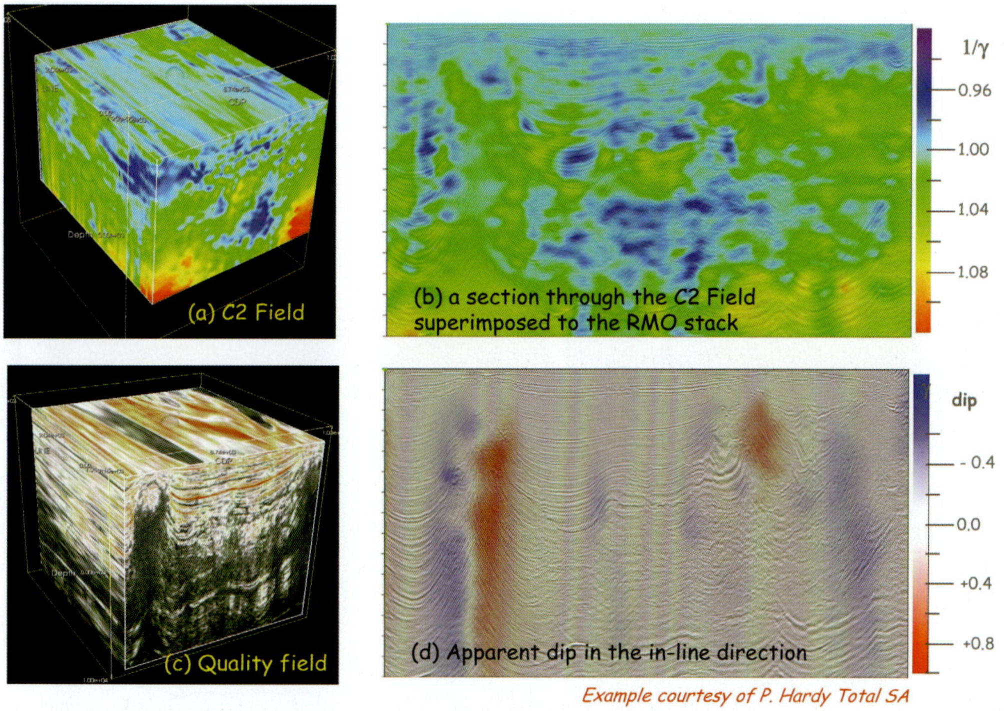

Figure 4.12 : Facet picking allows a high density estimates of several parameters in a full 3D sense. We display here real data examples of the C2 field expressed as 1/γ (a & b), the RMO stack (b), 'quality' parameter and dip and azimuth of the picked facets (c & d). All these parameters are used for high resolution grid tomography.

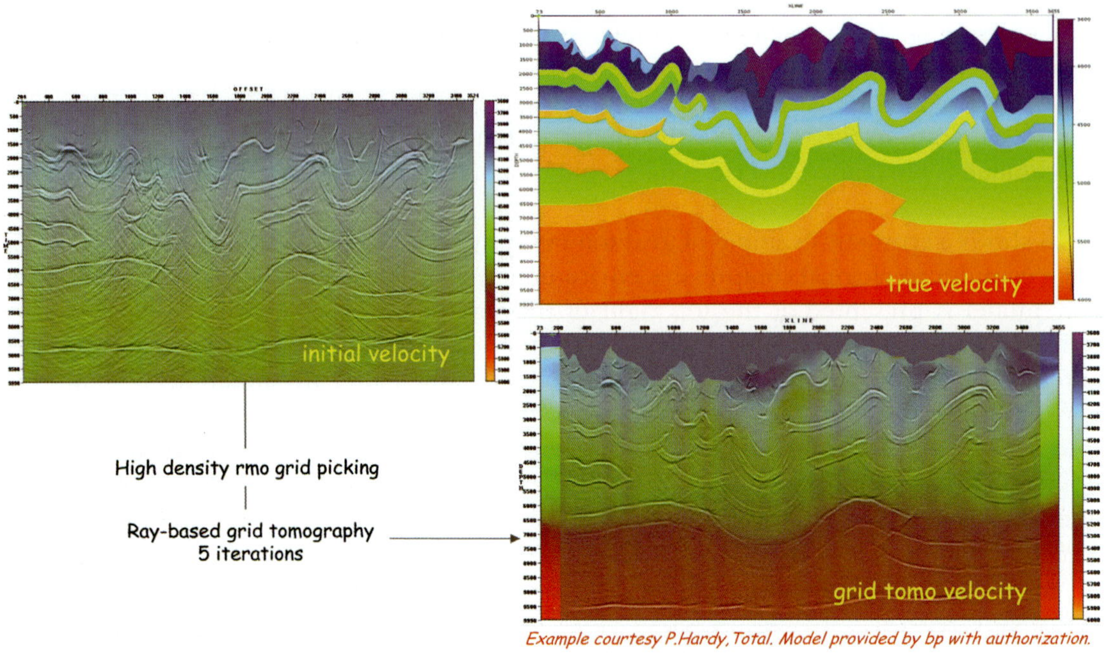

Figure 4.13: 2D synthetic example of grid-tomography in complex tectonics. Velocity models overlay the corresponding PreSDM result. Starting from a rough initial model (single V(z) function), grid tomography can recover the main features of the true velocity model, especially for shallow events (maximum offset = 3 km). Layers may be interpreted and introduced in a 'hybrid' model if necessary at this stage.

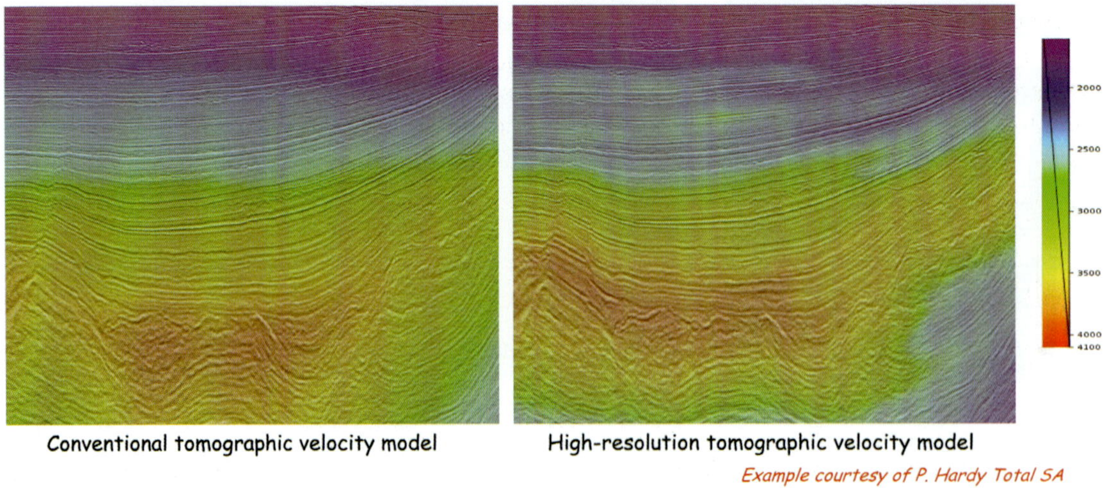

Figure 4.14: An example of grid tomography using the high density 'facet picks' from fig. 4.12. Note the increased resolving power brought by high density information.

The velocity elements obtained after grid tomography are already sufficient to obtain a decent image of reflectors. At this stage, it is advisable that interpreters become involved in the imaging process. If requested for interpretation and further refinement, layers could be introduced at this stage of the process, but at this stage only.

This is synthetic data and the reader should realize that imaging challenges are unfortunately often much tougher in real mountainous settings!

Fig. 4.14 is another example of grid tomography, now in 3D. It corresponds to the same data set and project as the one presented in fig. 4.12. The figure compares results of conventional tomography, on the left, with high-resolution grid tomography on the right, which results from the facet picks in fig. 4.12. The improvement in velocity resolution is clearly visible.

Velocity Model Building workstation

Fig. 4.15, courtesy of Western Geco, gives a more realistic vision of today's velocity model building (VMB) and tomography workstations than the cartoons presented so far. These workstations allow visual inspection, interpretation and picking of any results of an iteration of a PreSDM project: display of cross-section and constant depth sections of both the velocity model and the seismic data; display of any Common Image Gathers; display of well trajectories and markers, etc.

This facility used for QC purposes, *also allows interpreters to look at the data and to pick and update horizons*. In a salt tectonic environment, a picking facility is fundamental, since it is indispensable to *pick manually and very accurately the top and base of salt bodies*, such as the yellow one in fig. 4.15.

Note also the multi-dimensionality of CIGs in the figure: it is displayed not only as a function of offset, as mentioned so far, but also of azimuth.

Although visual inspection and human interpretive interaction in the imaging process is necessary, the measurement of flatness is performed automatically, as described above.

Constraints, well ties and regularization

The solution of tomographic inversion is generally highly non-unique: several possibly very different models can explain residual moveouts measured with the intrinsically limited accuracy and resolution at any given iteration of the tomographic process. As a consequence, constraints and any other piece of information must be added to the process on top of non-flatness indexes. They correspond to the second and third term of the cost function given at the top of fig. 4.04.
Interpretative or more 'quantifiable' or 'mathematically expressed' constraints are in the arsenal of the processor. We can list and discuss them as follows (non-exhaustive):

- direct involvement of interpreters;
- well-to-seismic tie;
- smoothness of the overall model or layer-specific;
- regularity of deep horizons;
- weights between various constraints;
- weights applied to input data.

Interpretation: an interpreter who selects one solution rather than another because it makes a more geologically sensible contribution brings into the system a first kind of constraint. An interpreter picking a top and base salt, as already mentioned, or considering whether a basement is faulted or not, or suspecting thrust faulting based on the regional geological context, are typical examples of such information in complex settings and/or poor seismic response (salt environment; foothills).

This requires *fast PreSDM and tomography* in order to be able *to test several scenarios of velocity models*. Kirchhoff PreSDM is fast compared with Wavefield Extrapolation techniques (chapter 6), but Beam migrations that are much faster than even Kirchhoff are now widely used for 'interactive' velocity model building. Their principle is discussed in chapter 5.

Fast algorithms but also appropriate workstations, as the one presented in fig. 4.15, are necessary for efficient interaction between interpreters and processors.

Well information: fig. 4.16 lists different pieces of *information that we may derive from wells* and use as constraints or direct measurements:

- *Depth markers*: they are crucial to perform the 'seismic-to-well tie' directly in depth while running tomography. In many projects, this necessitates introducing anisotropy parameters in the velocity model. The vertical tie controls mainly the δ Thomsen's parameters in layers. Using depth markers from wells means that the corresponding horizons have been identified on seismic and are selected in the layer-based or hybrid methods. This information is less straightforward to use directly in grid-tomography.
- A thorough analysis of *sonic logs and time-depth curves* from check-shot surveys from all wells available in the survey area gives a good estimate of acceptable ranges of velocity gradients k and V_0 velocities, if such a parameterization is selected (see fig. 1.19). For hybrid- and grid-tomography, these values may be used as the 'initial model value' and used as regularization terms during inversion, as suggested by the third term of C(**m**) at the top of fig. 4.04.
- Borehole seismics, and particularly *offset VSP*, offer the opportunity to assess anisotropic parameters (V_{vert}, ε and δ) around well(s) through traveltime tomography. Again, these values may be used as hard constraint (fixed in the model) or regularization terms (avoid too large a departure from it in the inversion iterations).
- More anecdotal, anisotropic parameters may be derived from sonic measurements in deviated wells or petro-elastic measurements on cores. In both cases, upscaling from log or lab measurement to the seismic scale remains a challenge! We discuss all these points again in chapter 8.

Model smoothness: the degree of smoothness of the velocity model is strong *a priori* information in the tomographic process. It is explicitly governed by the size of the grid interval. A coarse grid means a limited number of parameters to invert, hence a faster inversion process. The cost however, is that it impedes the assessment of small-scale heterogeneity, which may be of great importance, especially if it corresponds to *shallow anomalies*. Recall that shallow velocity anomalies act as 'blurring lenses' for deep reflectors. It is important to include them in the velocity model but unfortunately, not easy to estimate, as discussed in fig. 4.08.

Smarter and more 'geological' smoothing approaches are being used today. Three of them are presented in fig. 4.17.

Firstly, the model parameters $\{dm_i\}$ in equations (4.02) or (4.03) may be 'pre-conditioned' such that iso-velocity surfaces follow preferably seismic interfaces. The geological argument is that when the regional tectonic history is not too complicated (single tectonic phase, limited burial and uplift sequences, etc.), it is fair to assume that the velocity is constant along the layer. We may call it a 'dip-steered' constraint. Recall that in grid-tomography based on facet picks, the dips of the markers are picked and hence available to implement the constraint.

Secondly, other geological 'knowledge' is that in a fast subsidence environment, the compaction gradient is such that iso-velocity surfaces are quasi-parallel to the mud-line. We then introduce some pre-conditioning in the model parameterization to honour this rule as much a possible. One can also think of other rules, for instance the one sketched on the left of fig. 1.09 with a constant compaction gradient but a velocity controlled by the top layer depth. This is a 'compaction-driven constraint'.

Thirdly, poorly resolved shallow anomalies or complicated geological features may induce non-geological oscillations of deeper horizons. This is in a sense similar to issues faced with poor estimation of long wavelength static corrections. The migration velocity field may then be constrained such that a given horizon remains as 'straight' as possible during inversion. Here again, the judgement of the geologist and interpreter is essential.

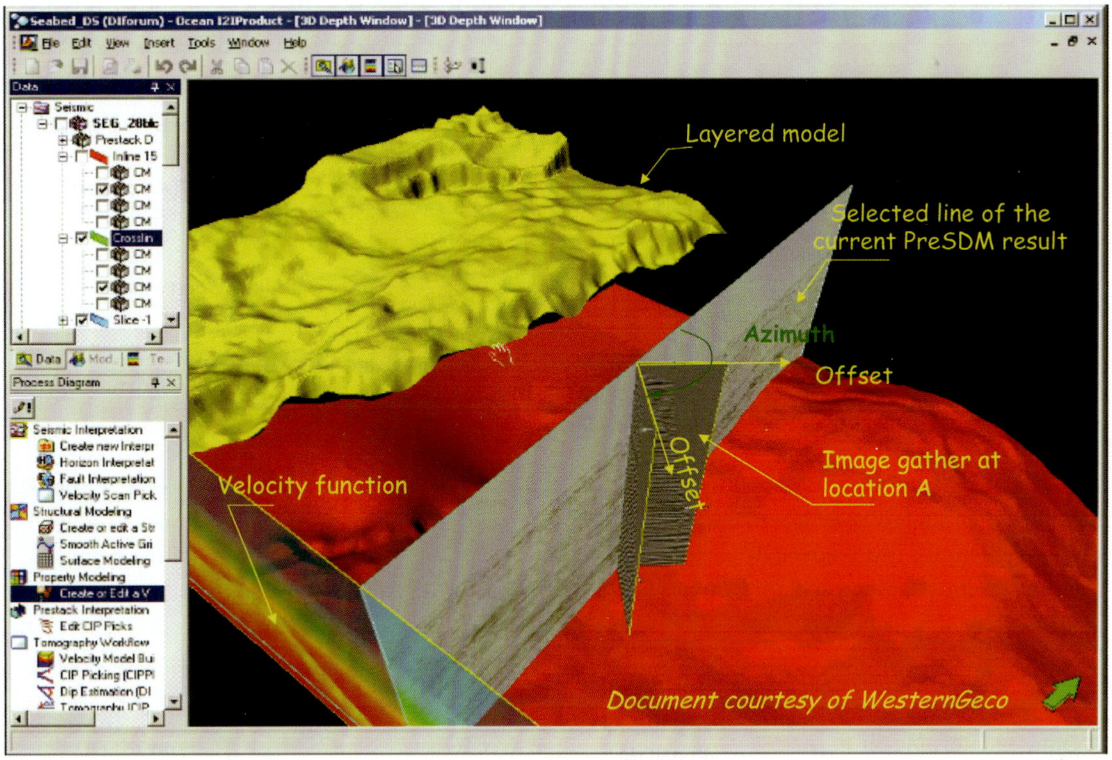

Figure 4.15: Example of 3D visualization of image gathers in PreSDM. For Wide Azimuth acquisition, CIGs are 3-dimensional objects as they depend on both offset and azimuth. Any image gathers can be interactively displayed for visual inspection, interpretation and QC purposes.

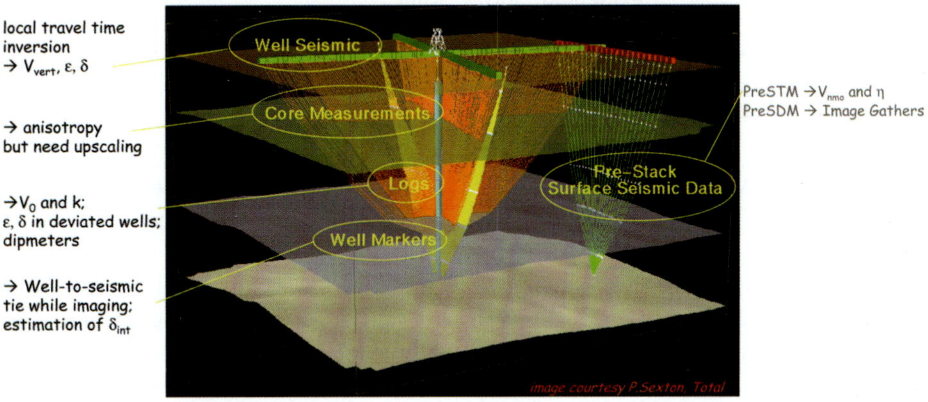

Figure 4.16: Different uses of well data and borehole seismic to constraint tomographic inversion.

This illustrates some ways to input geological knowledge into the inversion process but note that not all rules are so easy to implement in the set of equations (4.03), while keeping the numerical solver accurate and efficient.

Weights: the definition of the cost function C(**m**) if fig. 4.04 includes different terms, each multiplied by a scalar factor (α, β, ...). These factors should be considered as norms necessary to give to very different components of the equation equivalent influence in the minimization process.

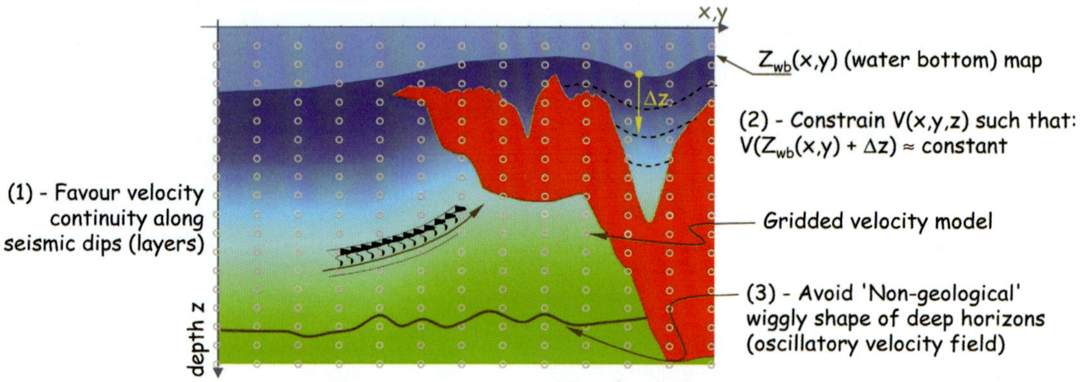

Figure 4.17: Three types of possible velocity constraints in tomography. (1) Favour lateral continuity of velocity parallel to layer interfaces. (2) Constraints may 'force' the iso-velocity contours to follow the water-bottom; this is specially useful in deep waters. (3) The overall smoothness of the velocity field may reduce lateral oscillations that yield un-geological wiggly horizons in depth.

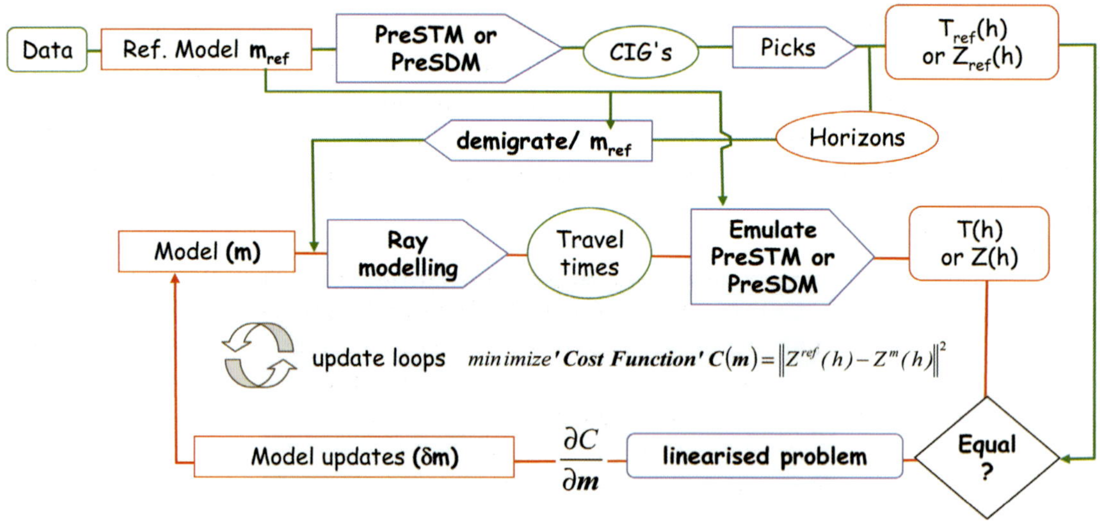

Figure 4.18: Principles of non-linear tomography. A first PreSTM or PreSDM is run with a reference model, producing CIG that will become the 'invariant' of the inversion. The optimization process consists in finding the model **m** that generates these invariant through ray modelling followed by a kinematic emulation of the PreSTM or PreSDM process using the reference model. Note that there are no expensive processes in the update loop.

For instance, the residual moveouts are in metres, which has nothing to compare with the model parameters $\{m_i\}$, which could be velocity, slowness or dimensionless anisotropic parameters. These factors must be set up before hand in the tomographic code. They can, however, be modified by the processor to give more weights to one or the other term in C(**m**). For instance, α may be increased or decreased, depending on the confidence the interpreter has in the well-to-seismic tie.

These factors have in fact more sophisticated forms. They can be expressed as a matrix acting on the ΔZ_j in order to weigh every measurement with its estimated 'quality', resulting, for instance, from the quality factor estimated in dense volumetric facet picks. Another physical approach is to include weights to the left-hand side of equations (4.03) to 'down-weight' the influence on shallow parameters in **m** of deep reflector measurements. This is because these have both a low signal/noise ratio and generally, a limited aperture. The measurement of their moveout is consequently less accurate, together with their respective gradients $\partial \Delta Z_j / \partial m_{ij}$.

In current practice, one starts with small values of the constraints (small α and β) to obtain as much of a 'data-driven solution' as possible. Constraints may be increased in further iterations to stabilize the process.

Clearly, tomographic velocity model building, although a very mathematical-looking process, is not a 'black box', and requires specific skills and experience.

Linear versus non-linear tomography

The linear tomography process of fig. 4.04 is expensive because it necessitates running a full pre-stack depth migration at each iteration of the minimization process of the cost function. In complex areas, the process may converge slowly. Consequently, all efforts are made to reduce either the cost of every iteration or the number of inner iterations.

One way to reduce the cost is to migrate only target areas or selected lines, or down to a limited depth only for the shallow layers. This facility is an *intrinsic benefit of Kirchhoff methods*, already mentioned for PreSTM.

An alternative method to reduce the turnaround time and cost of a project is 'non-linear inversion' (Adler *et al.*, 2008) illustrated in fig. 4.18. It benefits firstly from the possibility to *de-migrate* an horizon picked in a migrated domain to a zero-offset equivalent horizon and secondly, to emulate the kinematics of PreSTM (Reynaud and Robein, 1998) or PreSDM. The idea is then to PreSTM- or PreSDM-migrate the data with a reference model $\mathbf{m}_{ref}$ (this is more precisely a migration velocity field in the case of PreSTM).

The common image gathers produced by this first exercise become the 'invariant' target of the new optimization. A first velocity model is built; picked horizons are depth-migrated and added to the model; offset ray-tracing is carried out on these horizons and synthetic CIGs are computed with the reference model, simulating the initial exercise. If the current model **m** was correct, then the synthetic CIG should match the true reference ones. This is normally not the case, an optimization loop is then applied iteratively until convergence. The main difference with the linear route is that only one Pre-stack migration is performed during the tomographic exercise. Once the model has been optimized, the final PreSDM can be performed.

Migration operators in 3D

We have seen so far that in standard PreSDM Kirchhoff migration codes, data are sorted by common-offset volumes, which are processed independently, applying a diffraction summation principle over the aperture area. The aperture is defined in this case in the mid-point domain. In 3D, it is a surface, as sketched in fig. 3.19. We talk of apertures in X and Y. In theory, aperture should be equal in both directions to ensure that any dipping reflector is correctly imaged, whatever its azimuth.

As stated in chapter 3 for PreSTM, the concept of the 'mid-point' presented in fig. 3.03 extends in 3D to the concept of the 'bin'. The survey area to be migrated is divided into regular bins, generally squared. All traces that have their mid-point falling within the same bin are collected. This 'bin gather' is the equivalent of the mid-point gather in 2D. Traces in a bin can be sorted out as a function of their offset, but also azimuth. The process is called 'binning'.

Conversely, once traces are 'binned' one can select those having their offset falling within a given offset range, to produce the 'common-offset volume' that is the input to Common-offset Kirchhoff PreSDM.

Typical bin size ranges from 6.25 × 6.25 metres for high-resolution data sets, to 50 × 50 metres for low-frequency seismics. (note the link between spatial and temporal or vertical resolution).

A migrated trace is computed for each bin of a common-offset volume. It is made of image-points (or pixels) regularly spaced in depth with a spacing Δz, typically of 4 to 10 metres. Remember that each of these image-points is considered as a diffractor in Kirchhoff summation methods (chapter 2). The reflectivity at this point is computed for a given offset, as a weighted sum of samples along the corresponding diffraction surface.

In an *imaged-oriented implementation* of Kirchhoff PreSDM, the sequence of actions to image a point in the subsurface, for instance the point D of coordinates (x_D, y_D, z_D) in fig. 4.19, would be as follows:

- Sort data by common-offset range (we discussed the issues relating to this step in chapter 3).
- For a given offset 2h, repeat for all traces having their mid-point within the specified aperture around (x_D, y_D):
  - compute by *dynamic ray tracing* through the current velocity model, the traveltimes t_s and t_r respectively from the shot location to D and from D to the receiver location and the corresponding weights w. One such ray tracing is sketched in fig. 4.19;
  - take the sample at time $t = t_s + t_r$ of the trace;
  - weigh the sample with w and add the result to the reflectivity $r(x_D, y_D, z_D)$ in D, which builds up trace by trace; recall that the weight takes account, amongst other things, of the spherical divergence of the downward and upward propagation of the wave that illuminates D;
  - when all traces contributing to D have been processed, the last step is to 'scale' r, particularly but not exclusively to account for the number of traces having actually contributed to D. This number, sometimes referred to as the '*hit count*' (Audebert *et al.*, 2002), depends on the acquisition lay-out;
  - place the result at the corresponding depth z_D of the column '2h' of the Offset-CIG in (x_D, y_D);
- Repeat for all points D of the imaged trace: one trace of the Offset-CIG is now complete.
- Repeat this sequence for all offsets: the whole Offset-CIG is now computed.
- Repeat for all bins: the whole Kirchhoff summation is now completed.
- Use the CIG for tomography or residual RMO before the final stack.

In order to optimize data handling and overall code performance, actual implementations of Kirchhoff migration use a smarter sequence of actions than this one. Traveltimes and weights are pre-computed on a coarser grid than the imaged volume and stored on look-up tables. For all points D, tabulated values are interpolated 'on-the-fly' to obtain the appropriate value.

The ensemble of t and w values for one point D is the 'Kirchhoff operator', which we mentioned earlier, for instance, in fig. 3.08 for the 2D PreSTM case. By definition, only existing traces can contribute to the summation. The operator's completeness depends therefore on the actual acquisition lay-out (see also chapter 8, p.2).

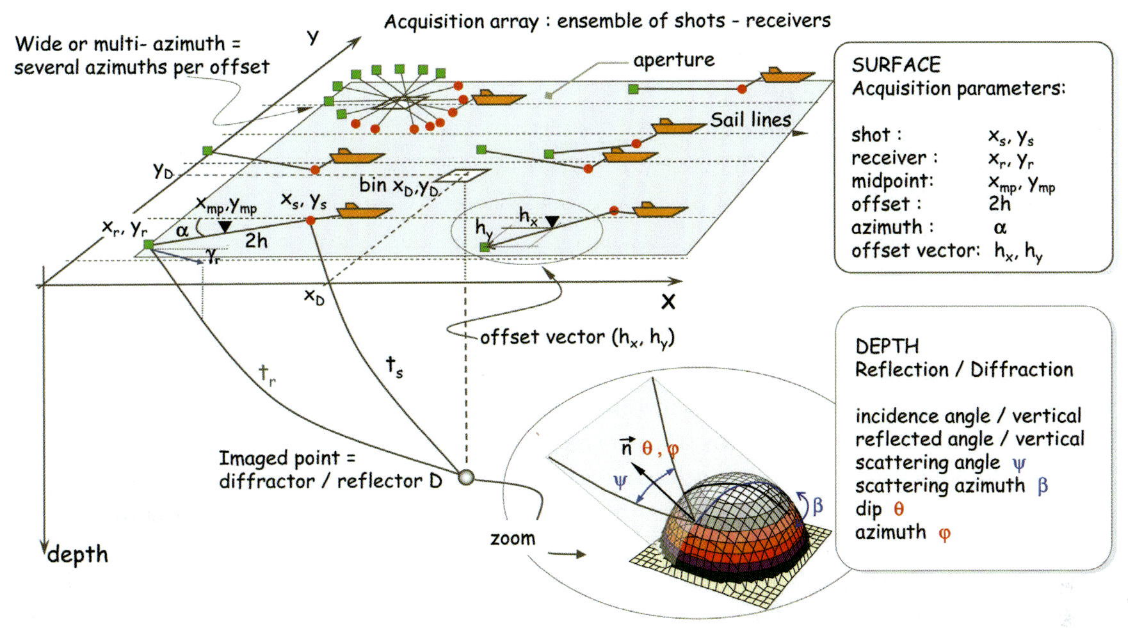

Figure 4.19: Relationship between surface acquisition parameters and reflection or diffraction angles in depth. Mapping [Midpoint - offset] ⇔ [dip θ -reflection angle ψ] is performed through ray tracing.

When dealing with 3D data sets, another trace attribute comes into the picture: the shot-receiver azimuth, denoted α in fig. 4.19. There can be traces with several azimuths for the same offset range in a bin. This is even what is looked for in 'wide-azimuth shooting'. An example is sketched for one bin in the figure.

Trace azimuth is taken care of when one computes the Kirchhoff operators, since we send rays to the actual shot and receiver locations. Azimuth may also be used in 3D tomography, although this requires storing the $r(x_D, y_D)$ not only as a function of offset, but also of azimuth, which makes the third dimension of the CIG. Azimuth is nonetheless not systematically considered when computing or displaying CIGs. There are two reasons for this:

- Firstly, the offset-azimuth CIG is a three-dimensional object, not so easy to manipulate; an example is given in fig. 4.15.
- Secondly, the offset-azimuth distribution is sufficiently dense only for acquisition of the wide azimuth type. In other cases, the three-dimensional CIG is poorly populated in the azimuth dimension.

Note that a popular alternative to the angle-azimuth coordinate system is the 'offset vector' (h_x, h_y). Its definition and relationship with offset and azimuth is sketched in fig. 4.19 and we will find it again in fig. 5.06.

Kirchhoff migration seen from the reflector's standpoint: reflection angle and other reflection-related CIGs

Common-offset Kirchhoff migration has practical advantages, particularly in the data sorting at the pre-processing stage. Unfortunately, offset is an acquisition parameter not related in a direct way to what actually happens at the reflection or diffraction point in depth, as illustrated in fig. 4.19.

The figure recalls the various acquisition parameters that characterize a specific trace of a 3D acquisition, namely:

- coordinates of the shot (x_s, y_s);
- coordinates of the receiver (x_r, y_r);
- coordinates of shot-receiver mid-point (x_{mp}, y_{mp});
- offset 2h (shot-receiver distance);
- shot-receiver azimuth α;
- offset vector (h_x, h_y).

These parameters are not independent and four of them, for instance {x_s, y_s, x_r, y_r} or { x_{mp}, y_{mp}, h, α} or again the more and more used combination { x_{mp}, y_{mp}, h_x, h_y}, suffice to fully locate and characterize the trace. For the sake of simplicity, we discard here elevations of shot and receiver that are crucial for land data, but land data processing goes beyond our scope, as it deserves a whole book in itself.

An event recorded at a time t of this trace corresponds to a reflection or a diffraction originated at a point D in the subsurface, which has its own 'local' properties:

- incident and reflected angles with respect to the vertical, which can be transformed into the more interesting ones below:
- the reflection angle ψ, which is also called 'opening' or 'aperture' or more classically now, the 'scattering' angle. This is the angle that controls the famous *'amplitude versus angle'* behaviour of the reflection in D, hence its importance. Recall that the normal $\vec{n}$ is the bisecting line of the incident and reflected ray only if the Earth is locally isotropic.
- azimuth β of the plane in which the reflection takes place is defined by the normal to the reflector and incident and reflected rays.

We can envisage several cases.
If D is an element of a reflector, the rays may be specular or not. Specular rays correspond to an actual reflection and one can then define the local reflector's parameters:

- reflector's dip θ
- reflector's azimuth φ

If D is a reflector's element but not specular for the rays, the corresponding reflected energy vanishes rapidly as the bisecting line of the incident and reflected rays depart from the actual normal to the reflector.

If D is a 'pure' diffraction point (isolated heterogeneity or short radius of curvature compared to the seismic wavelength), some energy is recorded for all dips and azimuths. This information is of interest to interpreters, since it may belong to fault elements, fracture corridors, karst features, pinch-outs, salt rugosity, etc. We saw in chapter 2 that such diffractions are present in seismic data sets. However, they are of an amplitude one or two orders of magnitude lower than typical specular reflections.

One can establish the relationship between these local parameters in depth and the acquisition ones by tracing rays between point D and the source and the receiver respectively, through the current velocity model.

Coming back to Kirchhoff PreSDM, we note that current practice is to regularize the data so that summation takes place on regularly spaced offsets within given apertures defined in the mid-points coordinates. It makes more physical sense however to perform the summation in the *local depth domain*, noting that the diffraction *summation principle* that we have applied so far to common offset data sets is *valid in fact for other common domains*, provided the weights are adapted to it. A natural and more physical one is thus the *scattering angle* ψ, and its azimuth β (Rousseau *et al.*, 2000; Ursin, 2004; etc.).

Fig. 4.20, similar to some extent to fig. 3.04, shows what would be the 'virtual acquisition' corresponding to a '*common reflection (or scattering) angle*' and variable hypothetical dip θ in a 2D case. The cartoon points out that a common reflection angle does not correspond to common offset. Computing the diffraction curve is a standard ray tracing exercise, but the diffraction summation implies a search for samples of traces that may be anywhere in the whole data set, as shown in fig. 4.21.

Common-scattering angle Kirchhoff migration (also called Common-Reflection Angle Migration or CRAM in short) consists thus in summing along diffraction curves for constant scattering angles and variable dips and azimuth, in a full 3D sense. By processing the data in this way, we create a new type of CIG: the 'Scattering Angle CIG'.
As for CIGs in the offset domain, the gathers must be flat to be stacked together.
There are clear benefits to Kirchhoff migration in the common-reflection angle domain:

- it *solves the multi-pathing issue*, because we scan reflections directly in depth;
- CIGs computed as functions of reflection angles are direct inputs to AVA analysis provided the Kirchhoff PreSDM code is 'amplitude preserving'. There is no longer any need for a poorly reliable offset-to-reflection angle transform, since this is done in a ray-based manner during the imaging process itself.

Fig. 4.22 is a real case comparison between Common-offset Kirchhoff PreSDM in panels (a) and (c) and Common Reflection Angle Kirchhoff Migration in panels (b) and (d)). The image improvement around the fault is clear when comparing panels (a) and (b). Panels (c) and (d) display a CIG and its corresponding Residual Move Out analysis (see fig. 3.22 for principles) for the Common-Offset and Common-Angle techniques, respectively.

Note on panel (d) the shape of the curve representing the maximum angle actually recorded, which also decreases as a function of depth. This reflects the physical limitation of resolution in velocity analysis and AVA, which generally decreases with depth.

As shown in fig. 4.19, summing along dip (and azimuth in 3D) at the reflection point in depth yields CIGs that are functions of both the scattering angle ψ and azimuth β. Again, the CIG is a

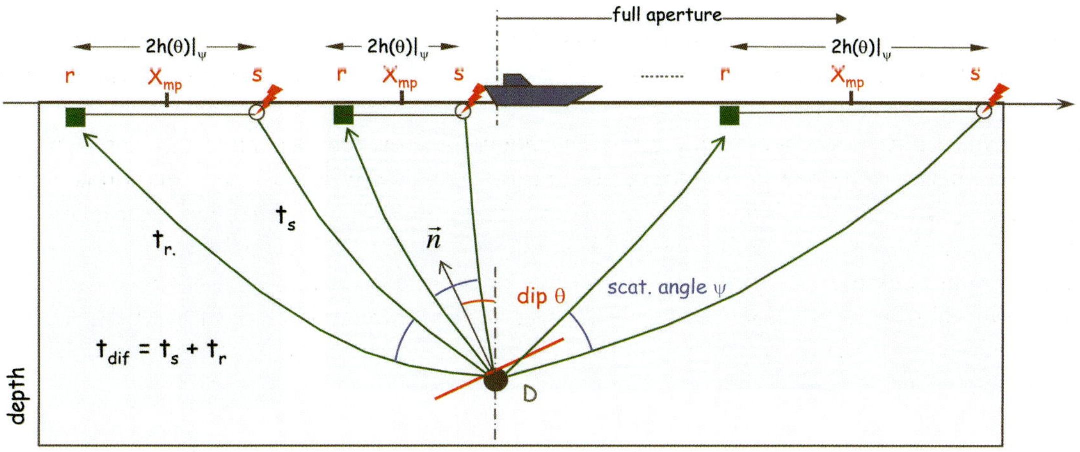

Figure 4.20: The diffraction point for a 2D 'common-reflection angle section'. Scanning dip for a constant scattering angle ψ is similar to scanning mid-points for a common-offset in fig. 3.09. If there is a dipping reflector in D, the corresponding rays will be 'specular' (i.e. honouring Snell's law) for the corresponding dip.

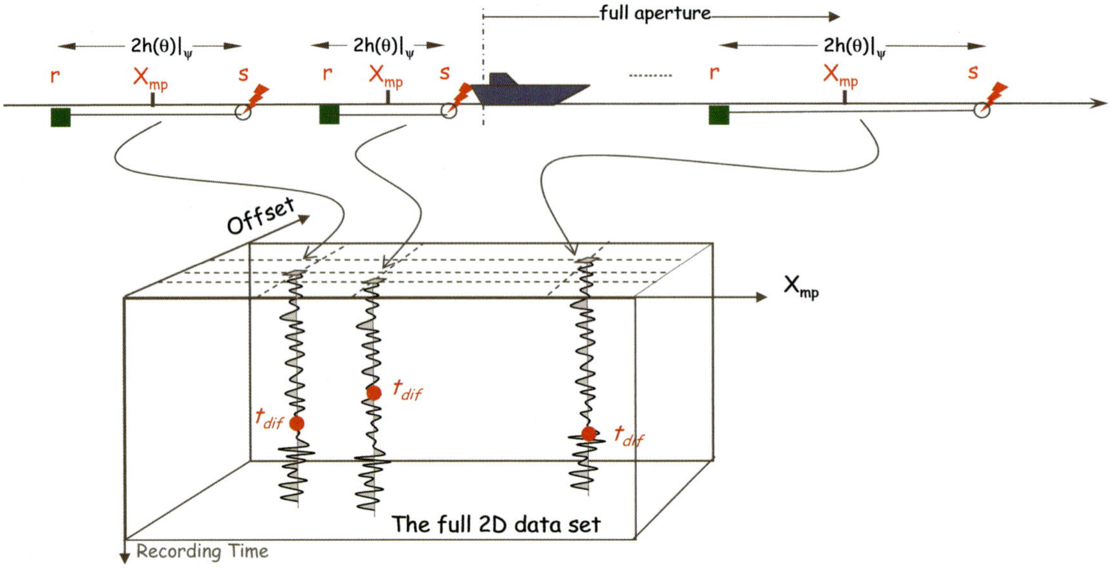

Figure 4.21: To get the samples along the Kirchhoff operator for a common-scattering angle gather requires some data manipulation. This allows however to compute Scattering Angle CIGs and other types of CIGs.

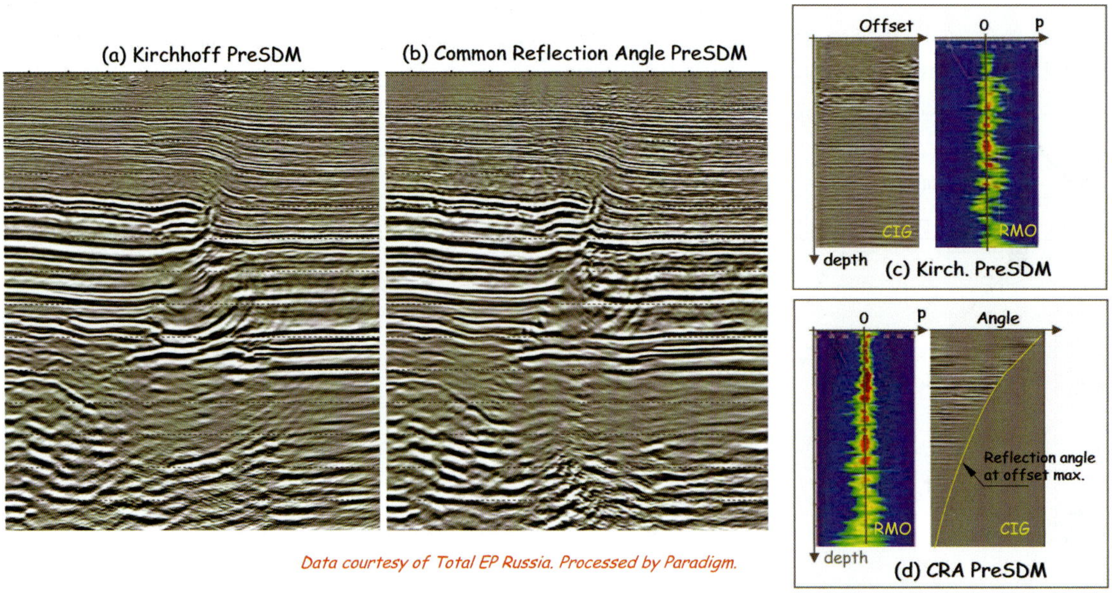

Figure 4.22: Example of 'Common Reflection Angle' PreSDM, compared with classical Kirchhoff PreSDM. The fault and the fault 'shadow zone' are better imaged with CRA migration. Note also the difference between the CIGs and the RMO panels.

3D object that should be visualized as such. Fig. 4.23 is an example of Paradigm's ES360 tool, which allows the computation and the display of various kinds of CIG. Here, we display Common Reflection (or Scattering) Angles CIGs, with the reflection azimuth β as the third dimension, either using a 3D viewer or as a juxtaposition of panels, where constant β panels are colour-coded.

Flattening these 3D gathers at each depth yields information on *azimuthal anisotropic velocity variations* (see chapter 8). Furthermore, if amplitudes are properly treated in the Kirchhoff migration process (especially when estimating the weights w in the Kirchhoff summations), these CIGs provide further information on Azimuthal Amplitude versus Angle. This is consequently a tool for *seismic reservoir characterization* as well as a 'pure' seismic imaging tool.

Diffraction imaging

Looking at the Kirchhoff summation from depth brings a new perspective to another imaging technology: diffraction imaging (Landa *et al.*, 1987; Landa and Keydar, 1998; etc.). There has been a renewed interest in this technology in recent years (Sava *et al.*, 2005; Taner *et al.*, 2006; Landa, 2007). As mentioned earlier, diffraction imaging may help to detect local heterogeneity that can be of interest, but it also opens new doors in migration velocity analysis (Reshef and Landa, 2009).

Looking again at fig. 4.20, instead of varying dip θ for a given scattering angle ψ, one can also vary the scattering angle for a given dip. The Kirchhoff summation will be performed in this case along scattering angles ψ for a fixed dip θ, but repeated for a series of dips (and azimuths in 3D, as usual).

The result of Kirchhoff summation along the scattering angle for one single dip is stored in the appropriate trace of a new type of CIG. When repeating the process for several dips and several depths at a given location, we obtain the Dip Angle CIG. As Audebert *et al.* (2002) and Reshef and Landa (2009) note, diffractions have a specific behaviour on a 'Dip Angle CIG'. Indeed, if D is an element of a reflector, the reflected energy on the CIG will present a concave 'bowl' of amplitude, vanishing as soon as the dip value departs from the actual dip. On the contrary, if D is an isolated diffractor, it will produce energy for all dips. It will then appear as a flat event on the Dip-Angle CIG,

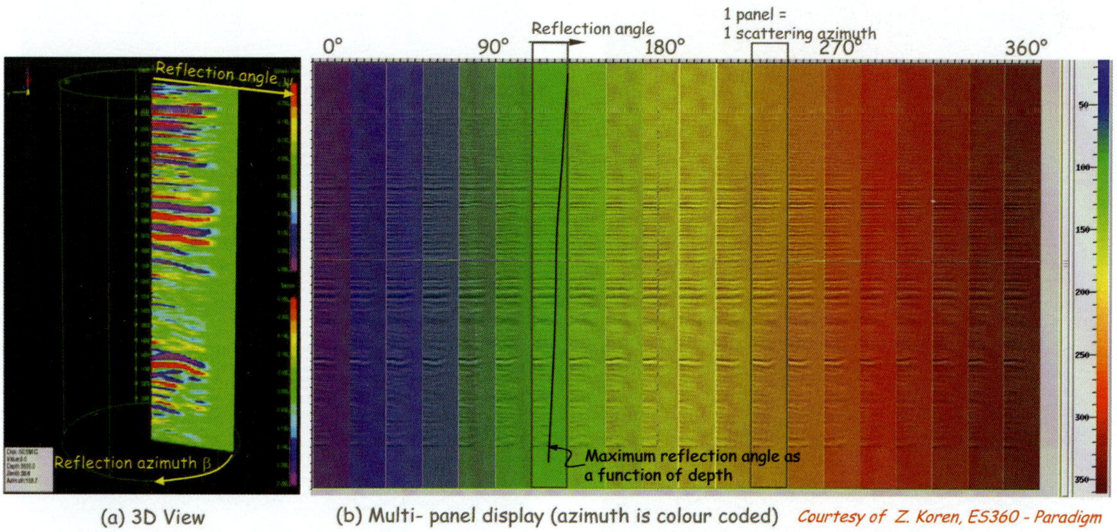

Figure 4.23: Example of 'Reflection (scattering) Angle and Azimuth' CIGs. (a) A 3D view of the CIG. The difference with the SA-CIG in fig. 4.15 is the reflection angle instead of offset dimension.
(b) A multi panel display: each panel corresponds to one reflection angle azimuth.

as long as the velocity model used for the Kirchhoff summation is correct. This would be a good criterion for velocity analysis. We can name two issues however, inherent to the process and topics of research:

- the weak amplitudes of diffracted energy compared to the reflected one; the question is then whether we should separate the two pieces of information and how;
- a given diffractor appears on many CIGs as a tilted event; it is a flat event only on the CIG just above it.

Fig. 4.24 illustrates a third potential benefit of diffraction imaging, i.e., the possibility to 'weigh' out in the Kirchhoff summation process the energy that may correspond to diffraction or diffuse noise or simply not 'specular'. The section on the left is a 'classical' common-angle Kirchhoff migration. On the section on the left, specular information is detected and weighted. The improvement in terms of signal -to- migration-noise ratio is clear.

Conversely, one may be interested in looking at the diffracted energy only and the corresponding imaged diffractors. As mentioned earlier, these may bring information on or enhance the image of geological features of interest, such as faults, shallow anomalies, karst, fracture corridors, etc. This opens a new domain to seismic imaging.

To conclude this section, we note that ray-based tomography based on Offset-CIG must be revisited for Common Angle Reflection Kirchhoff migration and the corresponding CIGs. This should not be an issue, since the principles of tomography remain unchanged and it is *a priori* simpler to implement. It is in fact easier to shoot rays from points in depth with a given starting angle than to solve the 'two-point ray tracing exercise' that consists in finding the specular ray that links a source point at the surface, a reflector in depth and a receiver at the surface. Because of its benefits and potential, common reflection angle Kirchhoff migration techniques are developing rapidly.

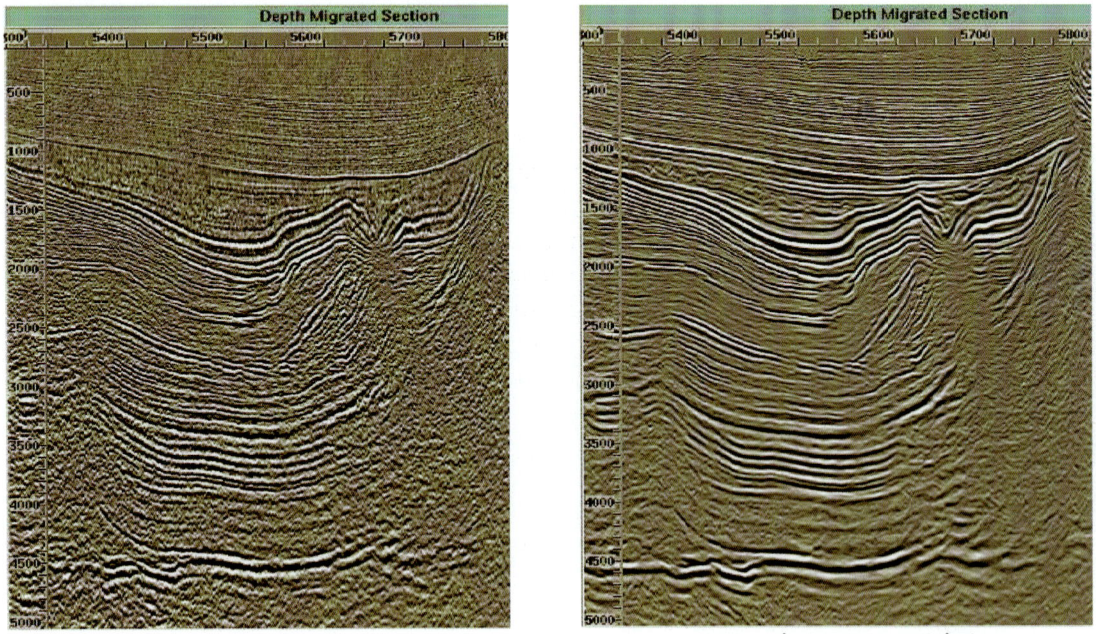

Figure 4.24: Example of signal-to-migration noise ratio enhancement obtained by weighing down non-specular information in a Common-Angle Kirchhoff Migration.

5 Ray-based Beam Migrations

This chapter 5, the second one dedicated to ray-based migrations, deals with two other classes of techniques. In the first part, we discuss the principles of offset-ray migration. For reasons that will become clear, all techniques start with a transformation of the data into the 'tau-p' domain. Some reminders of this transform used in many other applications and familiar to processors will be given next.

These preambles lead us to discuss first Gaussian Beam migrations, introduced and popularized particularly by (Hill 1990, 2001). We may qualify them as extended versions of 'data domain-oriented Kirchhoff migrations' (data domain- versus image domain-oriented methods are introduced in chapter 2 and particularly in fig. 2.08-a and b).

The second part of the chapter addresses the second class of migrations known as 'Beam steer', 'Controlled beam' or 'fast beam' migrations. As early as 1997, J. Sherwood and his group started revisiting beam migration (in Sherwood *et al.*, 2008) for PreSDM. The technique has proved to be successful and is now widely accepted and used by the industry. The various terminologies hide different implementations of concepts that are often similar, all clever extensions of the 'old' ray map migrations. The main ideas behind the different possible implementations are discussed in this chapter.

Concept of offset-ray migration of a reflection element from a shot gather

The concept of ray migration is introduced in chapter 2, fig. 2.04, for zero-offset data. The starting point is the relationship between the 'apparent velocity' or equivalently the 'time-dip' or 'ray parameter p' measured on data, the velocity at the surface and the emergence angle of the ray as the wave reaches the surface, discussed in chapter 1 and fig. 1.12.

In fig. 2.04, we presented the successive steps necessary to 'ray-migrate' an element of a zero-offset section through a given velocity model.

We extend it now to non-zero offset and the *migration of a reflection element from a shot gather*. Fig. 5.01 starts from forward modelling of fig. 1.12. We consider the reflection element corresponding to the ray traced in green in this figure. This element of the shot at x_s is defined by its receiver location (x_r), its reflection time t and ray parameter or time-dip p_r = dh/dt. The subscript 'r' (for receiver) specifies that the *emergence angle is measured at the receiver location*.

Knowing the velocity model V(x,y,z=0) at the surface, we can compute the emergence angle and initiate ray tracing. In the anisotropic case, phase (θ_{ph}) and group (ϕ_{gr}) angles must be computed jointly, noting that *p controls the phase angle* and not the group angle, as specified in red in the figure.

With the full anisotropic velocity field V(x,y,z) assumed to be known too, we can carry on ray tracing. This is generally performed stepwise, propagating the ray in depth, for instance by steps of constant time-intervals and honouring Snell's law at each step (fig. 1.11). In doing so, we *'back-propagate' the wave that reached the surface in x_r at time t.*

In the zero offset isotropic case of fig. 2.04, we knew that we had to back-propagate the ray during a time $t_{normal\_ray}/2$ in order to reach the reflector. The difference now with non-zero offset is that we know the total traveltime $t = t_s + t_r$ of an event recorded at receiver location R, but not the specific traveltime t_r from the *receiver to the reflection point in depth*, and t_r has no reason to be equal to t_s. The question is then where to stop back-propagation to locate the reflector?

There are different ways to address this issue and we present four of them below:

- In the shot domain, test several shooting directions p_s from the shot location.
- Sort the data to the common receiver domain and measure the ray parameter p_s at the shooting location on top of p_r.
- Measure the mid-point ray parameter p_{mp} in the common-offset domain and test various combinations of (p_s, p_r) pairs honouring p_{mp}.
- Measure the offset ray parameters p_h in the common-mid-point domain and test various combinations of (p_s, p_r) pairs honouring p_h.

The first solution is presented in fig. 5.01. The up-going ray, or 'receiver ray' reaching x_r is back-propagated during a time longer that the largest expected t_r. Similar to the receiver case, the down-going ray or 'source ray' that corresponds to the actual reflection in D has a shooting angle with its corresponding ray parameter denoted p_s. This ray is called the (source) *specular ray* because it corresponds to the actual reflection. The same relationship exists of course between shooting angle, velocity at the surface at the x_s location and ray parameter p_s.

Remember that our goal is to find the reflection point *D as the result of offset ray migration of the picked event*. The method consists in trying several different shooting angles, i.e., different p_s values. The corresponding rays are shot from x_s and propagated through the model, again honouring Snell's law at every velocity change. By construction, 'source ray' and 'receiver ray' intersect at the point D of the reflector. In 2D at least, all trial source rays will cross the receiver ray for a reasonable selection of p_s values. The multitude of source rays intersecting the single receiver ray gives rise to a multitude of possible reflection points. To resolve the ambiguity, we note that ray tracing provides us with travel-

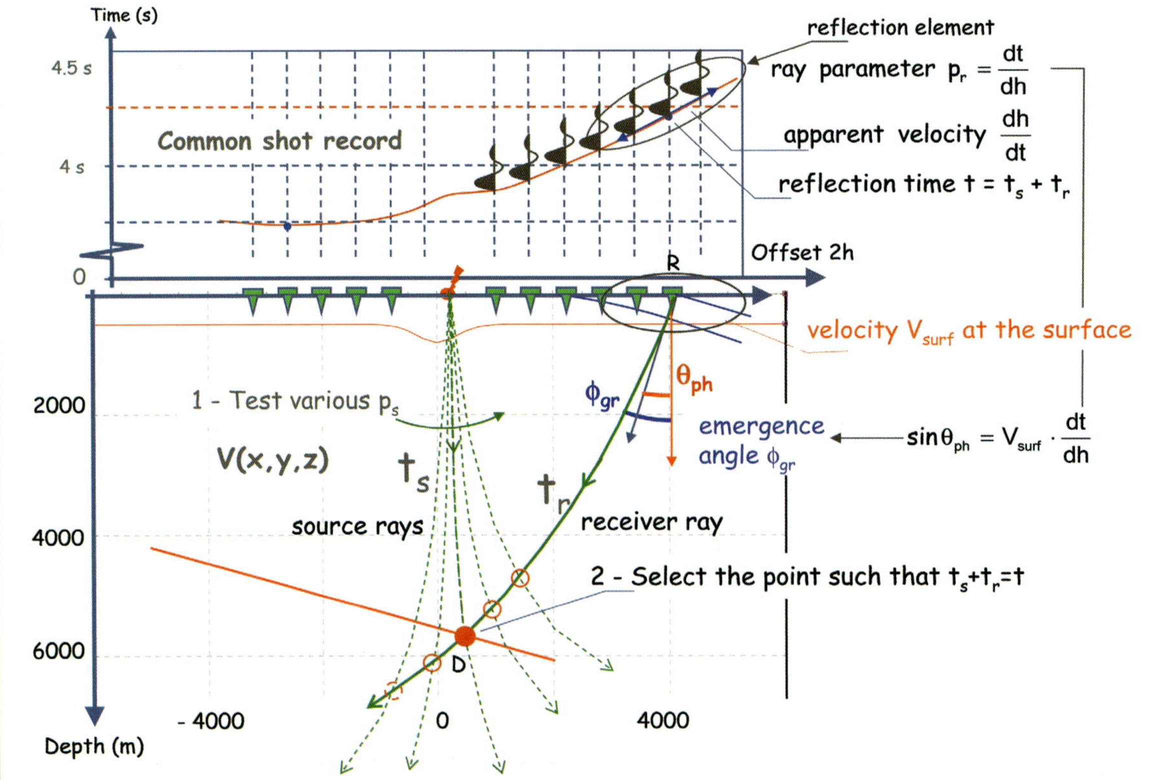

Figure 5.01: Offset-ray migration of a reflection element from a shot gather.

times t_s and t_r for both rays at their intersection. The 'ray-migrated' reflection point will therefore simply be the one such that $t_s + t_r = t$, t being the measured total traveltime of the event to migrate.

An alternative technique to this method is to measure p_s directly on the data, rather than testing hypothetical values. This requires sorting the data into common-receiver as illustrated in fig. 5.02. A common-receiver gather is the collection of all traces recorded at the same receiver location (not the same physical receiver in the marine case ...) but from various shots. The top of fig. 5.02 is a cartoon that shows how the gather may look in our synthetic example. Reflections can be picked on this gather, together with their reflection time and time-dip or ray parameter p_s. If we ensure that we *picked the same event* in both shot and receiver domains (they appear at the same reflection time t and same offset 2h), we can confidently trace the source ray with p_s and the receiver ray with p_r and *they should cross at the migrated location D*. We can check at this stage that the sum of the traveltimes of both rays from the surface down to this intersection D is equal to t. A discrepancy between t and $t_s + t_r$ indicates that the velocity model V is not correct or that p_r and/or p_s have not been picked accurately enough (... or that the picked event is not a primary one; we come back to this important comment later in the chapter). This is a way to express the over-determination of pre-stack ray migration. Conversely, finding V(x,y,z) such that p_s, p_r ,and traveltimes of all events picked in the seismic data set are compatible is the basis *of stereo-tomography (Lambaré, 2008), an alternative to the image gather tomography of chapter 4.*

This method requires adapted acquisition parameters: specifically the shot interval must be small enough to enable an accurate measurement of p_s. This may be a real issue, particularly in 3D. The method also necessitates data sortings, which may be cumbersome and time consuming.

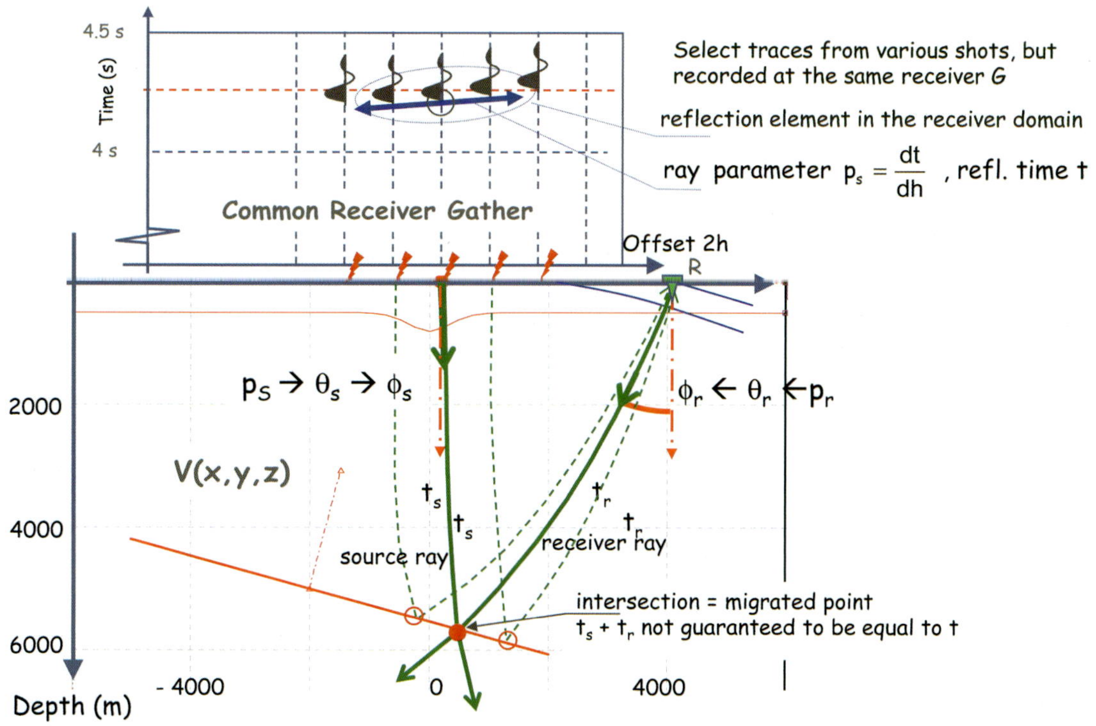

Figure 5.02: Offset-ray migration using a measurement of both p_r (in the common-shot domain) and p_s (in the common-receiver domain). This implies to pick the same event in both domains.

Offset-ray migration in the common-offset domain

The third option is widely used in beam migration and is performed in the common-offset domain. The method is illustrated in fig. 5.03. The cartoon shows the common-offset reflection element of our synthetic model and to what rays it corresponds. Apparent velocity of the reflection element can be picked as in the previous cases and is denoted p_{mp}, the p value at a mid-point location of a given common-offset data set. It can be shown that for a given reflection element, $p_{mp} = p_s + p_r$ where p_s and p_r are the ray parameters of the source ray and receiver ray respectively, of the corresponding *specular reflection*. We make use of this property to migrate the element in the way schematized in fig. 5.04. Now, both source-ray and receiver-ray are shot with hypothetical p values, but *in pairs such that* $p_s + p_r$ *are equal to the measured common-offset* p_h value. Pairs of rays are shot through the velocity model and for each pair of rays the intersection, if it exists, is a possible migrated location. The ambiguity between the various solutions is cleared up by taking once again the one that honours the total traveltime, i.e., such that $t_s + t_r = t$.

Offset-ray migration of elements picked in the mid-point gather domain

The fourth domain used in beam migration is the mid-point gather. The definition of a common mid-point gather and its geometry are given in fig. 3.03-a in the 2D case.

A cartoon of the mid-point gather at location x_{mp} for our synthetic model is shown at the top of fig. 5.05, as well as a reflection element in this domain. The corresponding rays are illustrated as usual in the model. The offset ray parameter of this element is denoted p_h. Parameters p_h , p_s and p_r are not independent, the relationship being $p_h = \pm (p_s\text{-}p_r)$. The minus sign comes from the fact that source and receiver 'move' in opposite directions when changing offset by an amount dh at constant mid-point location, and the ± sign from convention in the sign of offset.

Migration of the reflection element proceeds very much as for a common offset element in fig. 5.4, the only difference being that the pairs of rays shot must now honour the $p_h = \pm (p_s\text{-}p_r)$ relationship.

Issues in 3D offset ray migration

We have mentioned possible sources of error in offset ray-migration, making it such that the migrated point is erroneously located or not defined at all (no ray crossing). This is particularly true in 3D, where the rays may not be in the same planes. Fig. 5.06 shows in more detail some of the 3D geometrical aspects of ray shooting.

Firstly, the typical marine shot-receiver acquisition system is defined by four parameters: shot and receiver coordinates, or equivalently two mid-point coordinates, offset and azimuth. This latter information needs to be kept when building common-offset volumes and taken care of in ray-migration. For land or nodes data, elevation is another key parameter that we do not discuss in the framework of this book.

Secondly, the source ray and the receiver ray at the surface are each defined by three-dimensional vectors. In fig. 5.06, they are represented by the emitted and emergent *slowness vectors*. These vectors have the magnitude of the inverse of the velocity at the surface at their respective source and receiver locations, and the direction of the emission (source) and emergence (receiver) angles. Slowness vectors project onto the surface as 'ray parameter vectors' (see also fig. 1.11). The ray parameters in 3D are consequently characterized by two values each; classically their projections $p_{s,x}$, $p_{s,y}$, $p_{r,x}$, $p_{r,y}$ onto the x and y axis of the processing grid are used. These four parameters are necessary to send the rays in the 3D model and consequently have to be estimated from the data.

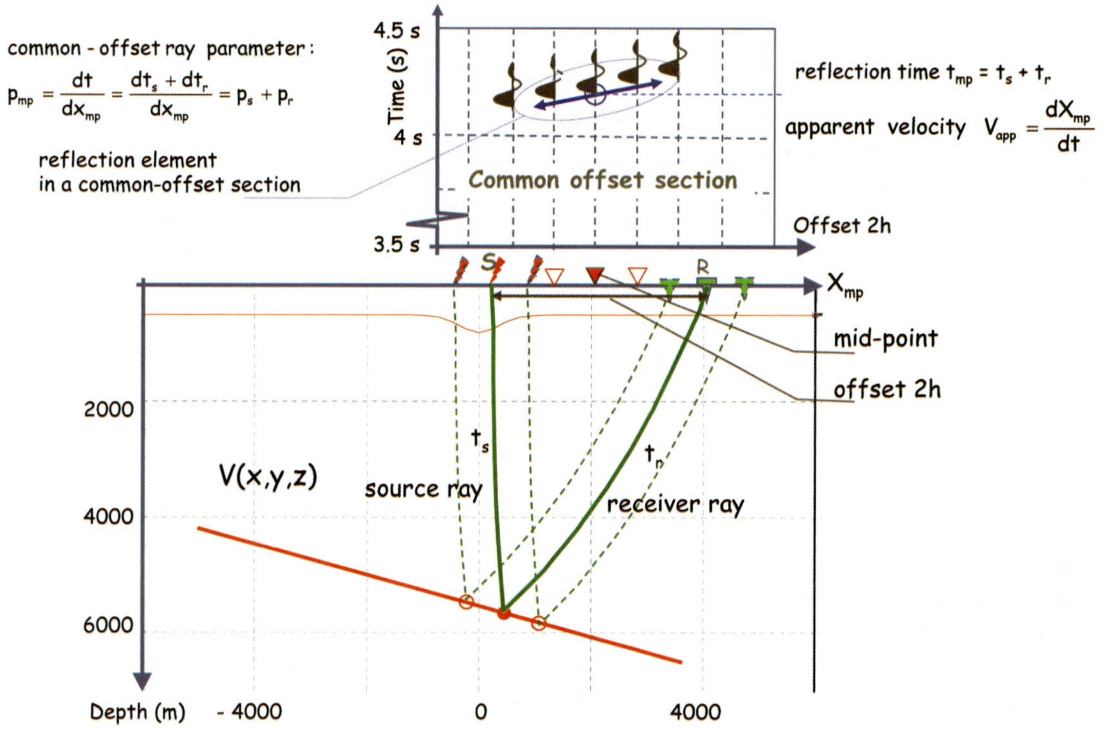

Figure 5.03: Ray parameters in the common-offset domain. The ray parameter p_{mp} of a reflection element at a mid-point location x_{mp} is the sum of the source and receiver ray parameters.

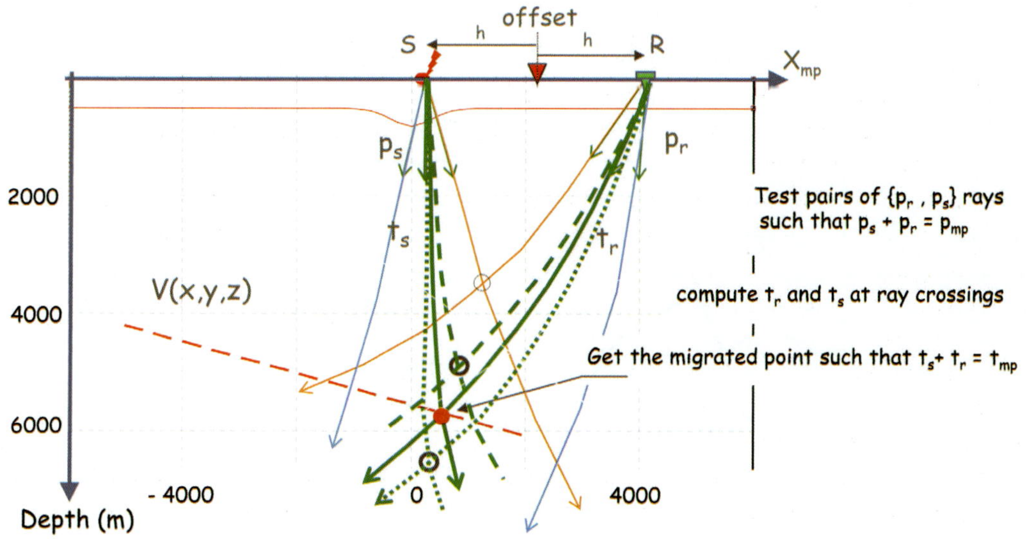

Figure 5.04: Offset-ray migration of an event picked on a common-offset section at location x_{mp} defined by its ray parameter p_{mp} and reflection time t_{mp}.

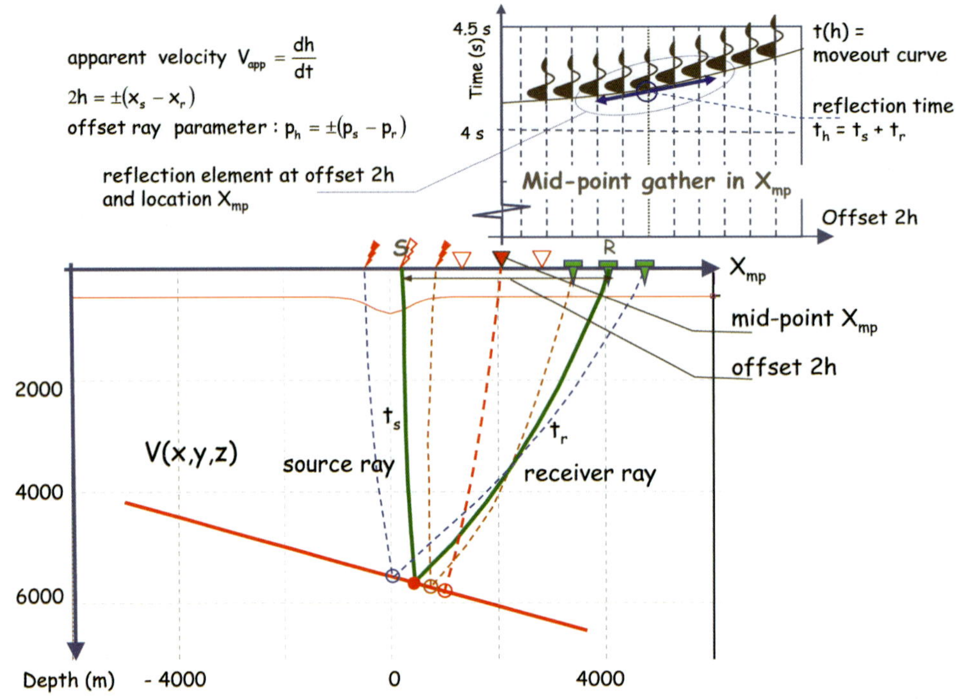

Figure 5.05: Ray parameters in the mid-point gather domain. The offset-ray parameter p_h of a reflection element at an offset 2h is the difference of the source and receiver ray parameters.

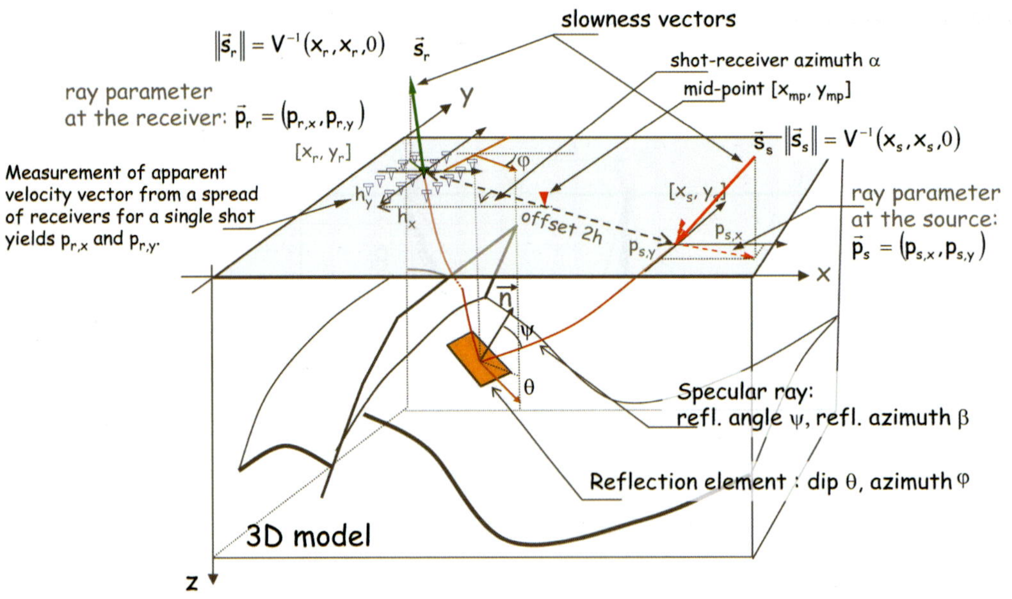

Figure 5.06: Ray geometry in 3D. A specular ray is defined by its source and receiver coordinates, offset and azimuth, vectorial ray parameters at source and receiver locations, reflection angles and azimuth at the reflector, dip and azimuth of the reflection element. All these parameters are interrelated through the velocity field.

In the shot domain, an extension to 3D of fig. 5.01 may be reasonably easy for 'wide-azimuth data', such as land, ocean bottom cables or nodes acquisition, as shown in fig. 5.05. For a given shot, receivers are spread over a grid around the central one currently processed. The measurement of apparent velocity on a gather of traces sorted by receivers aligned in the x direction yield $p_{r,x}$ and the same analysis for receivers aligned along the y direction gives $p_{r,x}$. Ray tracing can then be initiated in 3D for the receiver ray in the direction defined by $p_{r,x}$ and $p_{r,y}$. Similarly, pairs of ray parameters ($p_{s,x}$ and $p_{s,y}$) are hypothesized on the shot size and the source rays shot into the 3D velocity volume using these ray parameters.

Note as well in the figure that the shot-receiver offset 2h and azimuth α can be replaced by the 'offset vector' **2h**, which can be characterized by its two Cartesian coordinates h_x and h_y. Ray parameter in the mid-point domain (fig. 5.05) generalizes then in 3D as the pair (p_{hx} and p_{hy}), which measure the derivatives of arrival time as a function of h_x and h_y for a common mid-point.

Migration quality indicator

Clearly, there are no reasons why these various sets of rays may cross, hence the difficulty to estimate the hypothetical migration location. Various techniques are used in order to try and resolve this issue. A possible one, based on neighbourhood analysis, is presented in fig. 5.07. The figure is again in 2D for the sake of simplicity but extension to 3D is easy to imagine. In this example, traveltimes of both source-ray and receiver-ray are computed at regular depths and summed. The estimated migration location is taken as the barycentre of the ray-ends at the depth such that the sum of both traveltimes $t_s + t_r$ is equal to the reflection time t. The distance between the two end-points is a measurement of the 'accuracy' of the solution. If it is too large, the result may be simply discarded. This migration 'quality' indicator, also called a 'quality factor' or 'focusing factor', is used as a criterion of selection in fast beam migrations, as we see in the next paragraph.

Many other approaches are possible and used in practice. The point is made here to show practical issues and some of the limitations of ray-based migrations mainly related to:

- need for a precise and fast 3D ray tracing;
- accuracy of the measurement of 3D ray parameters;
- impact of acquisition lay-out on this accuracy;
- uncertainties in the migration velocity model.

The second point in this list is crucial. In all current applications of ray-based migrations, this measurement is made using some specific version of tau-p transform, which we now discuss.

Tau-p transforms

In order to migrate a reflection element, we need to measure its ray parameters first. As fig. 5.01 to 5.05 show, this consists in searching for *slopes of locally consistent events*. One classical way to do that is 'local slant stack' or linear local tau-p transform. Its principle is explained in fig. 5.08, with an application in the shot domain.

First, a 'nominal' offset and an 'offset window' of width W around it are selected. Then, for a given *time*, classically denoted tau (τ), and a given discrete *slope* parameter denoted p_j, we can sum samples within a dipping window of slope p, centred at τ and of time thickness Δt. For time τ, this summation is repeated for a whole range of discrete p_j values. The process is then repeated for a series of (possibly overlapping) tau windows at discrete τ_i values. Depending on implementations, Δt may reduce to a single sample and τ_i extend to every time sample of the data.

Each sum is placed in a 'tau-p panel' at the corresponding p_j abscissa and τ_i ordinate. This panel is the 'tau-p transform' of the offset window of the shot being analysed.

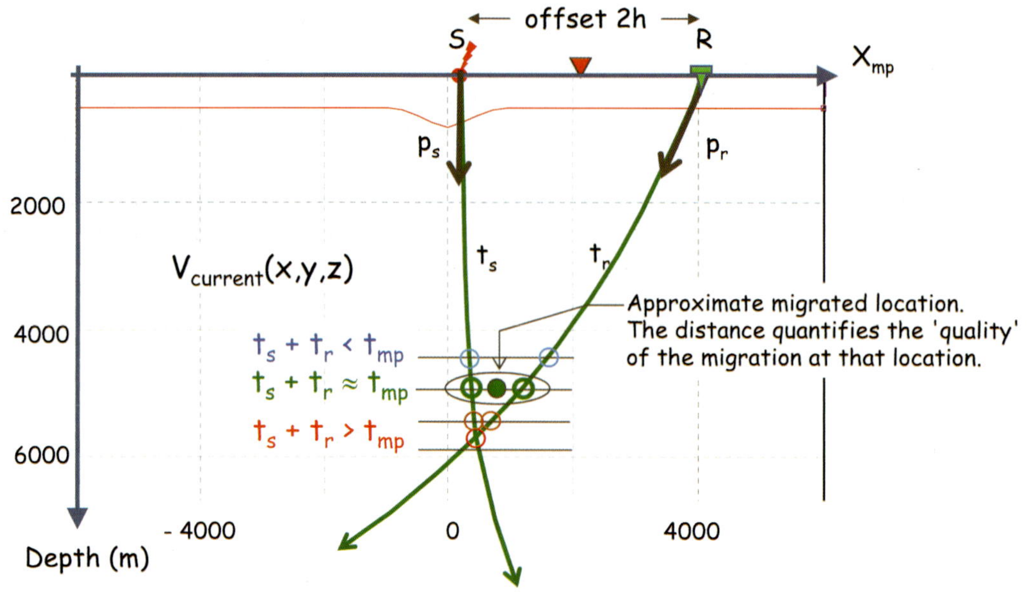

Figure 5.07: If the velocity field is not correct or ray parameters are inaccurate, rays may not cross, especially in 3D, or cross at a point such that t_s+t_r is different from t_{mp}. Approximate solution are to be found in a neighbourhood, the size of which gives an estimate of the 'quality' of the migration.

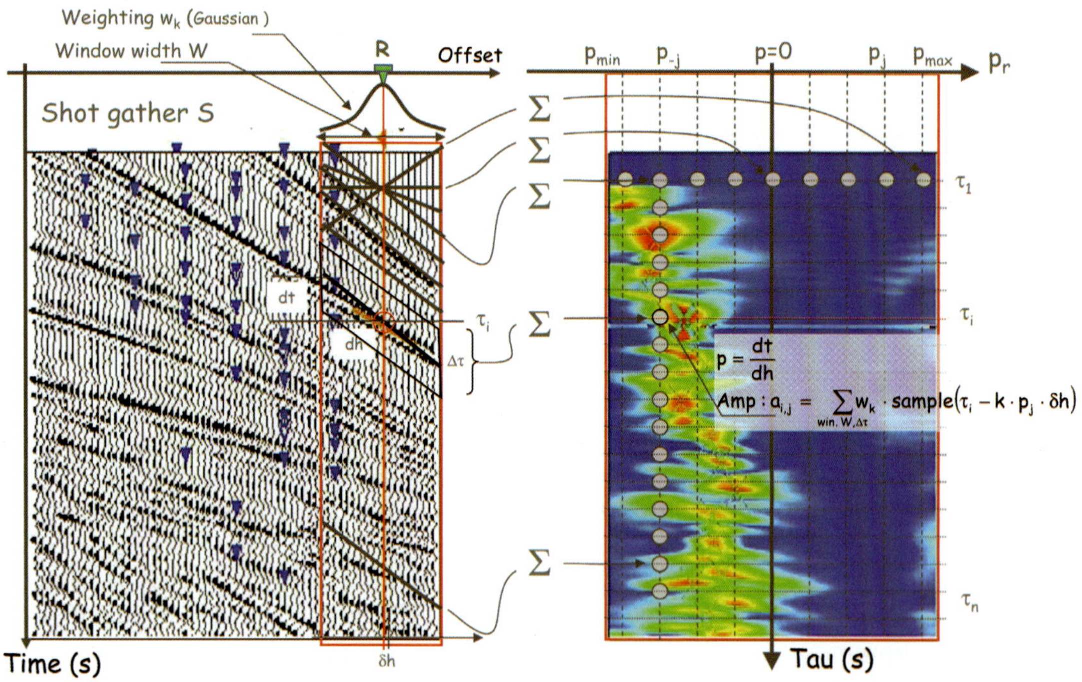

Figure 5.08: τ-p transform of shot gather. A shot is analysed window-wise. The amplitude of the transform of a window is the sum, possibly weighted, of samples in a dipping window. This transform yields the amplitude corresponding to the emergence angle of a reflection 'event' (beamlet).

The process can be slightly modified by weighting samples as a function of their distance to the central trace of the window. A weighting of the form of a Gaussian function, such as the one schematized in fig. 5.08, is systematically applied in Gaussian Beam migration in particular.

Peaks in the tau-p panel indicate the presence of locally coherent events. Such an event can be an element of a reflection somewhere in the subsurface, such as the one in cartoon 5.01. The abscissa of such a peak indicates the slope p of the corresponding event; its time τ is the time $t = t_s + t_r$ of the assumed corresponding reflection. These are precisely the pieces of information that we are looking for and that we need in order to migrate the event with one of the methods presented in fig. 5.01 or 5.02. In the following, we will discuss two types of migrations:

- Gaussian Beam migrations that migrate every single sample of the tau-p panel;
- other beam migrations that migrate only selected peaks of the panel.

One tau-p panel, as in fig. 5.08, corresponds to an offset window of a 2D shot only. The whole shot gather is analysed using such windows overlapping along the offset of the shot record, as shown in fig. 5.09. Thus, the tau-p transform of a 2D shot record is a 3D volume. The fact that offset windows overlap increases the probability that a given event that has some continuity with offsets is picked coherently from one tau-p panel to the next one. At the same time, we want neither excessive redundancy (cost) nor too strong lateral coherency enhancement. This is why weighting factors are generally chosen such that their sum along the offset direction remains close to one, as shown in fig. 5.09. In doing so, amplitudes of samples along a given event are maintained in the migration process that will follow. This is also a specific feature of Gaussian Beams.

Fig. 5.08 and 5.09 are drawn for 2D shots. If we deal with 3D shots, the offset window becomes an area such as the 'spread' of geophones sketched in fig. 5.06. In this case, the (weighted) summation of the 3D slant stack is *computed along a surface*, now defined by two p parameters, denoted p_x and p_y in this figure. The tau-p panel becomes a 'tau-p volume' function of these two coordinates, and the analysis of a 3D shot becomes a four-dimensional object! This point is made to emphasize the fact that modern processing implies handling multi-dimensional data sets: how these data sets are manipulated in the computer is a key practical issue in terms of efficiency, but it goes beyond the scope of this book to discuss it.

Gaussian Beam Migration principles

Gaussian Beam Migrations or GBM in short, can be applied either in the shot domain or in the offset domain. We use the shot domain GBM case to explain some of its principles.

The process starts with a tau-p transform of the shot gather, similar to the one presented in fig. 5.09 and is followed by the migration *of each element of this tau-p panel.*

For that purpose, we note that an event in this panel, characterized by its time τ, time-dip or ray parameter p, receiver location x_r , offset 2h, from the shot at coordinate x_S, would correspond to a reflection recorded by shot S and receiver R, at a reflection time $t_s + t_r = \tau$, and emerging at the surface with a ray parameter p_r. It can consequently be migrated through the current velocity model V(x,y,z) as described in fig. 5.01, remembering that its amplitude $a_{i,j}$ is the weighted average of seismic amplitudes along slope p_j within the window W.

'Gaussian beams', a kind of 'fat ray', allow us to take the *amplitude factor* related to geometrical spreading and the fact that it *comes from a window along the offset axis* into account.

Fig. 5.10 presents the concept of Gaussian beams and paraxial rays, another kind of 'fat ray'. Both are designed to describe the propagation of an element of wavefield, initially plane and laterally

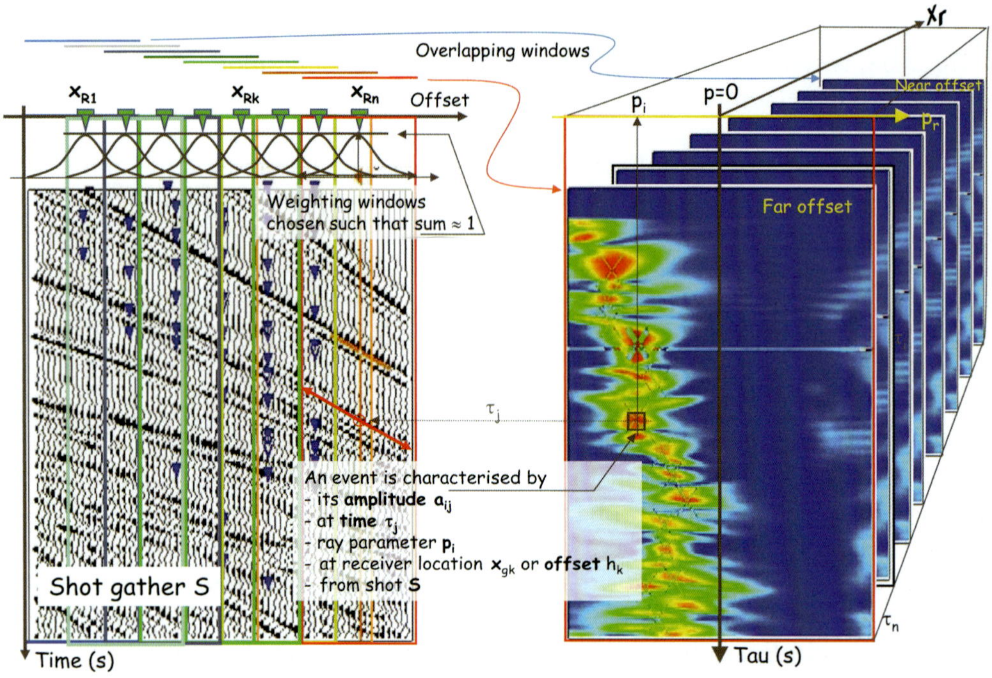

Figure 5.09: τ-p analysis of the whole shot. Each overlapping window corresponding to a nominal offset x_r is processed independently. The tau-p transform of a 2D shot is a three dimensional volume.

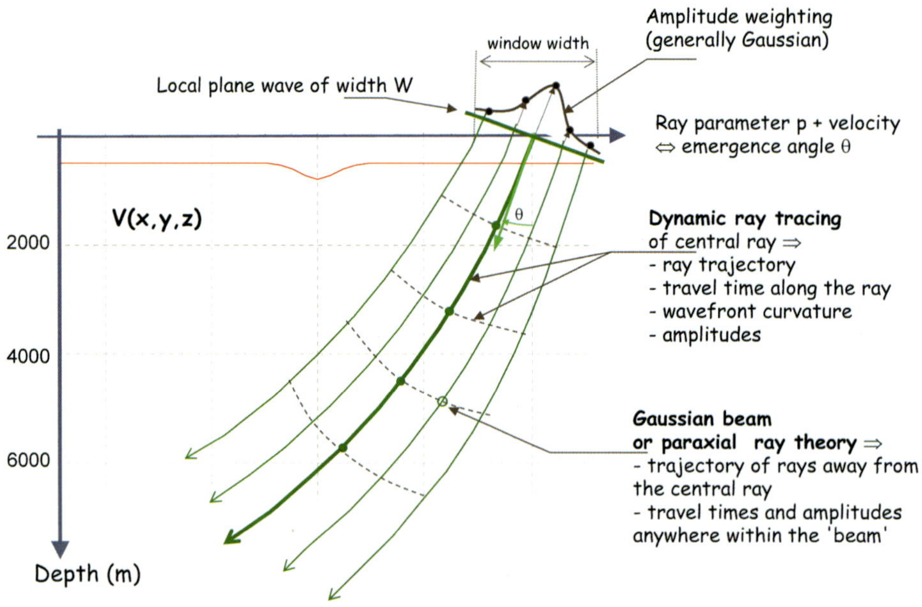

Figure 5.10: Paraxial rays and Gaussian Beams allow for the computation of 'fat rays' using dynamic ray tracing for a unique 'central' ray. It simulates the propagation of a local, weighted plane wave of a given width W and ray parameter p.

weighted, as it propagates through a medium defined by its velocity and anisotropy in every point. The goal is to emulate wavefield propagation (acoustic in our case), using ray theory and shooting only one ray, called the 'central ray'. The plane wave element may be tilted, its slope being characterized by its ray parameter p. Kinematic ray tracing, as described in fig. 2.04-b, provides the ray path only but 'dynamic' ray tracing (Cerveny, *et al.*, 1982) adds information on how the wavefront is deformed by refraction (wavefront curvature) in a heterogeneous medium and especially at velocity interfaces. This curvature quantifies the geometrical spreading, itself related to spherical divergence along the central ray. Gaussian beam (with its 'imaginary rays') and paraxial ray theories allow us to accurately approximate traveltimes and amplitudes of the waveform everywhere within a 'beam' around the central ray. In this way, the computation of a single ray, along with its associated paraxial rays, gives access to information about propagation in a much larger zone of the subsurface.

The possibility and benefits of using Gaussian beams for offset migration of events picked in the tau-p domain should become clear when comparing fig. 5.10 with fig. 5.01 for instance.

The Gaussian Beam Migration of one element of the tau-p panel of a common shot gather is thus schematized in fig. 5.11. In order to be consistent with the fact that we migrate points resulting from a *linear* tau-p analysis with a *constant width window* W, we use the same parameters in GBM and tau-p transform.

The sequence of actions to migrate a single point in the tau-p domain in a GB sense is an extension of the one in fig. 5.01. It could be the one sketched in fig. 5.11:

- pick the amplitude $Amp_{i,j}$ of an event in the tau-p domain defined by its shot coordinate x_s, the location x_r of the centre of the tau-p analysis window and its receiver ray parameter $p_{r,j}$, its time τ_i as sketched in fig. 5.09;
- compute the receiver central ray and various source central rays from x_r and x_s respectively, in order to define the migrated location D, such that $t_r + t_s = \tau_i$, as described in fig. 5.01. The central source ray is now known;
- compute the paraxial rays around both source and receiver central rays;
- combine both sets of paraxial rays to compute the locus of point within the intersection zone of the two 'fat rays' honouring also the relation $t_r + t_s = \tau_i$;
- compute also the corresponding amplitude factors at these points;
- these provide us with the locus where to 'spread' $Amp_{i,j}$, with the appropriate spherical divergence for source and receiver propagations and other amplitude correcting factors;
- place these amplitude values at the corresponding points in depth;
- The result is the *'migrated beamlet'* of the specific tau-p domain event, sketched in fig. 5.11.

Since paraxial or Gaussian beam rays keep track of the evolution of amplitudes along the central source and receiver rays and within the whole beam, amplitudes are taken into account in the GB migration process. However, the migration quality indicator defined in fig. 5.07 cannot be computed. In this sense, the Gaussian Beam offset migrations slightly differ from point-to-point migrations in fig. 5.01 or 5.04.

Clearly, all events in the p-column of the tau-p panel can be migrated in a similar way, as illustrated in fig. 5.12. The result is a 'migrated beam' of common-emergence angle.

In this figure we have also symbolically sketched part of the output migrated image grid. This is to recall that amplitude values from the 'migrated beamlets' need to be regularly digitized on this grid through some kind of interpolation. The size of the grid intervals (Δx, Δz) and the beam width W must be related.

The migration of the tau-p windows is naturally performed by 'Gaussian beam migration' in the way described above, amplitudes corresponding to all points of the tau-p panel, computing the migrated amplitudes on the output image grid and summing every individual beamlet's contribution. In doing so, we obtain the migration of the window around the current receiver location. The process needs to be repeated for all (overlapping) windows of the shot to obtain the 'migrated shot'. And finally the whole process is repeated for all shots to obtain the final GBM result. All these processes use the same and unique velocity and anisotropy model, of course.

GBM versus Kirchhoff migration: some comments

Fig. 5.13 illustrates the migration of *one single line of the tau-p domain* that represents the migration of an isochrone (only receiver Gaussian beams are represented). Gaussian beamlets are summed up along this isochrone line.

This result has similarities with fig. 2.08-a, where we saw that zero-offset migration of a single event can be performed by spreading the amplitude over an isochrone (circular in the homogeneous case). We called it 'data-oriented' implementation of Kirchhoff migration. However, there are several important differences between fig. 2.08-a and 5.13, which explain some of the potential benefits of GBM versus Kirchhoff migrations:

- The isochrone is not a circle because we deal with *non-zero offset* and possibly complex velocity models allowed by the use of ray tracing; this is true however for Kirchhoff PreSDM too.
- Amplitude along the isochrone is better handled. Firstly, the 'fat rays' introduce some degree of smoothing of the amplitude along the isochrone.
- Secondly, and more importantly, the slant stack process *enhances events that have some lateral continuity, with a maximum for their effective emergence angle* and reduces the amplitude of events away from the right ray parameter, or for non-coherent events (noise). It has consequently a degree of directivity that conventional Kirchhoff summation, which sends a given event to all possible directions with the same amplitude, as suggested by fig. 2.08-a, does not have. This *reduces the level of migration noise in GBM*.
- GBM solves the multi-pathing issue; this important point is valid for all beam migrations and is consequently addressed in more detail at the end of the section.

GBM key parameters

As a recapitulation of the paragraphs on GBM, we note that the window width of both tau-p transform and Gaussian beam migration impacts the migration results.

The window should not be too wide because:

- Gaussian beams provide reasonable approximations of events in the beam as long as the actual velocity field does not change too fast laterally away from the central ray and within the window W.
- It should comply with the *linear* tau-p analysis. Reflection events are generally curved, especially on a shot gather and can be approximated only locally by a straight line.

Choosing W too small:

- increases the probability of wavefront curvature to increase exaggeratedly.

The tau-p window width and 'thickness of fat rays' W, is then optimized as a function of the frequency content of the seismic data. The choice of W drives the number of windows per shot and equivalently the distance between the central line of two windows and the percentage of overlap.

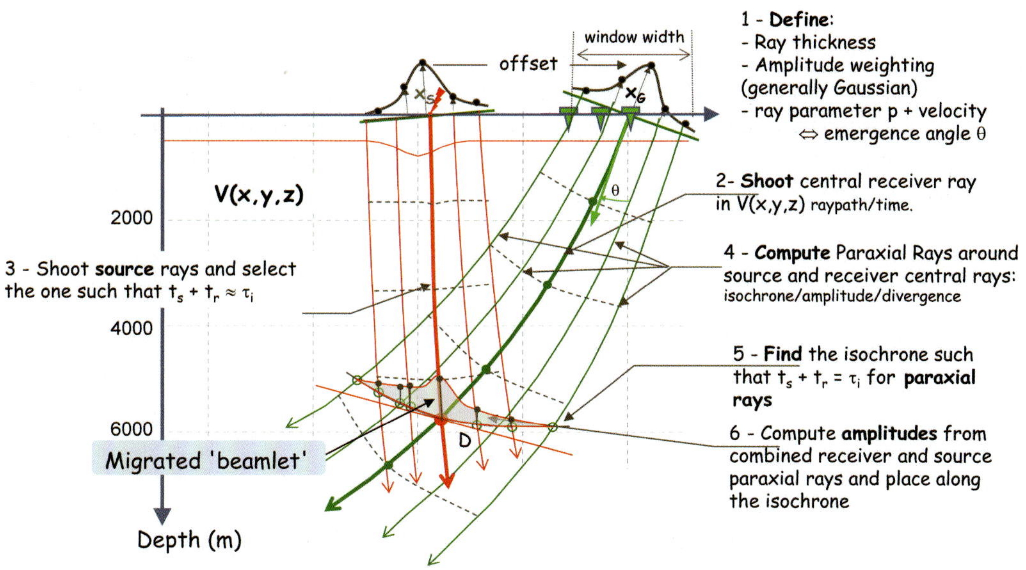

Figure 5.11: "Gaussian Beam Migration" of a single event in the shot tau-p domain. The event is defined by its reflection time τ_i, amplitude $a_{i,j}$ and ray parameter p_j. The source and receiver rays are computed as in fig. 5.01 or 5.03. 'Gaussian beam' or 'Paraxial rays' shot from both receiver and source provide the isochrone and weights to build the corresponding 'migrated beamlet'.

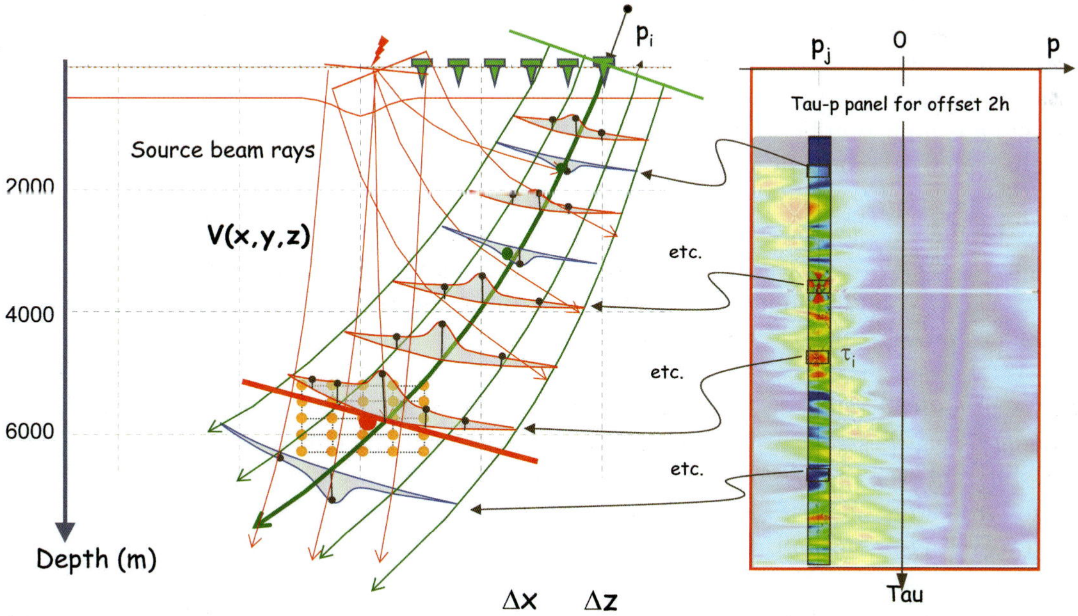

Figure 5.12: "Gaussian Beam Migration" of a p_j column of the shot tau-p domain. For each time τ_i, the source beam ray is computed and isochrones $t_r+t_s=\tau$ computed to build the corresponding 'migrated beamlet'. All migrated beamlets are summed together.

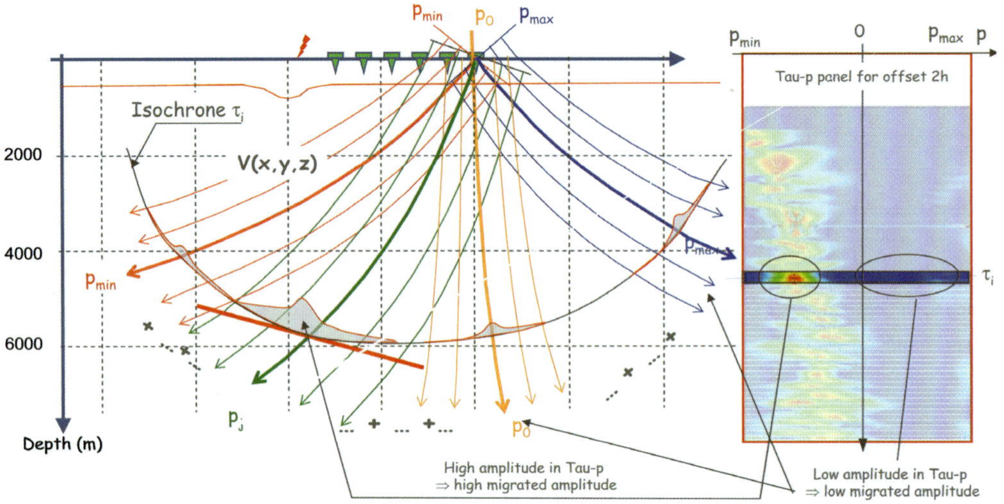

Figure 5.13: "Gaussian Beam Migration" of the whole tau-p panel. The operation of fig. 5.10 is repeated for all columns (all ray parameters) of the tau panel and all migrated beamlets are summed together. Note that the 'wavefront' corresponding to a single time has a 'directivity'. For the migration of a profile, this has to be repeated for all offsets, resulting in the migration of a shot; and then for all shots.

The other important parameters are:

- the weighting function in the window (typically Gaussian);
- the range of p values and especially their limits p_{min} and p_{max} ; these drive the maximum dip that may be migrated;
- the number of p values (typically around 50 in 2D);
- the p value interval, as a consequence of the two previous ones.

Fig. 5.14 is a comparison between 3D Kirchhoff migration (left panel) and Gaussian beam migration (right panel). The benefit in terms of signal-to-migration noise is apparent on the GBM result. On the contrary, some steeply dipping events appear more clearly on the Kirchhoff migration. This may be due to *selection of* p_{min} *and* p_{max} *in GBM that controls the maximum emergence angles and consequently, in many cases the maximum dips* that are effectively migrated.

The choice of W and number of p values is a trade-off between efficiency and frequency content of the data to process, ultimately related to lateral and vertical resolution.

So far, we have dealt with GBM in the shot domain. Many practical *applications of GBM are implemented in the common offset domain* (Hill, 2001). The principles are similar although the practicalities may be different. For the sake of avoiding redundant explanations, we will not detail them here. We choose instead to address ray-based migrations in the common offset domain for diplet migration in a following paragraph. The reader will easily infer from this how common-offset migration applies in principle to GBM too.

Controlled, beam-steer, wavelet-based, diplet-based and other 'fast' beam migrations

When looking at the tau-p panel in fig. 5.08, one realizes that the migration of all elements of this panel represents a lot of work done to little avail: we migrate either noise or zero amplitude data.

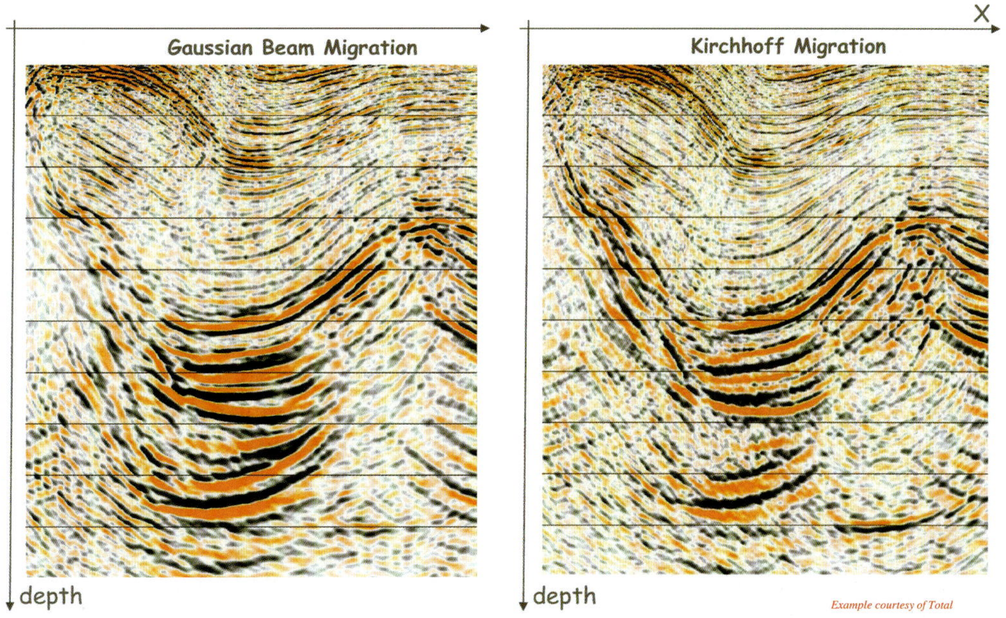

Figure 5.14: A comparison of a Gaussian Beam against Kirchhoff 3D migration. The signal/migration noise ratio is better with GBM, improving the continuity of events. Some steep dips are however better rendered with Kirchhoff.

It comes consequently as a natural idea to migrate only a (smart) selection of events; this speeds the process up to achieve 'fast beam migration'. However, practical implementations are not as simple as that.

As we have seen so far and as fig. 5.15 summarizes, the result of migration of a single event in the tau-p domain is a point in the image domain. This *point-to-point mapping* links via ray tracing through the current velocity and anisotropy model an event in the data domain, a point in the tau-p domain and a segment, in the depth migrated domain.

Point-to-point mapping is however not sufficient to build the migrated image. The end result of such a migration of a set of selected events would be strange looking, with horizon continuity in the migrated domain being lost. We consequently need ways to get around this issue and to migrate 'waveforms' instead of single points. The diversity of names in the literature for different implementations (controlled beam, fast beam, parsimonious, beam steer, diplet-based, etc.) reflects the variety of current approaches to achieve this goal. Gaussian beam in itself, as we have seen it, is one possible answer: a point in the tau-p domain migrates as a *beamlet*, as presented in fig. 5.11. The corresponding approach is called 'controlled beam migration' (Ting and Wang, 2008).

Fig. 5.15 suggests another way: a point in the tau-p domain corresponds in fact to a data window, called '*time wavelet*' in the figure. This time wavelet can be a base for the migrated waveform. Finally, we will see in a following paragraph yet another approach relying on an improved tau-p transform posed as an inverse problem.

'Controlled' beam migrations

'Controlled' beam migration (CBM in short) starts by the pick of the most relevant events in the tau-p panel. Selection criteria are common to the various beam migration implementations and are discussed at the end of the chapter.

The selection being done, we know how to migrate in a Gaussian Beam sense each individual point in the tau-p panel (fig. 5.11).

The key parameters are of course the selection criteria and the resulting number of picked events but important as well is the choice of the window width. Too small a value of W and the wavefront curvature becomes too high and continuity is lost; too large and the Gaussian beam will no longer be a good estimate of propagation away from the central ray. Recall that only the central ray 'sees' the actual velocity and anisotropy field. Moreover, the migrated beamlets being too wide will laterally 'smear' the data too much.

Fig. 5.16 shows a comparison between GBM and Controlled Beam Migration for the same case as in fig. 5.14. The further improvement in terms of signal to migration noise is clear. This is nevertheless at the cost of losing the character of seismic horizons and probably the reliability in relative amplitudes. All markers have been 'boosted' to a similar level. This result can be very useful and even indispensable for structural interpretation in complex areas, as illustrated, for instance, in the case of reprocessing of data in fig. 5.17 (courtesy of Total E&P UK Ltd).

Wavelet-based migration

In this paragraph, we deal with the migration of the 'time wavelet' mentioned in fig. 5.15. The whole process to migrate one selected point in the tau-p domain is described in fig. 5.18. As shown in fig. 5.15, the selected point of the tau-p domain of coordinates τ_0 and p_0 corresponds to

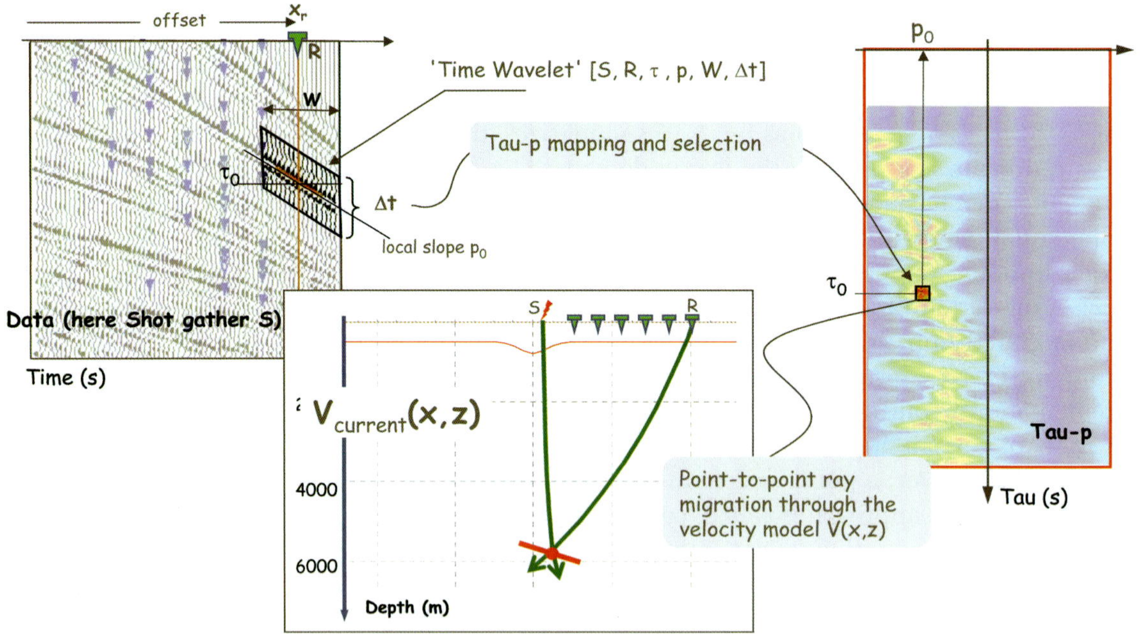

Figure 5.15: Point-to-point migration. Ray tracing through the current velocity and anisotropy model link the three domains: data, tau-p, and depth migrated. Point to point migration is however not sufficient to build the migrated image. The migrated point needs to be 'convolved' by a realistic waveform. Gaussian beam and time wavelets are two possible options.

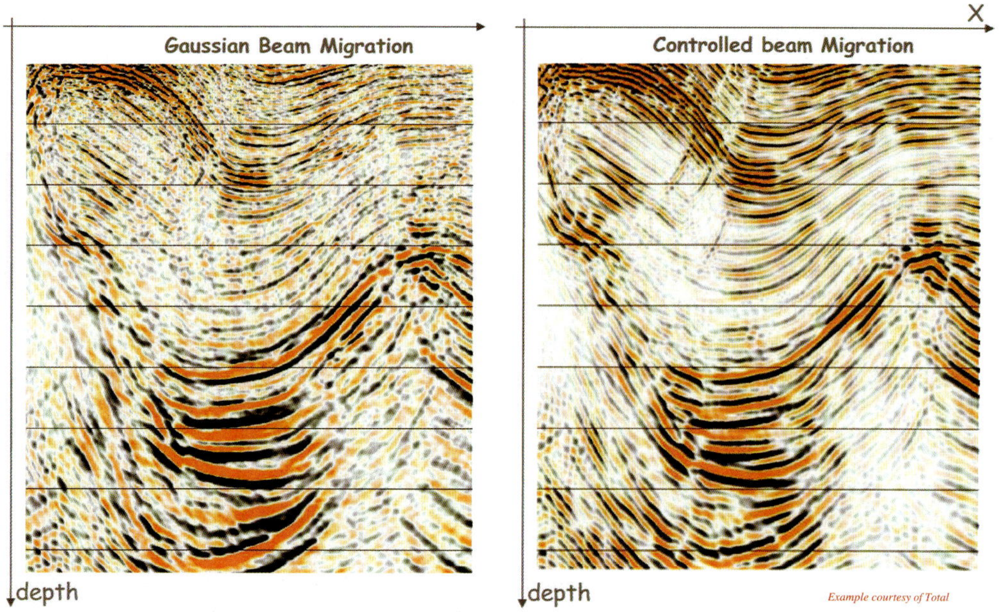

Figure 5.16: A comparison of a Controlled beam against Gaussian beam migration. The signal / migration noise ration is further improved with controlled beam, improving the continuity of events remove the sentence.

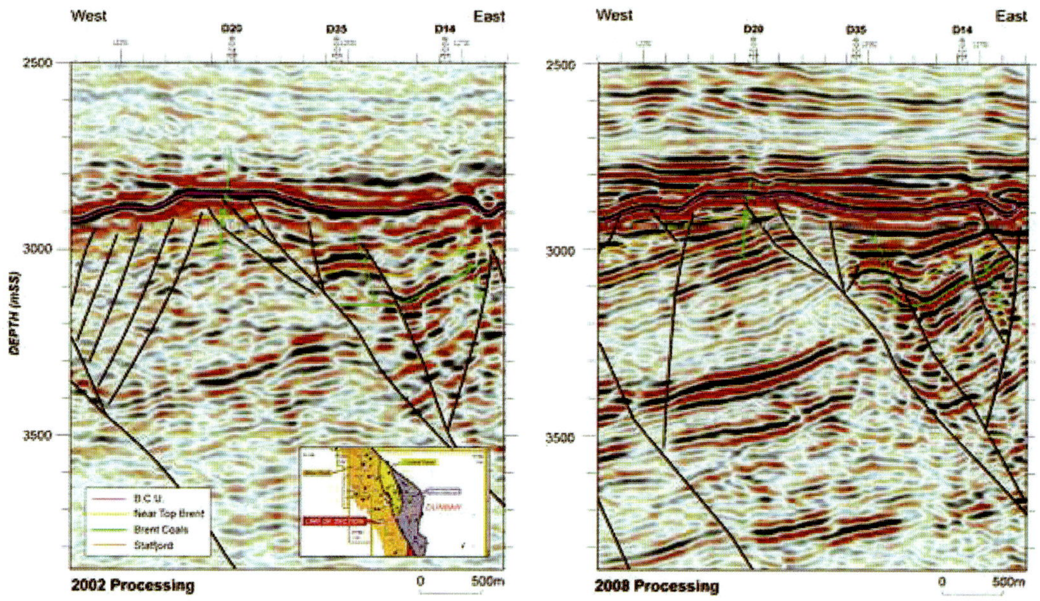

Figure 5.17: Comparison between PreSTM (2002 processing) and Controlled Beam Migration (re-processing 2008). Example courtesy of Total E&P UK Ltd from the Dunbar field. In (BenBrahim, 2008), with authorization. CGGVeritas' CBM re-processing allowed new structural interpretation of the area, enabling the definition of an intermediate panel which could be an additional target.

a time window of the analysed data set (here a common-shot gather), with space width W and time 'thickness' Δt. It corresponds to a local maximum value in the tau-p domain because the traces are reasonably well aligned and similar along the slope p_0 around the pivot point of coordinate x_r, τ_0. This is the *time wavelet* that we will use in the migration.

Next, we compute the migration location in the current velocity model V(x,y,z), using one of the techniques described in fig. 5.01 to 5.05, according to the domain in which the tau-p analysis has been performed. This yields the location at which the centre of the wavelet migrates. *Since we use rays explicitly, we know much more than its sole location,* as for instance and of particular interest, the incident and reflected angles at that reflection point, from which one can *derive the local dip and reflection angle* (taking of course local anisotropy into account as in fig.1.11).

Remember the added complexity when dealing with 3D data in real life, as hinted at in fig. 5.06.

The migrated wavelet will lie along the computed dip but before placing it there, we need to 'stretch' it firstly from time-thickness Δt to depth-thickness Δz; secondly, from time-width W to depth-width: this is 'beam spreading', an issue automatically solved by Gaussian beams in the case of CBM; thirdly, some implementations compensate for the natural stretch factor (similar to the 'old' moveout stretch factor).

As for Gaussian beam, the 'depth wavelet' needs to be interpolated onto the regular output image grid (fig. 5.12).

At this stage, it is classical to talk about 'wavelet properties'. Some are related to the data domain:

- Acquisition geometry: x_s, x_r, offset 2h, shot-receiver azimuth α, mid-point x_{mp}, etc.
- window size: width W, time-thickness Δt;

others to the tau-p domain:

- picked τ and p values;
- amplitude of the peak in the tau-p panel;
- traces coherency or semblance along the slope within the time-window;
- the event coherency and continuity in the multi-dimensional tau-p transform (for example, along the x_r direction in the three-dimensional tau-p transform of a shot in fig. 5.09)

and finally, the ones related to the migrated depth domain:

- of course, the migration location (x, y, z) of the centre of the wavelet;
- amplitudes compensation due to propagation (coming from dynamic ray tracing);
- incident, reflected and reflection angles and azimuths (resulting from ray tracing);
- dip and azimuth of the element of reflector;
- time-to-depth, beam spreading and offset stretch factors;
- migration 'quality' as derived from source and receiver ray distance (fig. 5.07).

Fig. 5.19 presents an example from (Sherwood *et al.*, 2008) that compares a wavelet-based beam migration with a traditional PreSTM. For the sake of comparison, beam migration is transformed to vertical time using the migration velocity model. The benefit in terms of signal-to-migration noise, imaging of steep dips and imaging below complex structure is clear. The technology used here, called by the authors BPSDM for beam pre-stack depth migration, is a 3D version of beam migration in the common-mid point gather (fig. 5.05), where the estimated ray parameters are a function of the Cartesian offset coordinates h_x, h_y (fig. 5.06).

Diplet-based migrations in the common-offset domain

In this paragraph, we introduce a new idea used in some implementations of ray-based migrations: the diplet decomposition (Peng, 2007), an extension of tau-p transform.

Let us start with the conventional tau-p transform applied to a common-offset section. As for shot domain analysis, the section is divided into overlapping windows of width W, with a pre-defined overlap factor. For each window, a τ-p transform is performed for a range of p and τ values.

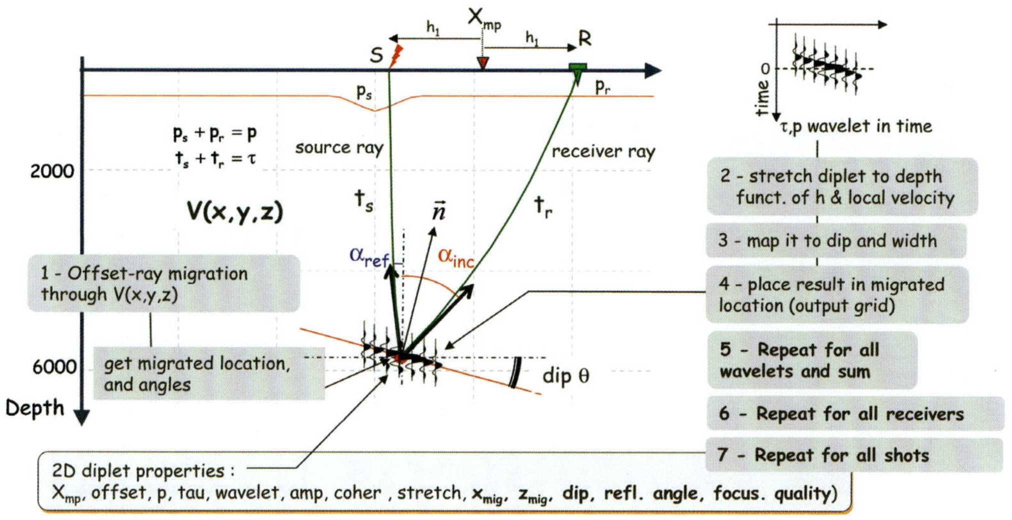

Figure 5.18: Wavelet or diplet migration. t and p values yield the migrated location of the selected wavelet in the velocity field (figures 5.03, 5.04). Since migration is ray-based, we know incident and reflected angles and consequently dip and reflection angles. The wavelet must be stretched to depth with its correct dip. The operation is repeated for all selected wavelets and results summed together after digitisation onto the output image grid.

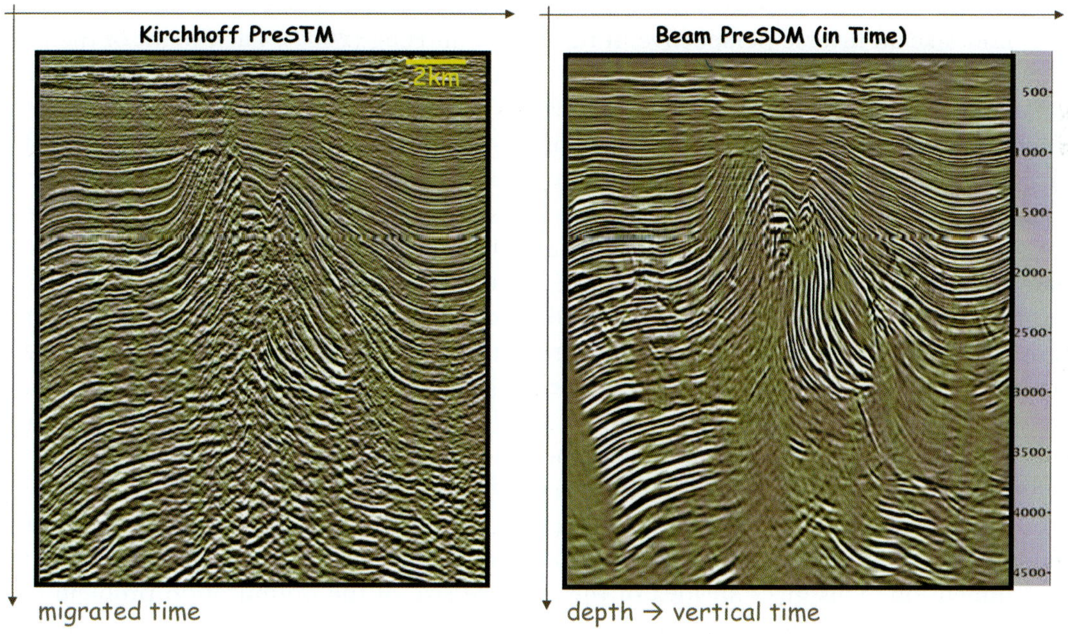

Figure 5.19: A comparison between Kirchhoff PreSTM and Beam PreSDM. Data courtesy of Devon Energy and Petroleum Geo-Services, in (Sherwood J. *et al.*, 2008) with authorization. Beam PreSDM has been transformed to vertical time for the sake of comparison.

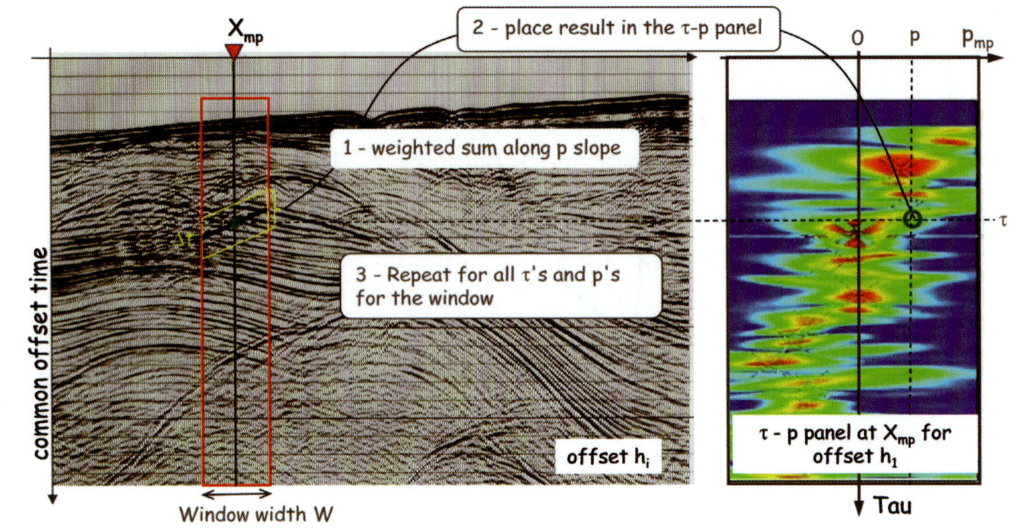

Figure 5.20: Principle of tau-p transform in the common-offset domain. The process is similar to the shot domain (figures 5.07 and 5.08). The common offset panel is divided into overlapping windows, each window having its own τ-p panel.

The process is presented in fig. 5.20 for one window of a common-offset section. As mentioned, the amplitude at a given τ-p point of the panel is the (weighted) sum of samples along the slope p, pivot (X_{mp}, t) and time-thickness Δt for the discrete mid-point location X_{mp}.

Let us now take an inverse perspective in fig. 5.21 as in Peng (2007). Each point of the panel of coordinate τ_i, p_j is now a set of two parameters: the classical amplitude $Amp_{i,j}$ plus the index k of the *wavelet shape taken from a catalogue* of wavelets with various frequency contents.

We define the '*forward problem*' of an inversion process as the construction of a 'diplet' made of wavelets Wav_k of amplitude $Amp_{i,j}$, juxtaposed at the trace interval of the section along a slope p_j, tapered by the same weights as the ones we used in the τ-p transform. This '*diplet*' is placed at time τ_i and coordinate X_{mp} on a 'synthetic common-offset section'. For the sake of completeness, a *stretch factor* may be optionally applied to the wavelets in order to take account of the actual offset being processed.

The objective is now to find the set of diplets defined by their parameters [$Amp_{i,j}$, Wav_k] for every τ_i , p_j, and all X_{mp} which build, when summed together, a synthetic common-offset section that looks as much as possible like the true one. This can be *cast as a large inverse problem*, where the unknowns are the two diplet parameters at every point of the [τ_i , p_j, X_{mp_l}] volume. The cost function may be the difference between the synthetic and the true common sections in a least-squares sense. Constraints and regularization may also be included to insure stability and convergence.

Note that in real life the common-offset 'section' is in fact a volume and the τ-p transform a four-dimensional object. And, if we also wish to consider multi-azimuth features, we need to add a further dimension! This gives an idea of the possible size of the optimization problem to solve for a large 3D survey. Remember moreover, that we have defined so far the process for one selected offset range. It has to be repeated of course for all offsets, possibly a few tens of them.

Once the inversion is completed, the next step consists in picking events in the τ-p volume, selecting the most relevant ones, as fig. 5.22 illustrates. Each selected diplet can then contribute to the final migrated imaged, in a way very similar to the one illustrated in fig. 5.18. Since we deal with events picked in the common-offset domain, the migration technique presented in fig. 5.04 would apply.

Events selection in Beam migration

As discussed throughout this chapter, an important step in beam migration is the selection of events that will eventually contribute to the final image. We can discriminate between initial selection of events in the tau-p domain and selection of events after their corresponding ray migration.
For the first class of events, the main criteria have already been mentioned as 'wavelet property', namely:

- events that give high amplitude peak values in the tau-p domain;
- events of weak energy but emerging from the background noise; being able to pick such events thanks to tau-p transform is an important benefit of beam methods;
- events with high coherency or semblance of traces along the slope within the time-window;
- the event with the highest coherency and continuity in the multi-dimensional tau-p transform;
- events with 'clean peaks' along the p dimension (for instance, an isolated noise burst on a trace will appear at constant tau with high amplitudes for all p values and should not be selected);

The selection criteria can act as a yes-or-no index or as a weighting in the final contribution of the event.

The first post-migration criterion is the migration 'quality' of fig. 5.07. The low migration quality may come from one or a combination of:

- an incorrect velocity model;
- poor measurement of the ray parameters in one or several of the domains where they are estimated;
- the fact that the picked event is a multiple instead of a primary.

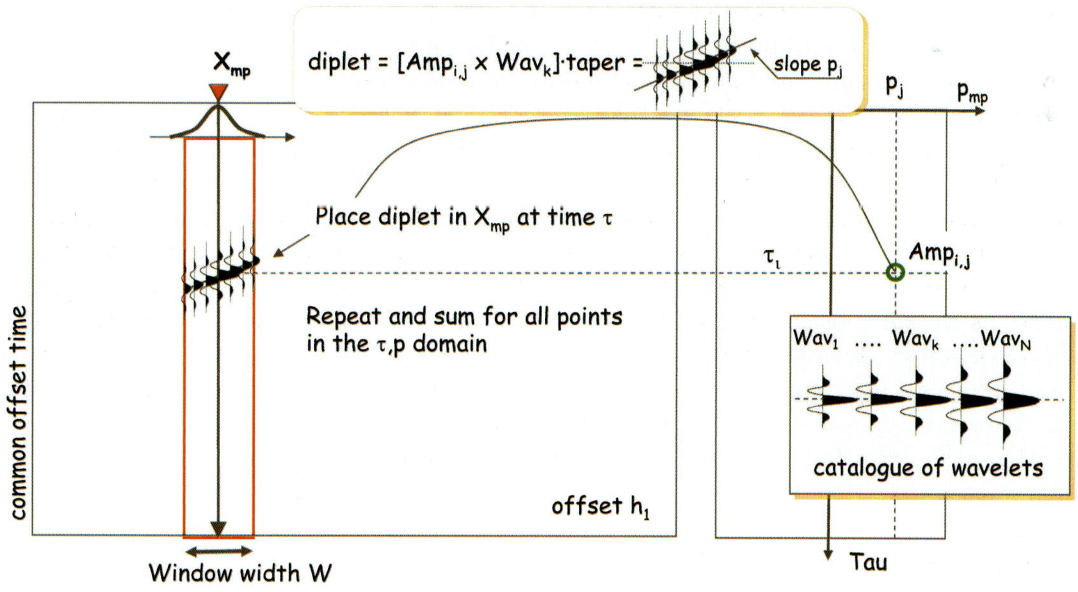

Figure 5.21: Diplet analysis in the common-offset: the forward problem. One point in the t-p domain of coordinates τ_i, p_j and amplitude $Amp_{i,j}$ transforms into a "diplet" in the common-offset domain characterised by its width W, slope p_j, wavelet shape w_k and taper.

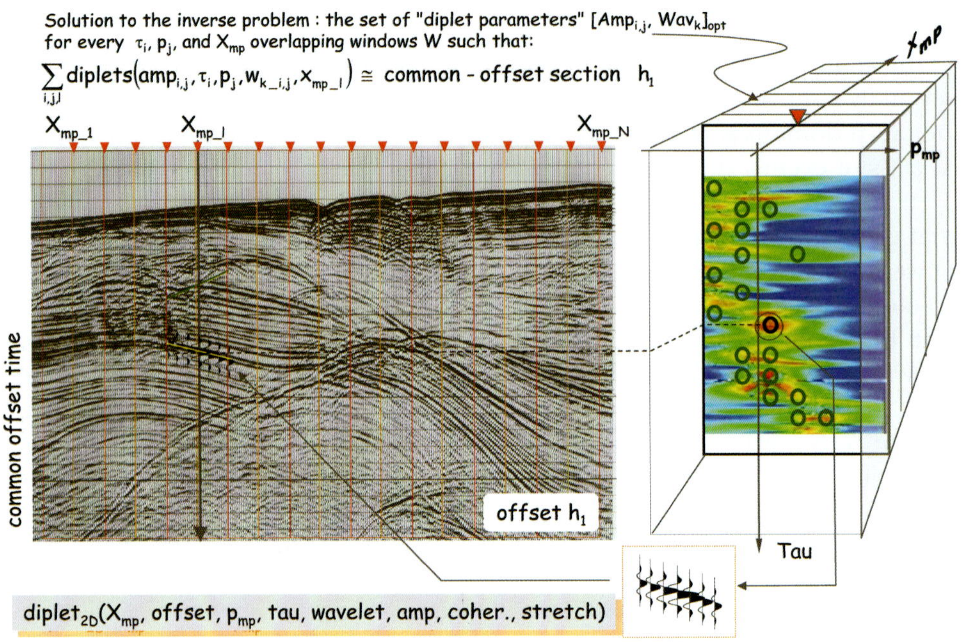

Figure 5.22: Selection of diplets in the transform volume. Each point in the 'τ-p volume' corresponds to a specific diplet, with its attached parameters. The most significant peaks in the volume (highest amplitudes, coherency, probability of being primaries events, etc.) are picked and selected for migration.

This third and most important point is featured in fig. 5.23, which shows schematically how the migration rays that assume primary reflection only, depart from the actual multiple reflection ones. This offers a *guide not to select events that could possibly be multiples*. In the figure, we assume that the velocity and ray parameters are correct. In real life this may not be the case, limiting the efficiency of this specific method for short period multiples. Other solutions are developed; this is currently a hot topic in the field of multiple attenuation in the depth domain. An example of 'demutipled' CIGs is presented in fig. 5.24.

Finally, a combination of rays and migration provides interpreters with new ways to actively participate in the imaging process. For instance, they can display and select rays and subsequently the corresponding reflector's images, going through a salt body only, illuminating a given fault (see also chapter 8, fig. 8.06), having more than one reflection (bouncing against a salt wall for instance - fig.8.07-), suspected to be multiples, etc. These parts of the images can be analysed in detail, flagged on the CIGs for visual inspection, included or not in the tomographic process or in the final image.

Offset- and Angle-Common Image Gathers and velocity analysis in beam migrations

When working in the common-offset domain, we are in exactly the same situation as in common-offset Kirchhoff migration: Common Image Gathers in the offset domain can be created straightforwardly at any selected location, before summing the migrated common-offset sections

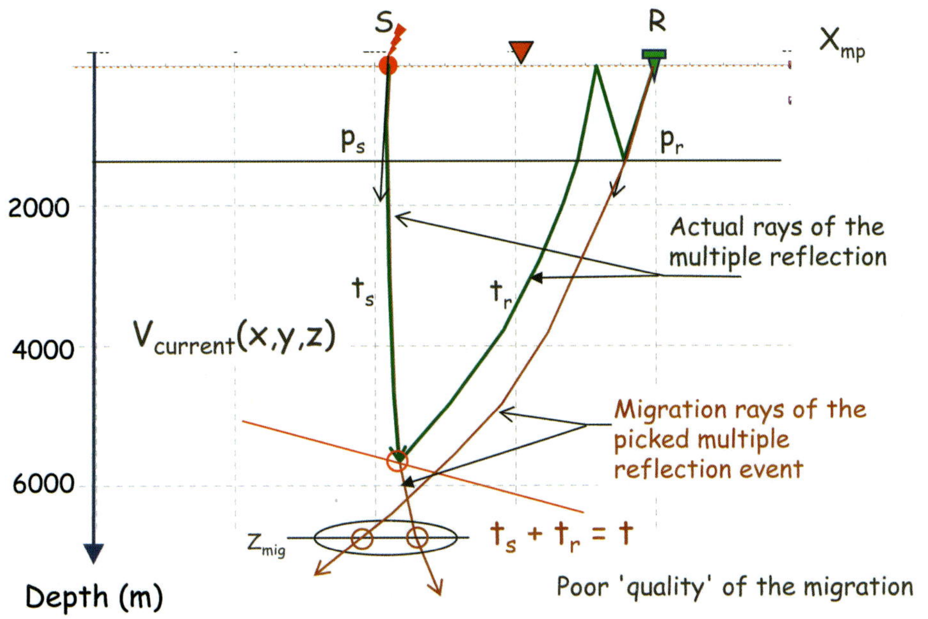

Figure 5.23: If a multiple reflection is picked in the tau-p domain, its migration rays (1) will not follow the true propagation rays and (2) be propagated in too long a time, resulting in a poor 'migration quality'. The event can then be rejected from the selection and hence from the final migrated image.

or volumes. This is similar to what is shown in fig. 4.02 and 4.06, for instance. Examples of Beam Migration offset CIGs are given in fig. 5.24.

In the case of shot point migration, it is trickier to obtain common image gathers, but the beauty of ray-based migration, is that we know the relationship between any event in the τ-p panel (data domain), its shot and receiver locations, offset, azimuth (acquisition domain) and migrated location, amplitude, dip and angles (image domain). We called it 'wavelet' or 'diplet' properties in fig. 5.18, but the 'migrated beamlet' in fig. 5.11 or the migrated event in fig. 5.15 have similar properties (at least the ones related to ray geometry).

In fig. 5.11, the migrated beamlet is naturally placed in the migrated shot panel. We also prepare a set of 'migrated offset' panels to which we assign nominal values of offsets. By construction, we know the offset that corresponds to the migrated beamlet and we can conceptually place it in the offset panel that has been assigned this actual offset of origin. All selected events of this receiver will add both to the migrated shot and to the migrated offset panel, but when moving to the next receiver, we carry on adding migrated beamlets to the migrated shot but start populating another migrated offset panel. At the end of the process (migration of all receivers of all shots), we have on the one hand the migrated section, the sum of all migrated shots, but also the various 'common-offset migrated sections', similar to the ones we get in common-offset Kirchhoff or beam migrations. From this set of common-offset migrated sections, we can easily extract the offset-CIG as needed. Fig. 5.24 is an example of a few such Common-Image gathers resulting from Beam PSDM (from Sherwood, 2008).

Clearly, this is an oversimplified description of the process, which implies some serious practical multidimensional binning and interpolation issues, especially in 3D and where azimuth adds

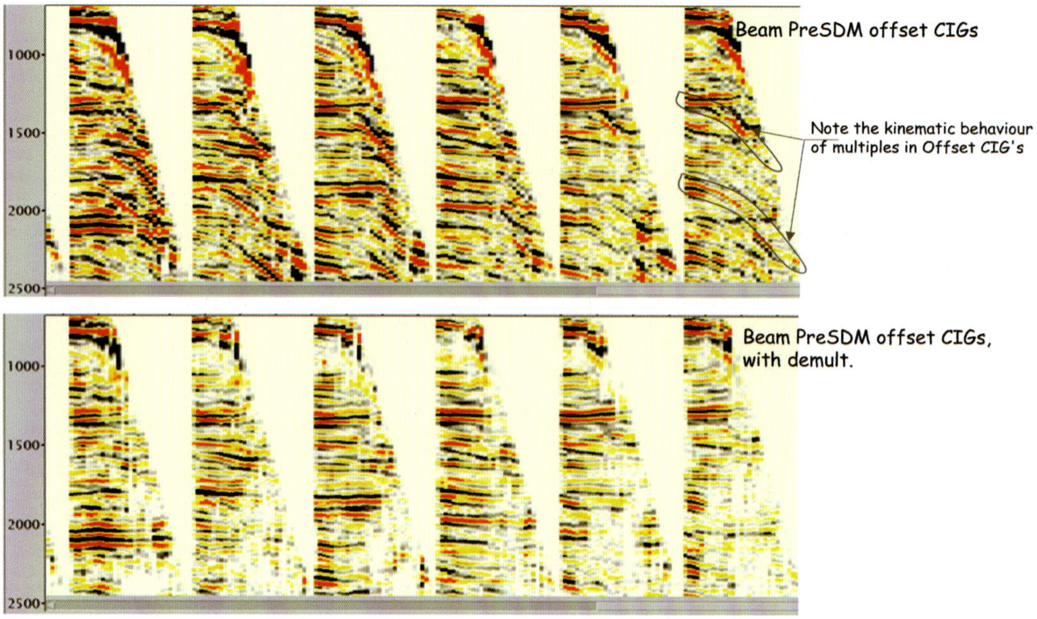

Figure 5.24: Beam PreSDM offers the possibility to discriminate between events that are clear primaries and events that could be multiples. These may not be selected for migration, allowing 'multiple removal in the depth domain'. North Sea Example of 'demultipled' offset CIGs, courtesy of Petroleum Geo-Services, in (Sherwood J. *et al.*, 2008) with authorization.

another level of dimensionality. There are two main reasons for that. Firstly, the actual offsets are different from the selected nominal ones and secondly, the migrated 'fat' beamlet *spreads over different offsets*. It shows in principle however, how to construct CIGs in the offset domain in shot-oriented beam migrations.

In chapter 4, we mentioned the benefits of building *CIGs in the reflection angle domain*, especially in terms of their physical meaning and possibility of direct, more reliable Amplitude Versus Angle analysis. What we have just described for offset can be extended to reflection angle. In this case, we prepare a set of 'migrated common-reflection angle panels' to which we assign nominal values of reflection angle. The reflection angle of each migrated beamlet or diplet is known as one of its properties. Thus, we can place the migrated beamlet in the appropriate 'common-reflection angle panels' (again some interpolation issues here!). As the migration progresses (more receiver traces and more shots), the various reflection angle panels populate and at the end of the migration, in parallel to the migrated shots or common-offsets, we obtain 'common reflection angle migrated sections', from which reflection-angle CIGs can easily be constructed. Note that we use both terms Reflection Angle or Scattering Angle CIG for the same object.

One should be cautious in the use of these CIGs for AVA as *amplitudes may not be properly handled* and consequently not reliably represent reflection coefficient in current applications of 'fast' beam migrations. Theoretical approaches of amplitude preserving GBM have recently been proposed (Gray and Bleistein, 2009) but not systematically applied in all implementations.

The reconstructed CIGs, whether coming from common-offset or common-shot implementations, in the offset domain or in the reflection angle domain, are *natural inputs to the ray-based tomography* addressed in chapter 4. They are also inputs to residual moveout analysis and correction to compensate for residual 'non flatness' and improve the final stack, as done in PreSTM or Kirchhoff PreSDM.

If the computing effort in Gaussian beam migration is of the same order of magnitude as Kirchhoff migrations, the other 'fast' ray-based migrations are, as the name suggests, much more efficient to run once the tau-p transform or the diplet decomposition of the input data has been completed. This characteristic offers the *advantage over Kirchhoff and GBM to be able to run many more iterations* of PreSDM, followed by ray tomography for the same project turnaround time; or to shorten this turnaround time if the geological setting requires fewer iterations to build the velocity model. This is one important advantage of these methods. Sherwood *et al.* (2008) mention the possibility to build a full 'salt' model in the Gulf of Mexico for a survey of 1000 km^2 in 4 to 6 weeks, a typical project including 8 to 10 iterations. Fast beam migrations are thus considered as very *valuable tools for fast model building*, especially in areas of complex geology and/or poor signal-to-noise ratio.

Advantages and limitations of ray-based beam migrations

We have presented in this chapter some of the techniques used in ray-beam migrations. We do not claim to be exhaustive, each implementation being generally a combination or an extension of them, with its own advantages and limitations.

The presentation of the various techniques is intended to give the reader a good understanding of the principles but overlooks many practical issues in 3D and in azimuth handling. Nevertheless, it should be clear from this chapter that:

- Ray-based beam migrations are pre-stack depth migration and as such are able to migrate events not accessible to Pre-stack Time Migration. This is illustrated by fig. 5.19.
- They must be considered as complementary tools to Kirchhoff and Wave Extrapolation technology but faster and cheaper.
- As such, they are particularly suited for 'structural imaging' and fast velocity model building in complex geological settings (see, e.g., Wang *et al.*, 2008; BenBrahim *et al.*, 2008; etc.).
- Ray tracing has no dip limitation; ray-based beam migrations can thus migrate very steep dips (fig. 5.16), overhanging reflectors via turning waves or steep fault planes (see, e.g., Dom Pham *et al.*, 2007).
- They can be implemented easily to process land data sets, including topography and multi- or wide-azimuth acquisition.
- They include ways not to image multiples (fig. 5.23).
- They are not limited by multiple ray-paths and caustic. This point is explained in fig. 5.25. Multi-pathing corresponds to multiple events in the tau-p domain. They will hence be automatically migrated as separate events in GBM and in other 'fast' beam methods, if the corresponding amplitudes are sufficient to be detected and selected.
- They easily handle any type of anisotropy in velocity models.
- Results can have an unusual appearance: reflectors look sometimes like the juxtaposition of segments, although this point is improving fast with most recent implementations and practice, especially in the way events are selected or weighted before migration.
- The general improvement in signal-to-noise ratio of the migrated reflectors is made at the cost of poor amplitude preservation.
- Ray-tracing has natural limitations in complex geological settings: application of Snell's law becomes unstable for large angles and sharp velocity contrasts.

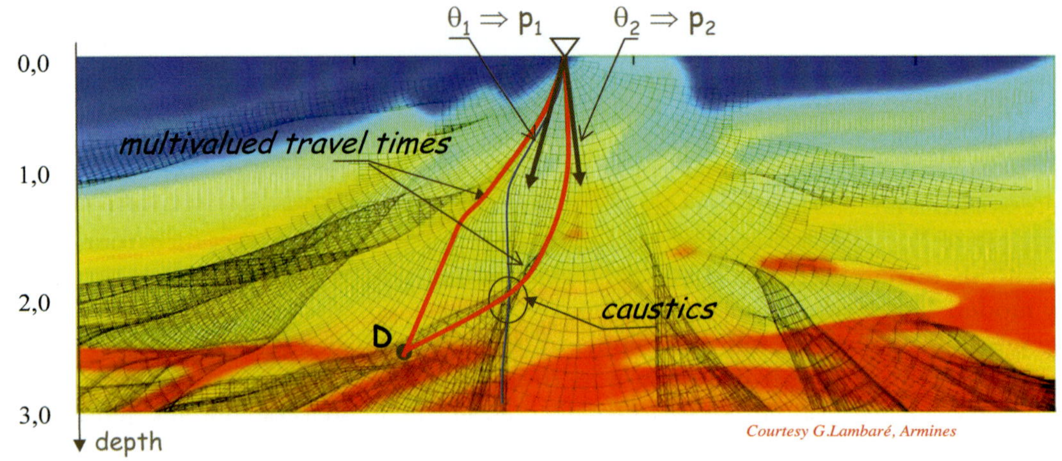

Figure 5.25 : Tau-p based techniques properly address the issue of multipathing. Multi-path events correspond to different ray parameter p values naturally handled separately when migrating the tau-p panel.

These three last points still make Wave Extrapolation-based methods valuable, even indispensable options for depth imaging. This is the topic of the next chapters.

6 Wavefield Extrapolation Migrations

After the two chapters dedicated to migration technologies that make use of ray theory for the simulation of wave propagation in the subsurface, we now address techniques using propagation of the 'full wavefield'. For this purpose, we draw on different kinds of approximations and numerical solutions of the scalar acoustic wave equation. These approaches are cast under the generic term of 'Wavefield Extrapolation' migrations or WE migrations in short.

In this chapter, we set the foundation for all methods: *the imaging principle*. We then address the numerous ways to extrapolate the wavefield, each corresponding to a specific method. Two classes of methods use *wavefield extrapolation in depth*: firstly, 'shot point migrations', also called 'shot profile migration' or simply 'shot migration', where shot records are processed one after the other and secondly, 'survey sinking' where the shots and receivers of a whole survey are recursively 'lowered' into the Earth. Extrapolation in depth necessitates the assumption that waves propagate in one direction only, either downwards for the wavefield initiated by the source, or upwards for the wavefield coming from reflectors. This assumption uses the so-called *one-way approximation* to the wave equation.

The next part of the chapter is dedicated to techniques where the wavefield is *extrapolated in time*, yielding today's quickly developing 'Reverse Time Migration' methods or 'RTM' in short. These methods do not assume one-way propagation; we call them *'two-way methods'*, with the benefits that we will mention.

In all WE methods, the construction of Common Image Gathers is not straightforward. As a consequence, Migration Velocity Model Building is a topic *per se*, still under development in many senses. We will address it in a subsequent chapter.

The imaging principle

Wave extrapolation-based imaging techniques started with Claerbout's 1971 paper 'Toward a unified theory of reflector mapping', where he stated his famous imaging principle: "*reflectors exist at points in the ground where the first arrival of the downgoing wave is time coincident with an upgoing wave*".

We implicitly stated this principle at the end of chapter 1 and in fig. 1.13. We met it again in chapter 5, for instance in fig. 5.01, where we claim that the migrated location of a picked reflection event is at the crossing point of the shot ray and receiver ray, at a time equal to $t_s + t_r$. In doing so, we *back-propagate* Claerbout's upgoing wave along the receiver ray and *forward-propagate* the downgoing wave along the source-ray to the point where they 'coincide', i.e., the reflector's location. Note that in fast beam techniques (chapter 5), we know that a reflection actually happened there because we individualized and picked it in the tau-p transformed data set. In all other techniques 'blind' applications of the imaging principle reveals whether or not there is effectively a reflector at any given point in the subsurface and what is its reflection coefficient.

In this chapter, we use the same principle in various ways no longer dealing with rays but with the full wavefield itself. Before getting into the details of these different methods, let us clarify once more what these different wavefields are exactly. For this we again use in fig. 6.01 the simple model of fig. 1.13.

As already mentioned, we distinguish the *'downgoing'* and the *'upgoing wavefields'*. The downgoing wavefield is the pressure or particle velocity field generated by the source that propagates downward into the subsurface; its ray-equivalent in chapter 5 is the source ray.

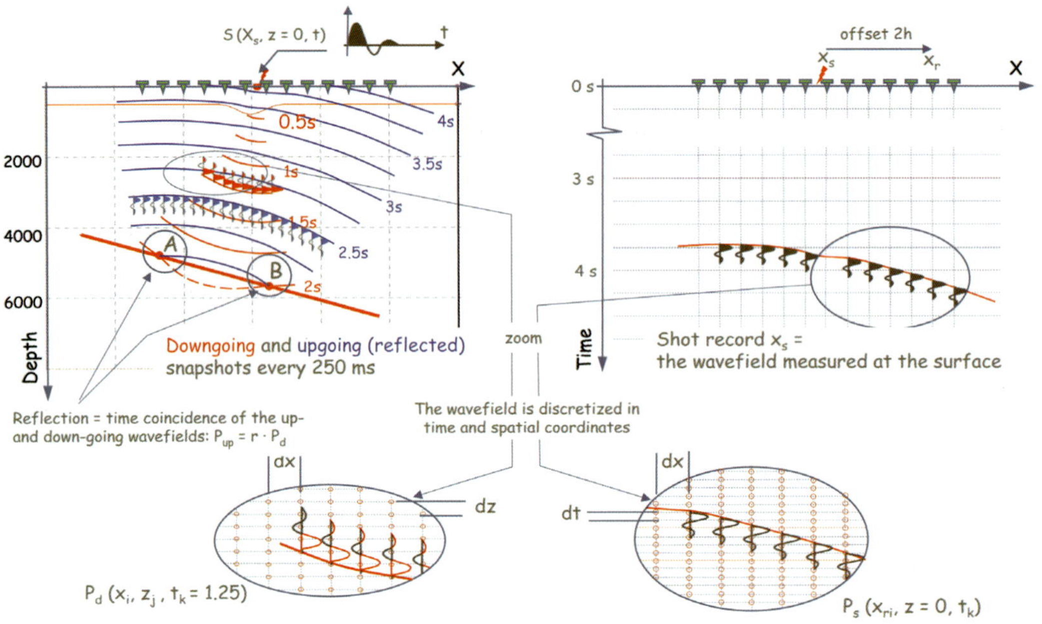

Figure 6.01: Shot record and wavefield. The recorded traces denoted $P_{up}(x_r, z{=}0, t)$ for the model in fig. 1.13 are discrete measurements at the surface of the upgoing wavefield resulting from reflections of the down-going wavefield induced by the shot in x_s. Snapshots are denoted $P(x_i,z_j,t_k{=}\text{constant})$.

The 'upgoing wavefield' corresponds to reflected energy, travelling upwards towards the surface where we record it; its ray-equivalent in chapter 5 is the receiver ray.

On the cartoon on the upper left-hand side of fig. 6.01, we represent in red the downgoing wavefield induced by a shot located in X_s, starting emission at time 0, with the source signature S(t). The wavefield is the value of the perturbation in pressure changes or particle velocity directly induced by the shot, in any location (x, z) in the subsurface as time t unfolds. Let us denote it $P_d(x, z, t)$. Note that for 2D seismics and for one shot, this field is a 3-dimensional function.

In fig. 6.01, we symbolically represent the wavefield at different times $P_d(x, z, t = \text{constant})$. For the sake of clarity of the figure, we limit the wavefield to its wavefront, apart from time t = 1.25 s, for which we represent the full 'time slice' of the wavefield $P_d(x, z, t = 1.25)$. Remember that we call such a 'frozen picture' of the state of the subsurface at a given time a *snapshot*.

In this simplistic model, only the area above the wavefront for a distance controlled by the signal duration is disturbed. The ground below has not yet been disturbed; the ground above has come back to equilibrium after the passing of the elastic disturbance.

When the downgoing wavefield encounters an impedance contrast, schematized by the red reflector in the figure, part of the energy is reflected, initiating the corresponding upgoing field, which we represent in blue and denote $P_{up}(x, z, t)$. As for the downgoing field, we represent a series of snapshots every 250 ms that we limit to wavefronts only, apart for the snapshot at time 2.5 s, which we can denote $P_{up}(x, z, t = 2.5)$.

Note that wavefields are limited to the *intrinsic frequency bandwidth limitation of the seismic signal*. For the sake of the demonstration, we have assumed only one reflector in the subsurface, which make the reflected or 'upgoing' wavefield particularly simple. Amplitudes are not realistic either in this cartoon and the following ones based on this model. A more 'realistic', although still synthetic, example of such a wavefield is presented in fig. 1.03, revealing particularly the imbrications of downgoing, upgoing, primary and *multiple reflections* wavefields: there are reflections everywhere in the subsurface, each contributing to the upgoing wavefield.

The upgoing wavefield propagates and is recorded by geophones or hydrophones at the surface *as a function of time*. The result is the 'shot record', which we can denote $P_{up}(x_r, z = 0, t)$, as mentioned earlier; a real example of such a shot record was given in fig. 1.05.

The imaging principle is obvious in fig. 6.01: there is a reflection where the downgoing (also called incident) and upgoing (or reflected) wavefields coincide in time and space. We can even say that the local reflection coefficient is the ratio between both wavefields' amplitudes.

The rules of the game in WE-based imaging are in simple terms:

- to *emulate the downgoing wavefield* $P_d(x, z, t)$ from the source S(t);
- to *reconstruct the full upgoing field* $P_{up}(x, z, t)$ back in time and everywhere in the subsurface from its measurement at the surface, i.e., the shot records $P_{up}(x_r, z = 0, t)$;
- to apply the *imaging principle, i.e., to find reflectors where and when the two wavefields are 'time coincident'*.

Computing the downgoing field knowing the source signal S(t) and the velocity model in which to propagate it is a pure modelling exercise. Reconstructing the history of the upgoing wavefield is no more complicated but less intuitive. The only evidence of it that we have is its measurement at the surface in the form of the shot record. The shot record is represented at the top right of fig. 6.01. We denote it $P_{up}(x_r, z = 0, t)$ to emphasize the fact that it is recorded at discrete locations x_r, at the surface z = 0 and as a function of propagation and recording time t (actually also digitized). Mathematicians call it the 'boundary condition' of the back-extrapolation exercise that we need to carry out to recover the full upgoing wavefield P_{up} at each time and location in the subsurface. This is very similar to the source signature being the boundary condition of the downgoing wavefield computation.

Computation of the down- and upgoing wavefields from information known or collected at the surface, gave the name 'wavefield extrapolation' or WE in short to this class of methods but of course, extrapolation whether up or down can *only be performed if one knows the velocity and anisotropy fields in the subsurface*. WE imaging does not escape this rule!

Note that wavefields are discretized functions in a computer. The time and space intervals Δt, Δx, Δy and Δz of the digitization grids impact the accuracy with which we simulate propagation. They are consequently important parameters in imaging, although some are already fixed when acquiring the data. For instance, Δx of the recorded shot is nothing else than the trace interval in the field (typically in the range 12.5 to 50 metres) and Δt is the sampling rate (typically 2 ms). The processors may however 'play' with these parameters by choosing different values for Δt, Δx, Δy and Δz, as they feel appropriate, provided that some basic sampling rules are honoured.

Imaging condition

We discuss in the following sections how to compute the two wavefields P_d and P_{up} but for the time being, let us assume we have got them and concentrate on what Claerbout meant by 'time coincidence' in his 1971 paper. Recall that our objective is to estimate reflection coefficients at every location in the subsurface.

For that purpose, we consider in fig. 6.02 the wavefields as could be recorded by a buried receiver at a given pixel A. A was selected as an element of a reflector. The downgoing wavefield is computed starting from time t = 0 and hits A after a propagation time t_s. Modelling allows us

$$r = \frac{P_{up}(x_A, z_A, t)}{P_d(x_A, z_A, t)} \ \forall t. \text{ Only possible when } P_d \neq 0 \Rightarrow r \approx \frac{\sum_t P_{up}(x_A, z_A, t) \cdot P_d(x_A, z_A, t)}{\sum_t P_d^2(x_A, z_A, t) + \varepsilon} \quad (r \equiv \text{scaled correlation coefficient})$$

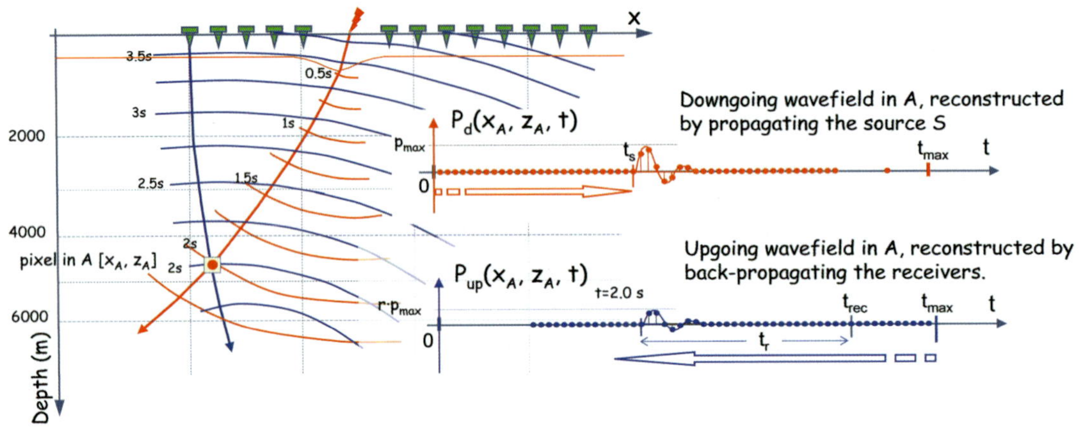

Figure 6.02: Imaging condition at the reflection point. The comparison of the downgoing and upgoing wavefields at any location (f.i. pixel A) allows us to estimate the local reflection coefficient r. One classical way to estimate r is to compute the correlation coefficient between the two fields at every location.

to reconstruct the history of the state of the Earth in A, schematized by the red series of recorded samples in the figure. Note as a side comment the relationship between wavefronts and rays that can be 'calibrated' in propagation time.

If the reflection coefficient in A is r, at every instant after t_s the reflected wavefield P_{up} is initiated with an amplitude equal to $r \cdot P_d$ and starts travelling upwards. The goal of the backward-extrapolation of the upgoing wavefield recorded at the surface is precisely to reconstruct this element $P_{up}(x_A, z_A, t)$ of the upgoing field. It will look like the record in blue in the figure. The blue arrow means that we reconstruct the history of the upgoing field 'seen' by A starting from the maximum recording time t_{max} backwards in time, until the zero shooting time. Note the different times that we encountered in ray-based migrations: the source ray time t_s, the receiver ray time t_r and the time t_{rec} at which the reflection in A has been recorded at the surface, sum of t_s and t_r.

The presence of a reflector in A is indeed revealed by the similarity between the two wavefields. They coincide at all times, differing only by a scaling factor equal to the local reflection coefficient r.

The reflection coefficient r could be estimated by taking the ratio between the two fields at any time around the moment when the reflection occurs, or an average of this ratio for all times but that would be possible only when the downgoing wavefield does not vanish. Recall also that fig. 6.02 is a cartoon for a very simplistic case (only one reflector). The wavefields are much more complicated in reality, as already mentioned. The most popular way to estimate r is then to take a scaled version of *the correlation coefficient of the two time series* corresponding to the down- and up-going wavefields in A. This is the formula given at the top of fig. 6.02, one simple version of the so-called *'imaging condition'*.

There are other ways to apply the imaging condition in order to improve the resolution of the result. This can be done however as a post-processing step of the final migration result.

Note in fig. 6.02 that the wavefields 'do not know' that they have met a reflector and their respective extrapolation carries on in depth and time. This is necessary for the downgoing propagation source, although transmissivity (hence amplitude) is poorly handled if there is no explicit velocity contrast at that location in the model.

The part of the reflected energy belonging to a specific reflector is also further back-propagated. This introduces 'migration noise', which is reduced when applying the imaging condition through the correlation process.

Depth versus time extrapolation

So far, we have shown how to obtain the reflection coefficient r in A. For this we needed to have the history of the wavefield at that specific location. We have thus to 'fill' the whole volumes $P_d(x, z, t)$ and $P_{up}(x, z, t)$, since we want the reflection coefficient everywhere, as schematized in fig. 6.03. There are two ways to fill these volume.

The first technique consists in the computation of the two fields at *successive depths*. We know the downgoing wavefield at the surface: it is the source signature S(t) in x_s; with the knowledge of the propagation velocity and anisotropy in the layer [0 – Δz], we have techniques at our disposal to estimate the downgoing wavefield $P_d(x, \Delta z, t)$ at depth Δz.

Similarly, we know the upgoing wavefield at the surface and, with the same velocity and anisotropy fields in the layer, we can backward extrapolate the wavefield and compute what it was at depth Δz: $P_{up}(x, \Delta z, t)$. Next, we apply the imaging condition at all desired points x_{im} at depth Δz. This gives us the Δz row of the migrated image. The x_{im} do not need to be the same as the x_r.

It suffices to repeat the three processes at all successive depths to recursively build the migrated image. Δz is called '*migration step*' and the technique is known as the 'one-way shot point migration', or OW-SP Migration (yet another acronym!). We develop it in the next paragraph. The migration step is an intrinsic parameter of recursive methods in depth.

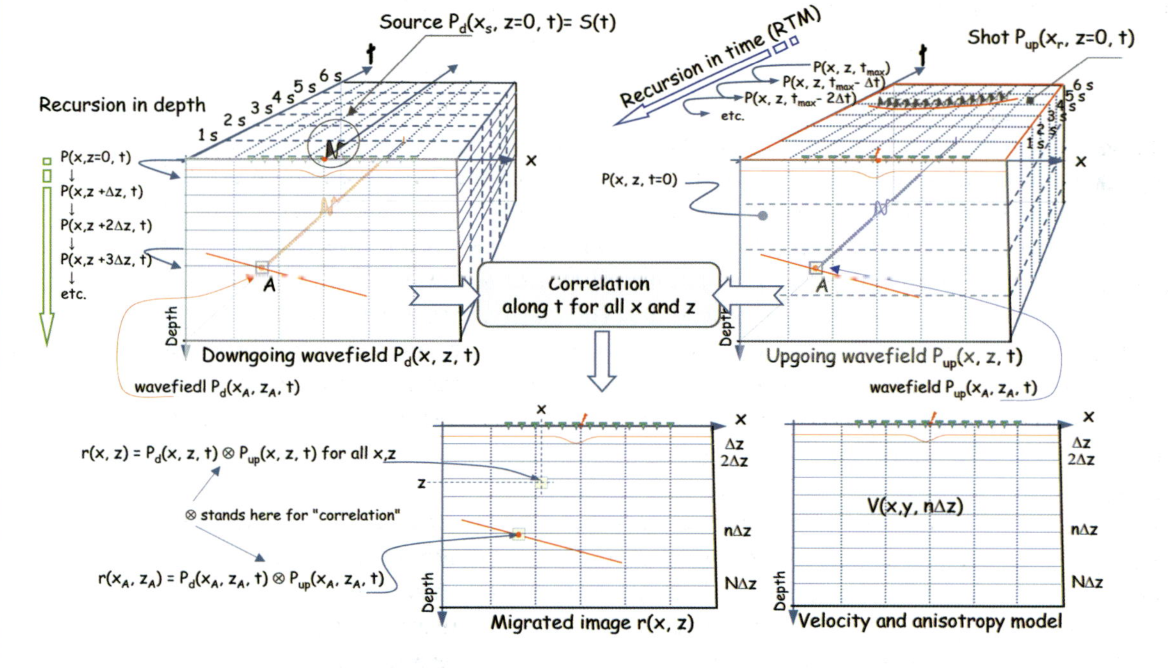

Figure 6.03: WE migration of a shot in 3 steps: compute the downgoing wavefield from the source; compute the upgoing wavefield from the recorded shot; apply the imaging condition (correlation coefficient along t) at all x, z locations. These different steps are actually performed recursively either in depth (one-way migrations; left upper panel) or in time (two-way RTM; right upper panel).

An alternative to the depth recursion is to 'fill' the two wavefield volumes time-wise. In this *time recursion* case, the downgoing wavefield is computed everywhere in space at different times Δt, $2\Delta t$, $3\Delta t$, etc. This is probably a more intuitive method than the depth recursion, as we compute the successive 'snapshots' of the downgoing wave emitted by the source: we reproduce the 'movie' of the propagation. It is again exactly the modelling exercise of fig. 1.03. At the beginning of the time recursion, only the top part of the (x, z) domain is excited but the whole space is filled progressively, as time steps go on.

On the upgoing wavefield side, on the contrary, things are again trickier as we start from the maximum recording time t_{max}. At that time, we know the wavefield at the surface only (first initial condition) but with the velocity and anisotropy field defined everywhere and the wavefield at the surface at time $t_{max} - \Delta t$ (second initial condition), we can reconstruct what the wavefield was in depth at time $t_{max} - \Delta t$.

The recursion is carried out by time steps Δt, from t_{max} until zero time, at which moment the upgoing wavefield volume is full. The imaging condition can now, and only now, be applied everywhere in space, since we have the wavefield time series for all points such as A in fig. 6.03.

Note an important difference between depth and time recursion. In the former case, the imaging condition is applied at each nth depth step and we need to keep in memory successively only the 'slices' $P_d(x, z = n\Delta z, t)$ and $P_{up}(x, z = n\Delta z, t)$, in order to apply the imaging condition and proceed with the next iteration. In time recursion, we receive all the migrated image at once, but only at the end of the recursion. This implies a clever management of data if one does not want to keep in memory the two complete wavefields to apply the imaging condition. In 3D these volumes are huge and multi-dimensional and it is just practical to do so with current computing power. Hence, the necessity of 'tricks', which we discuss in the paragraph dedicated to Reverse Time Migration, which is based on this concept of time recursion.

Principles of one-way shot migration

In this paragraph we go into more detail about how depth recursion works for the so-called 'one-way shot point migration'. Firstly, we note for the moment that the technique is applied shot by shot. We will discuss multi-shot migrations later.

Most practical cases deal with *3D shots*. A 3D shot record corresponds to receivers spread over an area, as sketched, for instance, in fig. 5.06 (land acquisition, multi-streamer boats, nodes, ocean bottom cables, etc.). For the sake of simplicity and the clarity of the figures, we assume a 2D layout.

As said earlier, the recursion is based on the computation of the down- and upgoing wavefields at successive constant depths. As an example, fig. 6.04 features a downgoing wavefield, simplified to wavefronts, at different times (snapshots) on the left-hand side panel. On the right-hand side panel we sketch the downgoing wavefields that would have been recorded at three different depths: 0, 2 and 5 km.

At zero depth, it is the source signature S(t). It is limited to location x_s. At deeper depths the wavefield expands over space: it reaches more and more points in the subsurface as time increases. On the right of fig. 6.04, the three wavefields are superimposed on the same panel, but each represents in fact a different horizontal slice of the wavefield volume featured on the top-left of fig. 6.03.

Similarly, fig. 6.05 presents on the left different snapshots and the hypothetical depth at which receivers would be placed to 'record' the upgoing fields. The reconstructed upgoing wavefields at different depths 2, 4 and 5 km are sketched on the right-hand side of the figure. At depth 0, it is the actual shot record (also shown as the zero-depth slice of the upgoing wavefield volume at the top right of fig. 6.03). Note how the reflected wavefield 'shrinks' and moves towards shorter times when we 'focus it back' towards the reflector that generated it.

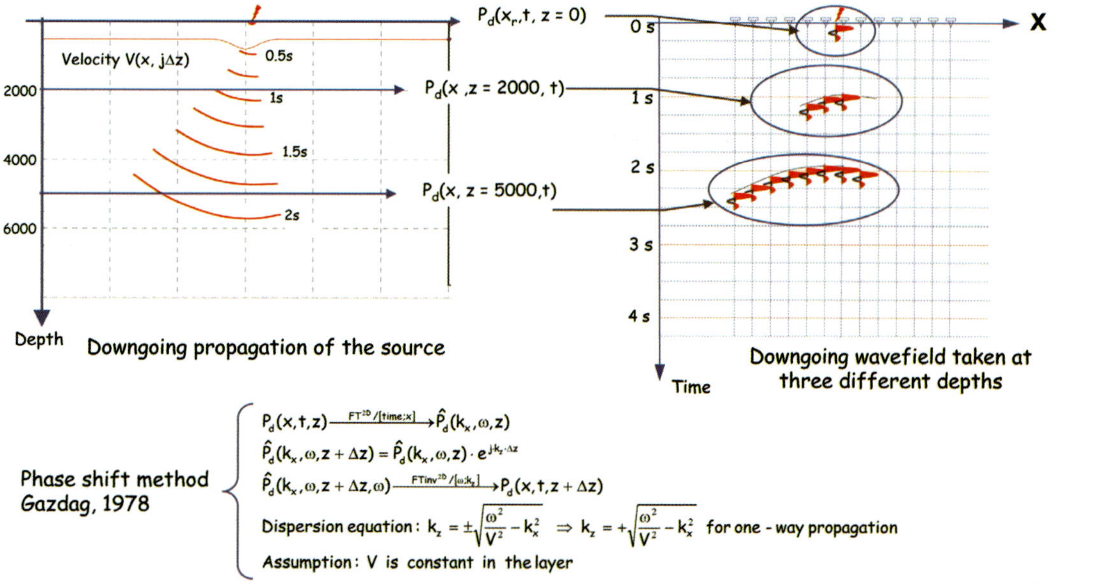

Figure 6.04: One Way Shot point migration. The source is downward propagated through the velocity model at depths Δz, 2Δz, 3Δz, etc. Here the wavefields at depths 0, 2 and 5 km, superimposed on the same diagram, correspond to three depth slices of the upper right volume in fig. 6.03. Phase-shift is a simple method to compute the wavefield at depth z+Δz from the wavefield at depth z if the velocity field is laterally constant at each depth step.

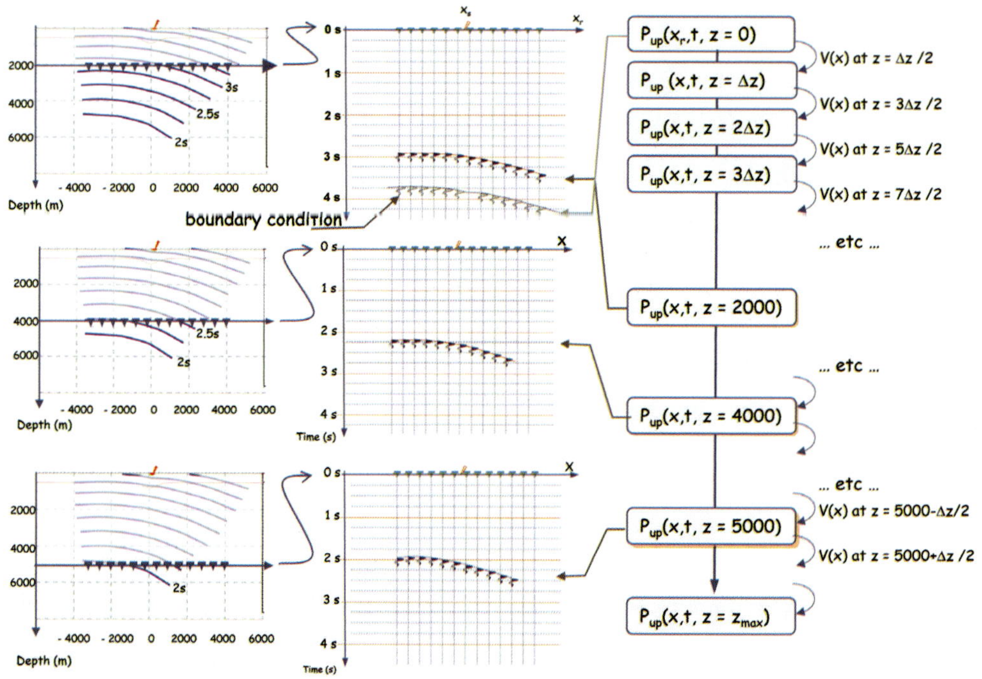

Figure 6.05: One Way Shot point migration. Same as fig. 6.04, but now for the upgoing wavefield reconstructed successively at depths Δz, 2Δz, etc. Only wavefields at 0, 2, 4 and 5 km are shown.

As sketched in the figure, the wavefields are computed recursively by small depth steps Δz, starting from the 'boundary conditions' at the surface, which are respectively the source and the shot record. The extrapolation is called 'one-way' because at each depth step, we assume that the waves propagated only downwards from the source (downgoing wavefield) and upward from the reflectors (upgoing wavefield). Δz is typically chosen in the range 5 to 10 metres, depending on the geological setting (smaller for complex areas) and the wavelength (related to the frequency content) of the seismics (smaller for higher resolution data sets).

We come back in a later paragraph on how a wavefield at depth $z + \Delta z$ is computed from the wavefield at depth z, but for the time being, we consider again in fig. 6.06 the imaging condition, which is applied at each depth. For this purpose, we have superimposed on this figure the down- and upgoing wavefields at depth 5 km. These are the ones that ought to be correlated in the application of the imaging principle. A small panel in the middle right of the figure shows the *correlation as a function of time shifts* of the two fields (we consider this panel again in chapter 7).

The imaging condition requires that we look for time coincidence, i.e., that we take the zero-lag value of this function for each trace. In information theory, this is the correlation coefficient of the two time series; one needs, however, to scale it by the downgoing wavefield to extract the r value (a stabilizing term ε is added in practice to avoid numerical division by zero).

This reflection coefficient r, a function of location x, can now be copied to make the row r(x, z=5000) of the migrated image at depth 5000 m.

In our simplistic model, there is only one dipping reflector that intersects that depth at a location x =-2000 (fig. 6.01) and we would expect to find a non-zero value at that location only. However, the band-limited nature of seismic waves makes it such that the reflection coefficient is spread over a distance related to the dip of the reflector and the dominant seismic wavelength. For a horizontal reflector, r would be evenly spread over the whole depth (at least the part illuminated by

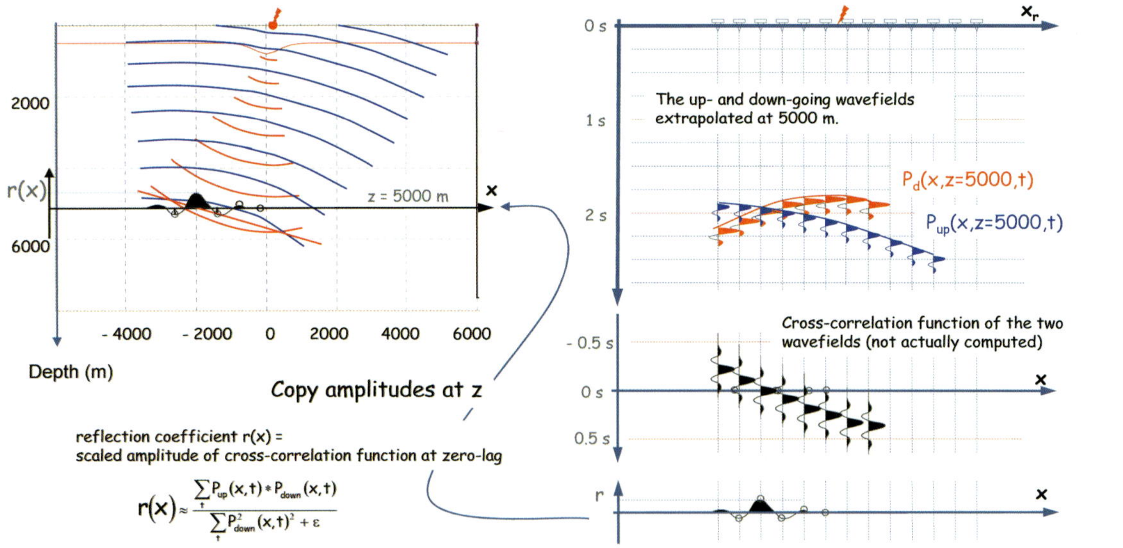

Figure 6.06: One Way Shot point migration: the imaging condition applied here at 5000 m depth. The back propagated upgoing wavefield (in blue) is correlated with the downgoing propagated source (in red). If a reflection occurs somewhere at that depth, the two fields coincide in time. The reflection coefficient at any point x at that depth is thus given by their zero-time lag correlation coefficient. Only a band limited version of the reflection coefficient can be estimated by seismic methods.

this specific shot, see later), but with an amplitude depending on the wavelet shape. The full vertical wavelet shape is reproduced when processing the different successive depths: we find again the well-known evidence of seismic limited resolving power, and the necessary link between the depth step Δz and the seismic bandwidth!

Fig. 6.07 recapitulates the different steps of the recursive process that leads to the 'one-way-migrated shot'. The top two rows of the figure are the upgoing and downgoing wavefields reconstructed recursively at successive depths 0, Δz, 2Δz, etc. (only depths 0, 2, 4 and 5 km are represented).

The recursion goes as follows:

- start with the zero depth; generally no signal has been recorded yet and anyway, the source has just started emitting; so the correlation is nil at all locations;
- compute the downgoing wavefield from the source at depth Δz; the migration step Δz is small as mentioned and the assumption that the velocity and anisotropy fields are vertically constant in the layer [0 – Δz] is valid. Nonetheless, velocity and anisotropy can be highly variable laterally; we denote it V(x, Δz/2);
- compute the upgoing wavefield from the shot record at the same depth Δz and with the same velocity and anisotropy function;
- apply the imaging condition, i.e., compute the reflection coefficient by cross-correlation of the time series (as in fig. 6.03 or 6.06) at each *desired x location*; note that the selected x values where we want to build the migrated image generally differ from the receivers locations x_r;
- populate the Δz row of the migrated image r(x, z = Δz) with this result;

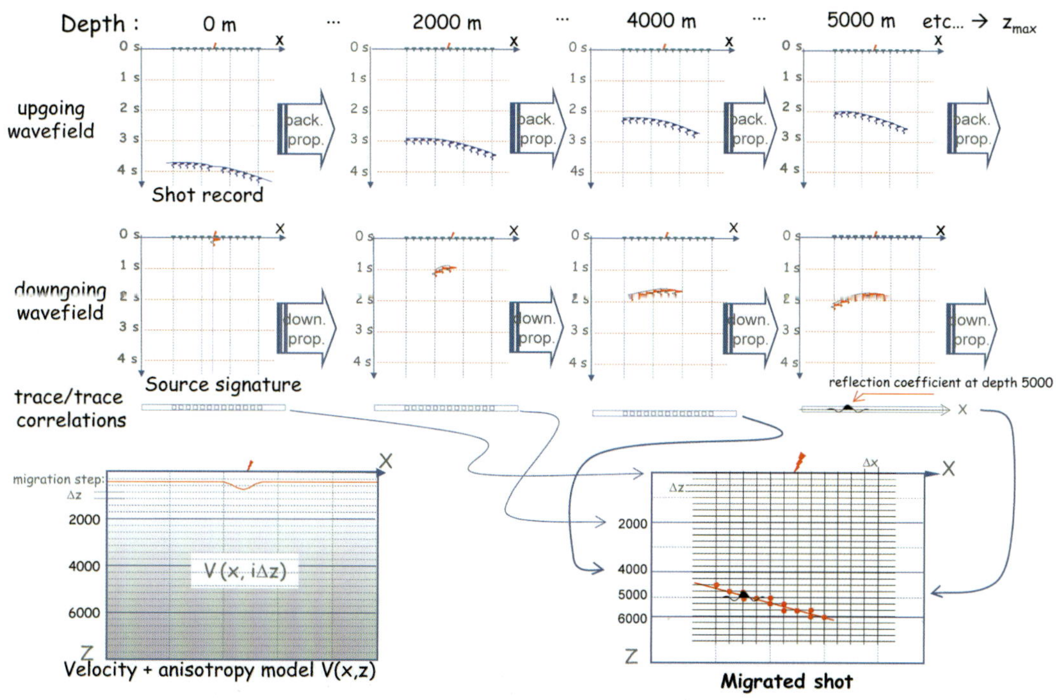

Figure 6.07: The shot point migration workflow. At each depth 0, Δz, 2·Δz, etc. we compute the up- and downgoing wavefields through the velocity layer V (x, iΔz). The two fields are correlated trace per trace and the result placed at the corresponding depth, building in this way the migrated shot.

- the $P_d(x, \Delta z, t)$ and $P_{up}(x, \Delta z, t)$ now become the new 'boundary conditions' from which we can repeat the operations above to go from depth Δz to depth $2\Delta z$ through a new velocity function $V(x, 3\Delta z/2)$ and compute $P_d(x, 2\Delta z, t)$ and $P_{up}(x, 2\Delta z, t)$;
- applying the imaging condition at that depth allows us to populate the row 5Uz of the migrated shot $r(x, z = 2\Delta z)$;
- and of course, the recursion carries on depth step by depth step through the entire velocity and anisotropy model, until we reach the desired maximum depth z_{max} of the migrated image $r(x, z)$.

The challenge in this process is to efficiently and accurately compute the two wavefields at successive depths $z + \Delta z$ from wavefields at depths z. This is addressed in a following paragraph.

Fig. 6.08 is a cartoon of what the result of such a migration of a single shot would be for our simple one-reflector model. Only the part of the reflector that has been illuminated by the shot-receivers lay-out can be imaged. This is highlighted in the figure by drawing on top of the result, the rays and truncated wavefronts belonging to the data actually recorded.

We do not know *a priori* what part of the Earth will be illuminated by a single shot. For steeply dipping and deep events, this may be kilometres from where the shot and the receivers are placed. It is thus necessary to allow the down- and upgoing wavefields to develop in a large area.

This is an important practical issue, related to the notion of *aperture* that we encountered in chapters 4 and 5 for ray-based techniques. Ideally, we would leave the possibility for the data to migrate in the whole image domain (full aperture), but in practice, we face memory space limita-

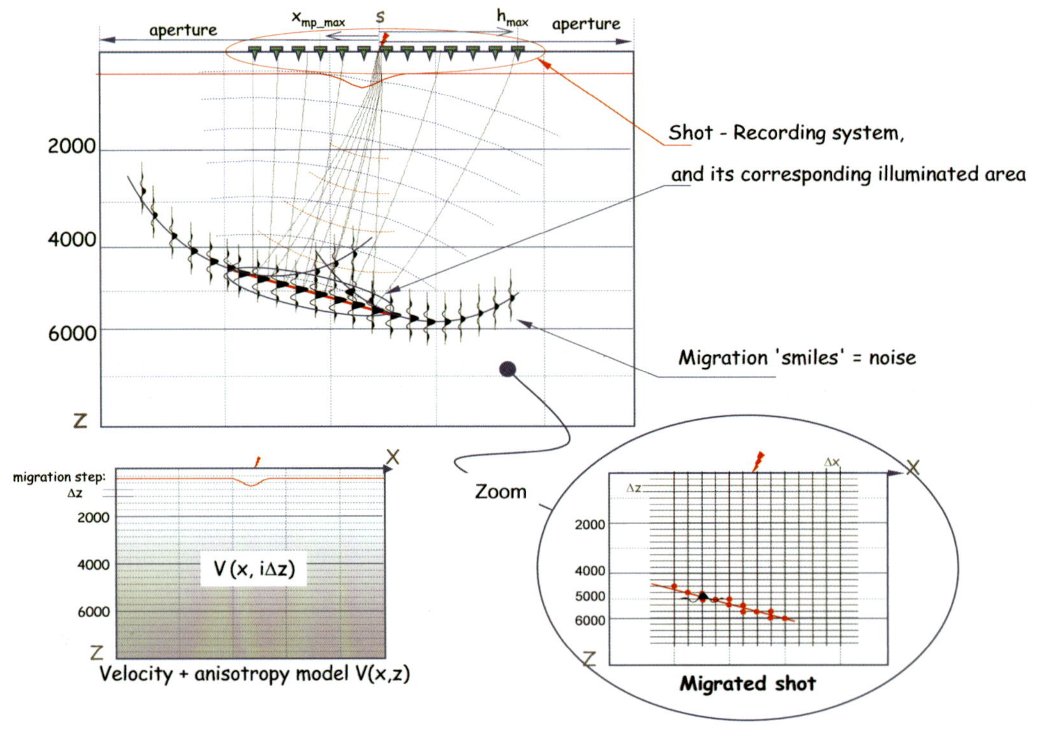

Figure 6.08: Cartoon of the migrated shot of the synthetic data. Only the section of the reflector illuminated by the system shot-receivers is imaged. Migration has edge effects in the shape of 'migration smiles'. This noise will largely stack out when summing the result of the different shots together.

tions (even more so for RTM, discussed later). Processors thus define the '*shot aperture*'. An example is sketched in fig. 6.08 as a fixed distance beyond either the last active receiver or the last mid-point of the shot being processed. Note that too small an aperture may lead to the non imaging of steep, deep reflectors in complex geological settings. Too large an aperture impacts the computing cost. As for ray-based migration, the *shot aperture is a surface* in 3D. This parameter can be tested at the start of an imaging project.

The migrated shot image is digitized with a trace interval of Δx, Δy (generally equal) on the order of 6.25 to 50 metres; it may differ from the nominal interval of the receivers during acquisition, as already mentioned. In current imaging technology, vertical sampling of the final image is also the migration step Δz, in the order of 5 to 10 metres, hence, no vertical interpolation is needed, as was the case in the past.

Some migration noise, often called '*migration smiles*', appears at the edge of the illuminated section of the reflectors. This is an intrinsic effect of shot migration. If the reflector in the subsurface would actually be limited to the illuminated segment, then the edges of the segment would induce diffractions that would be recorded when we acquire the data in the field. Migration would focus these diffractions to yield sharp ends at the reflector segment edges (see fig. 2.08-b). Since the reflector is not limited in space, the diffraction does not exist in the data and the migration process results conversely in 'smiles', the equivalent of the circles in fig. 2.08-b.

These migration edge effects partially sum destructively when we stack the result of several neighbouring shots, provided the shots are sufficiently close together. If this not the case, this migration noise will be high. This has to be taken into account in survey design, knowing also that it can be further reduced by 'muting' the 'Shot –CIG', as discussed in the next paragraph.

Handling all shots

In the process described so far, every shot is migrated independently though using of course the same velocity and anisotropy model. This is illustrated in fig. 6.09. Each shot illuminates different elements of the reflectors in the subsurface. It is the sum of the contribution of all migrated shots that yields the final migrated image.

If shots are sufficiently close together, the same reflector element will be illuminated several times and appear consequently on different shots. This is nothing other than the usual concept of '*multi-fold coverage*', which yields improved signal-to-noise and primary-to-multiple reflections ratios, provided the velocity model is correct. We note in passing two types of issues for 'true amplitude' imaging:

- A given reflector element is illuminated by different shots with different angles of incidence. Summing them together yields a kind of 'average' reflectivity that depends on the actual incidence angles themselves.
- Moreover, in order to obtain a proper average, we should count the number of times a given element appears in the summation and divide the sum by this number; this requires an analysis, feasible in ray-based migrations (chapter 5) but that remains an issue for WE methods.

On the positive side, the migration edge effects illustrated in fig. 6.08 are different for all shots and will hopefully cancel out, as already mentioned, thus improving the signal-to-migration noise ratio.

The concept of Common Image Gather holds for shot migration as it does for Common Offset migration. This is also shown in fig. 6.09. In effect, we can collect the traces from migrated shots having the same location before they are stacked, as for instance the one marked 'CIG' in the figure.

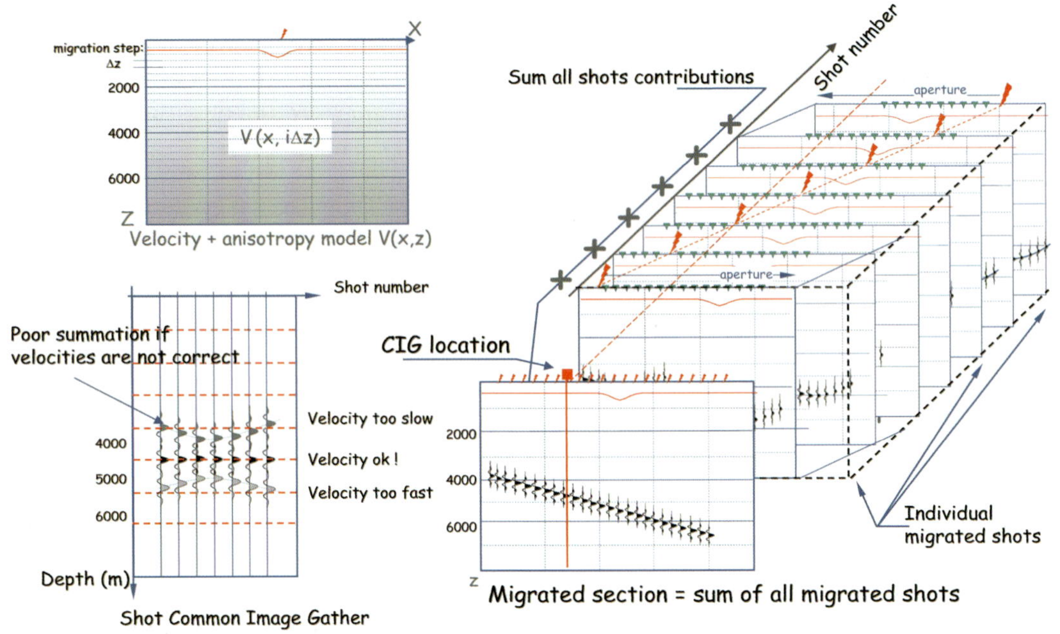

Figure 6.09: Schematics of the one way shot point migration sequence. Each shot of the survey is processed independently through the unique velocity model as in fig. 6.07. The migration is the sum of the contribution of all shots. 'Shot number CIGs' may be computed for cosmetic processing (noise muting) before summing the migrated shots.

Such a collection of traces is called a 'Shot Number CIG' (one also finds the term Shot Index CIG, or SI-CIG). It is used for two things:

- to quality-control the data;
- to process the data and particularly to apply an appropriate muting in order to further reduce the edge effect and any other noises.

It is important to note that a reflection appearing on a CIG may come from any receiver and it is not straightforward to find out which one it is. This is even more difficult when dealing with 3D shots as in most practical situations. This is because in the upgoing wavefield extrapolation we mix signals coming from all receivers, losing the azimuth and offset dimension and the physical relationship between a reflection and its acquisition (shot ↔ receiver) system. This is an important limitation of WE methods, as compared to ray-based techniques. It explains why, contrary to the Offset CIGs, the Shot Index CIGs cannot be used easily in practice for tomography. Migration velocity analysis (or MVA) in WE migrations is done in a different way, which we explore in a later chapter.

Wavefield extrapolation in hybrid Fourier + space domains

In this paragraph we come back to figures 6.04 and 6.05 and discuss in a more detailed manner the important steps in OW Shot point migration that permit the derivation of the wavefields at depth $z + \Delta z$ from the wavefields at depth z.

There are several techniques for this. The most efficient one was described by Gazdag in 1978 and is known as the '*phase shift method*'. Its efficiency comes from the use of *Fourier transform* but the counterpart is that it is only valid where no lateral variations in velocity occur.

The benefit comes from the fact that after Fourier transform the data are decomposed in a series of n frequency panels $\omega_{i,\ i=1,n}$. Each frequency panel is processed independently from the others as an application of the linearity principle. This reduces for each frequency the dimensionality of the propagation problem and makes it a natural and efficient domain for code parallelization.

Let us recast the extrapolation problem with Gazdag's idea, first for the downgoing wavefield. The recursion is initiated at depth 0 (fig. 6.04), where we know the downgoing wavefield $P_d(x, 0, t)$. We know also the velocity $V(\Delta z/2)$ and we assume for the time being that it is laterally constant. We want to derive the wavefield at depth Δz , i.e., to extrapolate the wavefield while honouring the wave equation (1.5) mentioned in chapter 1. For this we recall that the wave equation has a specific expression in the Fourier domain, called a 'dispersion equation', which is written:

$$k_x^2 + k_z^2 - \frac{\omega^2}{V^2} = 0 \qquad (6.01)$$

k_x, k_z and ω are the respective space wavenumbers and radial frequencies corresponding to the variables x, z and t in the Fourier domain; recall that $\omega = 2\pi f$, with f being frequency. When deriving equation (6.01), one *assumes that the velocity field is constant* in any places where the propagation takes place. This is the limitation of Gazdag's method that we mentioned earlier. It has subsequently been completed by correction terms or methods to overcome this theoretical constraint, while keeping some of the benefits of the Fourier domain.

An example is the '*phase shift plus interpolation*' method (Gazdag and Squazzero, 1984), which applies the phase shift method for several velocities covering the range of lateral variability found in the layer. Next, the wavefield at a given location x is interpolated from the different phase shifts based on the local velocity. A wealth of other more accurate methods have been proposed since then. The reader interested in technical details will find all the mathematical developments of these often tedious methods, in the literature. Biondi (2006) gives a very clear overview, addressing moreover the specificity of 3D data sets.

Our goal is not to be exhaustive but simply to present a popular method, the Fourier Finite Difference (or FFD), which is an example of 'hybrid domain methods'. This should give the reader a sense of the different steps needed and the related necessary computing efforts. We do so for its 2D implementation only, again for the sake of simplicity.

This technique was first proposed by Ristow and Rühl (1994). They suggest to decompose the extrapolation into the effect of a constant background velocity equal to the lowest velocity in the layer and denoted 'reference velocity' V_{ref}, plus corrections that should be limited and account for the true velocity function $V(x)$. The extrapolation at each depth is applied in three steps:

- A simple phase shift with velocity V_{ref} very efficiently performed in the $[\omega, k_x]$ domain.
- Appropriate corrections to compensate for errors related only to vertical propagation when using V_{ref} instead of $V(x)$. These operations are applied in the $[\omega, x]$ domain.
- A third correction similar to a 'residual wavefield propagation' also applied in the $[\omega, x]$ domain and that accounts for non vertical propagation. The steeper the dip to migrate accurately, the more computer intensive these corrections become.

The sequence of operations is detailed in fig. 6.10 as follows:

- 2D Fourier transform the 'slice' z = 0 of the downgoing wavefield $P_d(x, t, z = 0)$; this is the initial condition equal to the source signature. It is often a simple zero-phase band pass filter at the x_s location:

$$P_d(x,t,z=0) \xrightarrow{FT^{2D}} \hat{P}_d(k_x,\omega,z=0)$$

- Make use of a well-known property of Fourier transforms to express the field $P_d(x, \Delta z, t)$ at depth Δz, as a function of the field at depth 0:

  $\hat{P}_d(k_x, \omega, z+\Delta z) = \hat{P}_d(k_x, \omega, z) \cdot e^{j \cdot k_z \cdot \Delta z}$
- In this equation, the imaginary exponential acts additively on the phase of $\hat{P}_d$ only; hence the term *'phase shift'* for the method.
- Estimate the unknown k_z term in the phase shift expression using the dispersion equation (6.01). Equation (6.01) has however two solutions, which come from the physics of propagation: in any medium, the waves propagate both upwards and downwards. $k_z = \pm\sqrt{\frac{\omega^2}{V^2} - k_x^2}$
- We want the wave to propagate downward to reach the depth z + Uz, which is achieved by *selecting one solution only*. The choice of sign is a question of convention but the selection in itself yields a *'one-way'* solution to our downwards propagation problem.
- We now need to account for the lateral variations in velocity V(x). The k_x domain is not appropriate for this and one returns thus to the [ω, x] domain, via a 1D inverse Fourier. The principle of the corrections proposed by Ristow and Rühl can be explained by returning to the dispersion equation and developing the k_z quantity as a Padé's expansion, which is written as follows after some algebra:

$$k_z = +\frac{\omega}{V(x)}\sqrt{1 - k_x^2 \cdot \frac{V^2(x)}{\omega^2}} \approx \frac{\omega}{V_{ref}} + \omega \cdot \left(\frac{1}{V(x)} - \frac{1}{V_{ref}}\right) - \frac{2\omega \cdot \left(V(x) - V_{ref}\right) \cdot k_x^2}{4\omega^2 - \left(V^2(x) + V(x)V_{ref} + V_{ref}^2\right) \cdot k_x^2} \qquad (6.02)$$

There are three terms in the right-hand side of this equation, each corresponding to a specific correction applied sequentially.

- The first term in the expansion corresponds to a simple phase shift with velocity V_{ref}. This is the one that we efficiently applied in the [ω, k_x] domain. This results in the computation of grid 1 sketched on the right of fig. 6.10. Note that we deal with matrices or grids of complex numbers in the Fourier domain. They are represented by their amplitude and phase components in the figure.
- The second term, referred to by Claerbout (1985) as the *'thin lens term'* compensates for errors in vertical propagation due to the use of V_{ref} instead of V(x). It is again a phase shift but applied now in the time frequency domain only and varying laterally:

$$\Phi_2(\omega, x) = \Phi_1(\omega, x) + \omega \cdot \left(\frac{1}{V(x)} - \frac{1}{V_{ref}}\right)$$

The results are grids number 2 in the figure.

- The third term is more complex. Since it is space variant, it needs to be transformed back into a new differential equation in x in the [ω, x] domain, recalling that $-k_x^2$ is the Fourier transform of the operator $\frac{\partial^2}{\partial x^2}$.

Differential equations are classically solved numerically by transforming them into 'finite difference' equations (hence the acronym FD in the name of the method). We give an example of such a scheme in fig. 6.19 when we discuss RTM. In RTM, we deal with 'explicit schemes': one point in the extrapolated domain is a linear combination of several points in the different previous panels.

In the case of One-Way depth extrapolation, the application of Finite Differences leads to a system of equations where three points in the extrapolated domain $\hat{P}_{d\_final}(k_x, \omega, \Delta z)$ are a linear combination of n points in the approximate field $\hat{P}_{d\_approx}(k_x, \omega, \Delta z)$. We then talk about an 'implicit scheme',

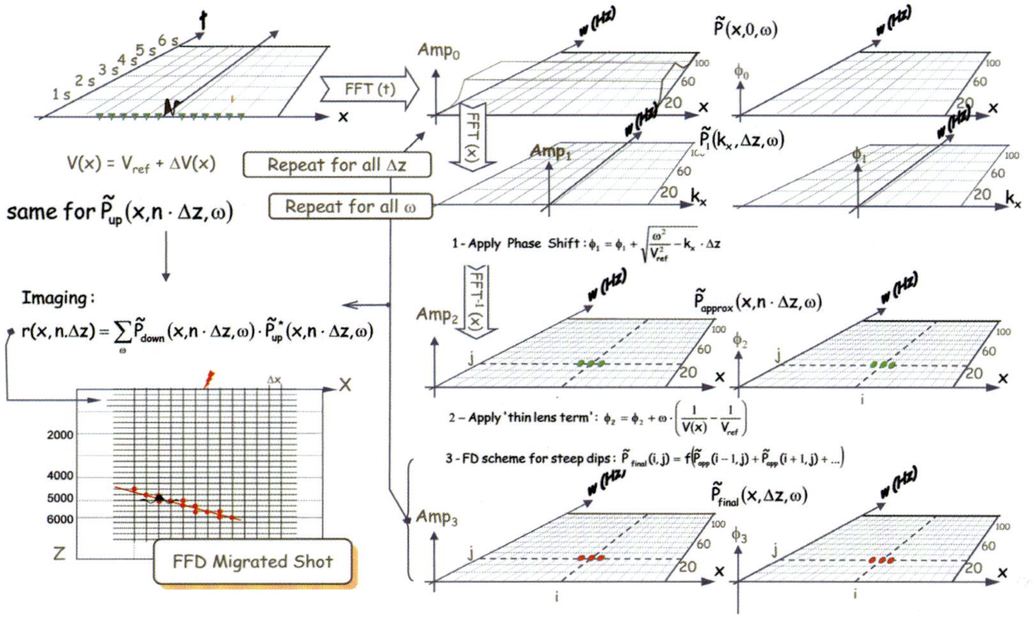

Figure 6.10: Fourier Finite Difference scheme. The process works in the frequency domain ω, but goes back and forth to the (ω, k_x) and (ω, x) domain, filling successively grids 1, 2 and 3. An 'implicit scheme' is used: 3 points of the final grid are jointly function of n points of grid 2.

which necessitates the solution of a set of simultaneous equations. Its resolution goes through the inversion of tri-diagonal matrices. This is an exercise that numerical analysts know how to solve efficiently, but nevertheless requires extra computing efforts, since the size of the matrix to invert is large, especially in 3D. We develop further a Finite Diffrence scheme in fig. 6.19 in the frame of RTM.

- In parallel to the downgoing wavefield, we apply exactly the same process to the upgoing wavefield, starting with the initial condition that is now the Fourier transform of the shot record and selecting the – sign in the one way dispersion equation. This is to account for the fact that we 'reverse the time' of the propagation.

- When this is done, we are in a position to apply the imaging principle. The correlation of the down- and upgoing wavefields, being a correlation in the time domain, becomes a product in the frequency domain:

  $\hat{r}(x,\omega) = \hat{P}_d(x,\omega,\Delta z)\cdot\hat{P}^*_{up}(x,\omega,\Delta z)$ (* denotes complex conjugate)

This yields the value of the reflection coefficient at depth Δz but for one frequency only. We then have to repeat the whole series of operations for all frequencies within the useful seismic bandwidth. Each frequency is generally treated on a different processor, but still properly selecting the frequency range allows computer time and memory savings.

- The reflection coefficient computation requires only the zero-time lag of the cross-correlation of the two fields (fig. 6.06), which is expressed as the following sum over frequencies (a property of Fourier transform):

  $$r(x,\Delta z) = \sum_{\omega}\hat{P}_d(x,\omega,\Delta z)\cdot\hat{P}^*_{up}(x,\omega,\Delta z)$$

At this point, we have the reflection coefficient for depth Δz and there is no need to run a full inverse Fourier transform. On the contrary, the two fields $\hat{P}^*_{up}(k_x, \omega, \Delta z)$ and $\hat{P}_d(k_x, \omega, \Delta z)$ are ready to become the initial conditions of the next step in the recursion.

- The whole series of actions described above is repeated again, now using the velocity function in the second layer. This yields the reflection coefficient $r(x, 2\Delta z)$ at depth $2\Delta z$.

This is repeated recursively, depth step by depth step, until one reaches the final depth z_{max}. As a typical order of magnitude, the set of operations sketched on the right column of fig. 6.10, which includes two 2D Fourier transforms and the inversion of a large matrix in the 3D case, is repeated for 100 frequencies for the two fields and for 1000 to 2000 depth steps, i.e., possibly $4 \cdot 10^5$ times! If this is computer intensive, there are on the other hand, no serious memory issue.

Fig. 6.11 illustrates the benefit of FFD over a simple phase shift approach for a 2D synthetic example. In the FFD case, horizons appear at the right depth and steeply dipping events are reasonably well imaged. This is not the case of course for phase shift, where the model does not include lateral velocity variations.

Delayed- and encoded-shot migrations

Wavefield extrapolation methods are expensive and many techniques are developed to try and make the process more efficient and reduce the migration project's cycle time and cost. One idea consists in the simultaneous migration of a combination of several shots. The principle of combining shots is not new (Berkhout, 1992; Rietfeld, 1995). These authors introduced the concept of 'areal shots technology': instead of having one point source, one generates a 'thought experiment' with several sources spread over a given area and triggered, one possibly delayed with respect to the others. This areal source generates a downgoing wavefield, which is the combination of the wavefields individually generated by each single source and similarly, the reflected upgoing wavefield and the recorded field is the same combination of each individual record.

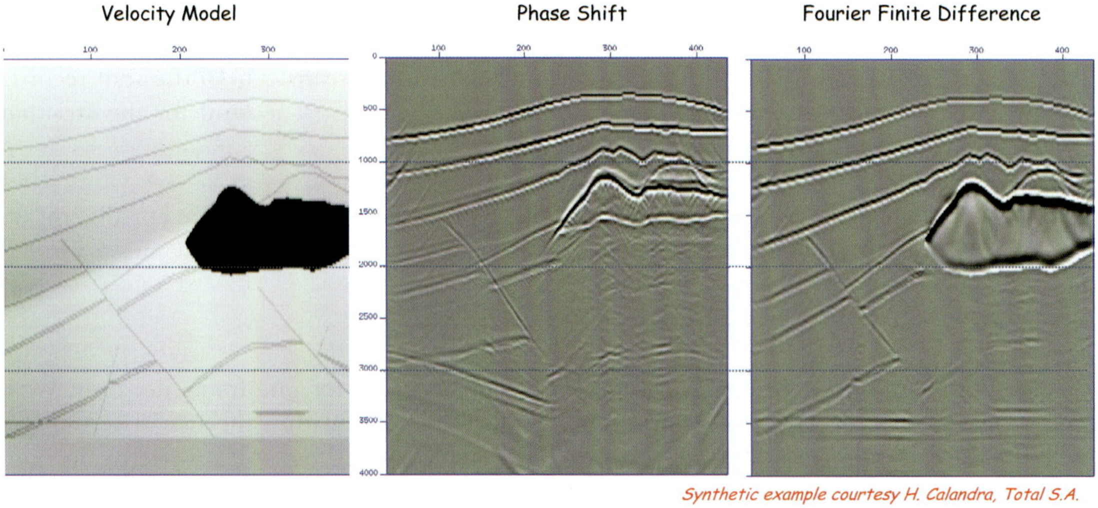

Figure 6.11: One-way shot point migration. Comparison between phase shift and Fourier Finite Difference (FFD) methods (Ristow and Rühl, 1994). FFD methods properly handle steep dips and strong lateral velocity variations, with an acceptable computing efficiency.

This concept, based on the linearity property of seismic experiment, is used in shot point migration in two ways: 'plane-wave or delayed shots' on the one hand (Whitmore, 1995; Duquet *et al.*, 2001; Liu *et al.*, 2002; Zhang *et al.*, 2005; Liu *et al.*, 2006) and 'encoding-shot' or 'modulated-shot' techniques on the other hand (Romero *et al.*, 2000; Soubaras, 2006; Zhang and Sun, 2009).

The idea of *delayed shots* is illustrated in fig. 6.12. If we shoot different shots one after the other with a delay Δt and if the distance between shots Δs is small enough in terms of seismic wavelength to avoid aliasing issues, the combination of each individual wavefield generates the equivalent of a *plane wave source* tilted with a 'take off' angle θ, or equivalently a ray parameter p_{source}. We obtain the response of the Earth to that 'areal source' similarly by summing the N shot records, each delayed by the time $n\Delta t$, according to its location at coordinate $n\Delta s$. In doing so, we obtain for one particular Δt (or take-off angle θ or ray parameter p_{source}) one 'macro-shot record'.

The delayed-shot migration technique consists then in generating such plane wave source emissions with a given propagation direction or '*take-off angle*' θ, to construct the corresponding upgoing wavefield as it would have been recorded by all the receivers at the surface and to migrate these combined records as a unique seismic experiment and finally, to repeat the operation for a series of M take-off angles θ_i.

The only differences between the migration of a single shot and a macro-shot is that for the latter, the downgoing wavefields boundary condition is now made of the series of delayed source signatures spread over the entire line length and the boundary conditions of the upgoing wavefield is the sum of the actual corresponding shot records, each delayed by the appropriate amount $n\Delta t$, according to its corresponding source. The recorded upgoing wavefield covers the whole line of receivers as well. These macro-shots are large volume data sets, especially in 3D.
To recapitulate, the delayed shots process proceeds as follows (fig. 6.12):

- select a series of M values of take-off angles θ_i (or Δt_i) ranging from $-\theta_{max}$ to $+\theta_{max}$; this maximum angle is related to the maximum dip that will be migrated;
- for each θ_i, create the corresponding macro-shot record;
- migrate the macro shot record with a shot migration technique;
- repeat for all M values of θ and stack all 'migrated macro-shots' to obtain the final migrated image.

The following considerations come to mind when applying this process in practice.
Firstly, although wavefield extrapolation is a linear process, the migration of the sum of the N delayed shots is not equal to the sum of the migration of each shot. This is because of the imaging condition. In the simultaneous source experiment, we indeed obtain the sum of the contribution of each correlation of all downgoing wavefields from the ith shot with its corresponding i^{th} upgoing wavefield, but we also sum all the unfortunately non-zero contribution of correlations of all the N-1 other upgoing wavefields. The effect of this '*cross-talk noise*' is however attenuated when repeating the migration for several θ_ι values and stacking the different results.

An increase in computational efficiency of the overall migration process, which is the main target of delayed-shot migration, is consequently achieved if:

- one can use fewer combinations of shots, i.e., fewer plane wave directions θ_ι than individual shots (smaller number of migrations) for an equivalent quality of the migrated image;
- and this benefit is not offset by the cost of creating and migrating the combined macro-shots.

The cost of a single macro-shot migration is related to the size of the down- and upgoing wavefields, which increases with the length of the line (number N of shots to migrate together) and θ. As an example, for a survey with a shot interval of 50 m, a surface velocity of 1500 m/s and a line 20 km long,

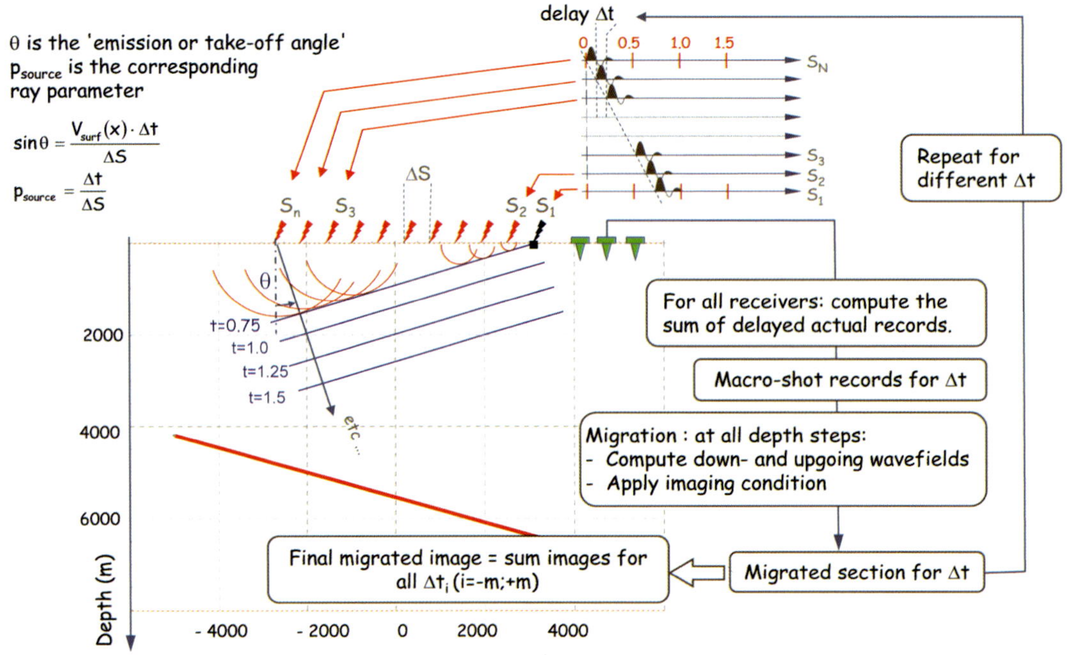

Figure 6.12: The thought experiment and processing sequence in delayed-shot migration. Plane wave emission is simulated by delaying each shot by a quantity Δt. The corresponding 'areal source' and shot record are computed. Single shot migration is applied to this thought experiment. The process is repeated for all selected value of Δt and the individual results summed together.

the maximum 'virtual recording time' would be increased by 9.4 s for a take off angle θ_{max} close to 45° and 13 s for a θ_{max} of 90°, thus more than doubling the size of the $P_d(x, z, t)$ and $P_{up}(x, z, t)$ volumes. This may have limited impact when dealing with frequency-based approaches such as the FFD, since a delay in the time domain is equivalent to a phase shift in the frequency domain. For techniques working in the time domain such as RTM, this becomes a real issue.

The selection of the proper series of θ_ι values (or equivalently delays Δt) is critical. Two parameters are to be defined: the maximum take-off angle θ_{max} and the number M of θ_ι. Take too small a value of M and the result is noisy and we lose information; select too large a value of M and we lose the benefit of the shot combination.

Moreover, in 3D surveys the 'plane wave' source is really a plane defined by a dip θ, as in fig. 6.10, but also a series of azimuths α, the number of parameters M being the product of both.

The selection of the [θ] and [α] range is thus important, as it impacts directly and linearly the efficiency of the method. Etgen (2005) lists the number of variables that may affect this selection: shot spacing Δs, seismic frequency bandwidth, cable length, line length, recording time, velocity gradient, depth of target, etc. This author also emphasizes the impact of proper data regularization.

Note that the maximum take-off angle is related but not equal to the maximum migrated dip. A strong velocity gradient for instance may lead to turning waves, i.e., the recording of reflections against overhanging reflectors (dip > 90°) for small take-off angles (<30°). Hence, the two last parameters in Etgen's list.

Generally speaking, the industry uses delayed-shot with rules of thumb to choose M. The reported average efficiency improvement of delayed-shot techniques is of a factor 2 to 3 with respect to single shot migration, with a number of θ values in the order of one hundred to a few hundreds

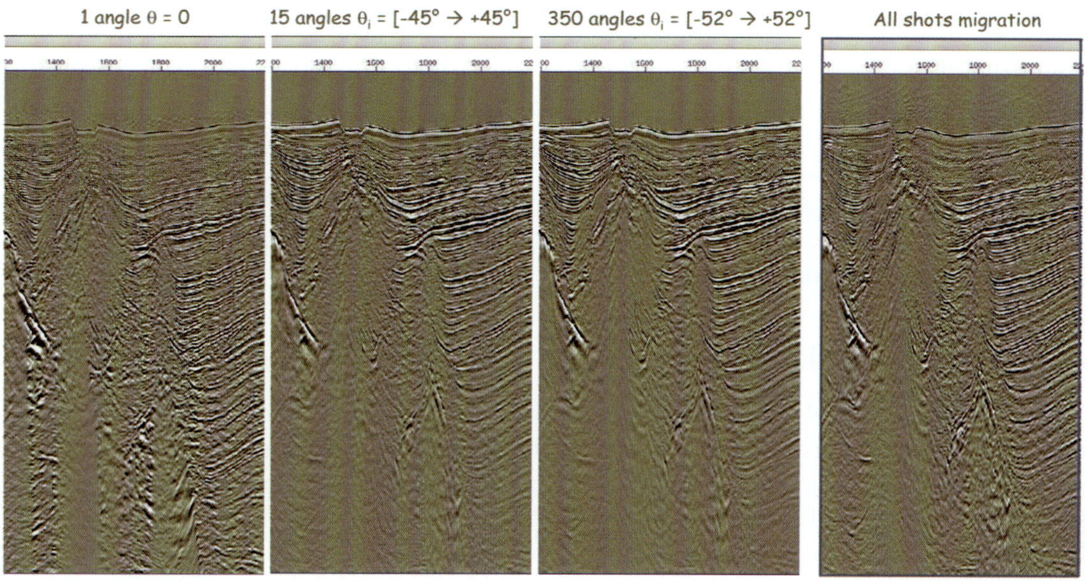

Example courtesy Total HGRG Team with authorization from TGS-Nopec

Figure 6.13: Example of delayed (plane wave) shot migration compared to shot point migration of all shots. For this specific acquisition, a very good image quality is reached surprisingly already with 15 plane waves only.

for 2D types of surveys. The selection of M is not easy anyway. Fig. 6.13, a test courtesy of Total's HGRG on data from TGS-Nopec, compares the result of shot-point migration using all shots (right-hand side panel) with the results of migration with different θ_{max} and M pairs of parameters. In the context of this example, a surprisingly low number of plane waves (15) yields an acceptable result, whereas the value of M estimated beforehand was 350!

As mentioned, the delayed shots technique, when considered in the frequency domain, appears as a linear phase shift encoding of the individual shots. Romero *et al.* (2000) consider other possible *phase encodings*, including random and 'chirp' ones. The goal is still to reduce the cross-talk noise due to the correlation at the imaging step of a downgoing wavefield with all the upgoing ones that do not correspond to it in the combined macro-shots. Soubaras (2006) shows that with M=N, one can always find a proper phase encoding in the frequency domain such that the two approaches are strictly identical. Such a case does not reduce the project's cost, but the author suggests ways to use this encoding in a smart way to reduce the number of composite shots to migrate.

Other techniques have been suggested. Among them is 'Harmonic Source encoding' proposed by Zhang at al (2007). In this case, shots are not delayed but encoded in amplitude by a cosine function of the shot location and a sine function of both time and shot location. The macro-shots consequently have approximately the same size as the individual shots, thus making the approach interesting for RTM. The scanning parameter is no longer a delay or take-off angle but the wavelength of the cosine and sine functions. Encoded techniques are used in the industry and often preferred to delayed-shots nowadays.

Before summing the contribution of each macro-shot, one can place the data in a 'Common Image Gather' of yet of another type! For the delayed shots, the CIG is now a function of the take off angle θ at a given location. We could call it 'Take Off Angle Common Image Gather'. As mentioned for 3D acquisition, this object is a three-dimensional object as the take off angle is composed of a dip θ and an azimuth α. These new CIG may be used similarly to the Shot CIG for QC purposes and processed for migration noise muting.

As a comment to conclude this paragraph we note that the delayed- and encoded-shot techniques described in this section are dedicated to one-way shot migration. They can however be used for any type of WE migration in the shot domain. Delayed shot is nevertheles not practical for Reverse Time Migration, which we discuss later.

Concept of survey sinking

The survey sinking methods introduced by Claerbout (1985) is another way to migrate all shots together. The process is illustrated in fig. 6.14. It starts from the basic observation that at zero time the downgoing and upgoing wavefields coincide where the source and the receiver themselves coincide, i.e., at zero-offset, and this is true for every shot.

Making use of this trivial remark, we can start building the migrated image by a simple copy of the zero-time and zero-offset row of the data set, provided that it has been 'divided' by the source signature. The 'division' process is known in fact as 'signature deconvolution' in conventional seismic pre-processing. This is also what we addressed in figures 6.02 and 6.06, with the correlation replacing the division in the 'imaging condition' to obtain the reflection coefficient.

The first copying step does not lead us very far since in practice, we have no recording at zero-offset. The trick is however that we can compute the whole data set we would have recorded at a depth Δz and again apply the imaging principle to it. In doing so, we look now for the presence of reflectors at time zero and offset zero of the new data sets at depth Δz. If we repeat the process further in depth, depth-step by depth-step until a predefined depth z_{max} (actually limited by the recording time and velocities), we will have 'imaged' reflectors at all depths and hence constructed the depth migrated image.

The cartoons on the left of fig. 6.14 represent a simple fictitious model and its corresponding 2D data set volume. The latter is sorted by mid-point, offset and time (refer to fig. 3.03 for data sorting). These data are supposed to be 'signature-deconvolved'. The migrated section is sketched on the top right of the figure. The yellow line is simply a copy of the zero-offset, zero-time row of the data volume, in the application of the imaging principle.

Next, we assume to know the propagation velocity and anisotropy in the layer [0 – Δz]. Using a wavefield extrapolation technique, we can compute the *data set that we would have recorded with all sources and all geophones now at depth* Δz. In doing so, we have 'extrapolated' all sources and all receivers to this new depth.

The new data volume, traditionally called '*residual data*' with respect to depth Δz, is schematically represented in the middle of fig. 6.14. Some offset rays drawn in the model at the top right of the figure illustrate the behaviour of a reflector as the measurement plane 'sinks' into the subsurface. We are now again in a position to apply the imaging principle by copying the zero-time, zero-offset data of the new data volume (the green line in the figure). It goes to the Δz depth of the migrated section that has become the new 'zero' reference depth.

The process is recursively repeated. A new data volume simulating full acquisition at depth 2Δz is derived from the first residual data volume by wavefield extrapolation through a velocity model assumed to be known in the layer [Δz – 2Δz] and we apply the imaging principle again to obtain the migrated row at depth 2Δz.

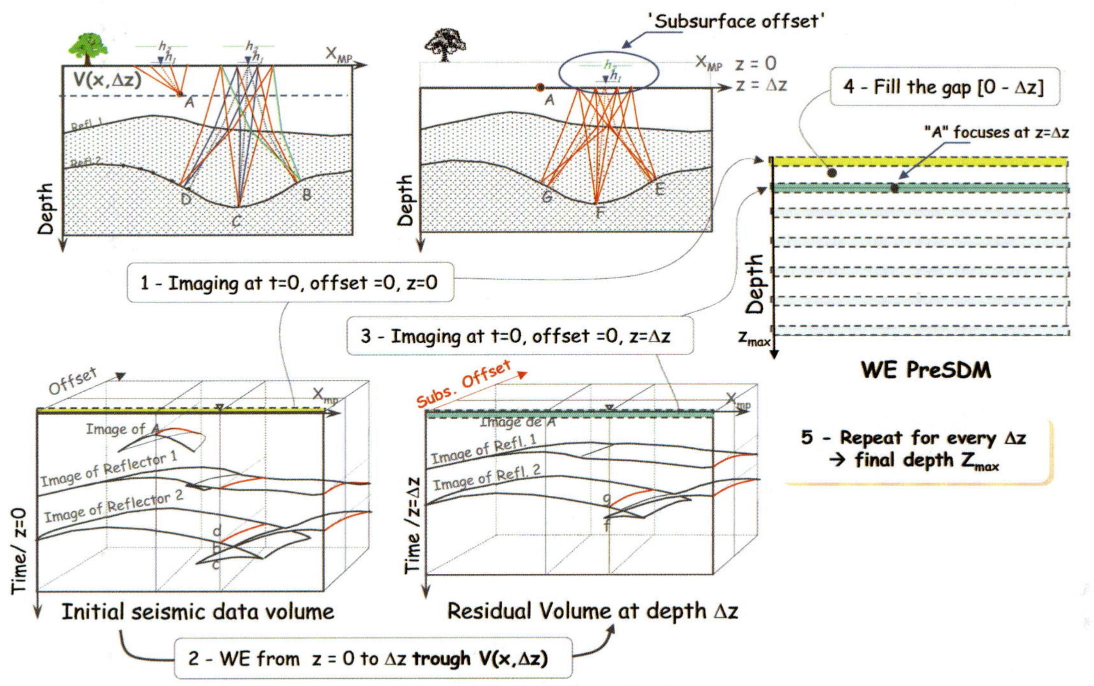

Figure 6.14: Survey sinking concept. The imaging principle states that all events at time 0 and offset 0 at recording depth is imaged at that depth. The imaging principle can be used recursively by 'sinking' the survey step by step Δz in depth through the velocity model V(x, i·Δz).

And so on, we 'sink' the acquisition, compute the residual volumes, apply the imaging principle and build in this way the migrated section depth step by depth step through the velocity model. Let us note some important points featured in figures 6.14 and 6.15:

- The 'initial data volume' in fig. 6.14 is the measured upgoing wavefield at depth z = 0, $P_{up}(x_{mp}, 2h, z = 0, t)$. Now that we are dealing with the whole data set of a 2D survey, it is a function of four variables: the shot and receiver coordinates x_s and x_r or equivalently, mid-point x_{mp} and offset 2h, time and measurement surface z=0.
- a diffraction point A at depth Δz appears as a pseudo hyperboloid surface in the data domain with respect to z = 0. It has 'focused' in the data set with respect to depth Δz and is consequently imaged at that depth by application of the imaging principle. The data have 'moved towards a shorter time', but also shorter offsets!
- The offset dimension of the 'sunken' survey is called '*subsurface offset*'; we find this term again, although with a different meaning in the further sections.
- By the very principle of the method, the migration velocity and anisotropy fields are gridded at the same vertical depth interval as the migration step Δz, thus enabling easy variations with depth. Any lateral variations can be handled simply by allowing lateral variations of the velocity at each depth step.
- The vertical grid size of the final depth image is generally smaller than the migration step Δz (typically 5 metres against 10 to 20 metres), thus necessitating some 'infill' between the rows resulting from the direct application of the imaging principle (as, for instance, between the yellow and green lines in the figure). This may be done by extending the imaging principle to zero-offset and small times such that $\Delta t(x_{mp}) = \Delta z/2 \cdot V(x_{mp})$, where $V(x_{mp})$ is the laterally varying velocity in the layer.

Fig. 6.15 is an extension to the survey sinking case of the phase shift method. Note the increased dimensionality and the fact that the k_z vertical wavenumber term is now the sum of two square roots, one for the shots and one for the receiver. The wavefield extrapolation that we perform in two separate operations (shots and receivers) is combined here in a single but more complex one. The survey sinking is often referred to as the '*DSR extrapolation method*'.

For 3D surveys, the data volume may be sorted according to five dimensions, which makes it difficult to sketch! These are, for instance, the four shot and receiver coordinates plus the recording time; or alternatively, the two mid-point coordinates, the shot-receiver offset and azimuth and the time. The application of the 'simple' phase shift becomes tedious and even more so all practical applications, such as phase shift plus correction or FFD methods.

The only current industrial application of the survey sinking technique makes the assumption that traces have all the same azimuth. They are called for this reason '*Common Azimuth Migrations*' or 'CAM' in short. There is, however, a further assumption made (Biondi, 2006), namely that at each

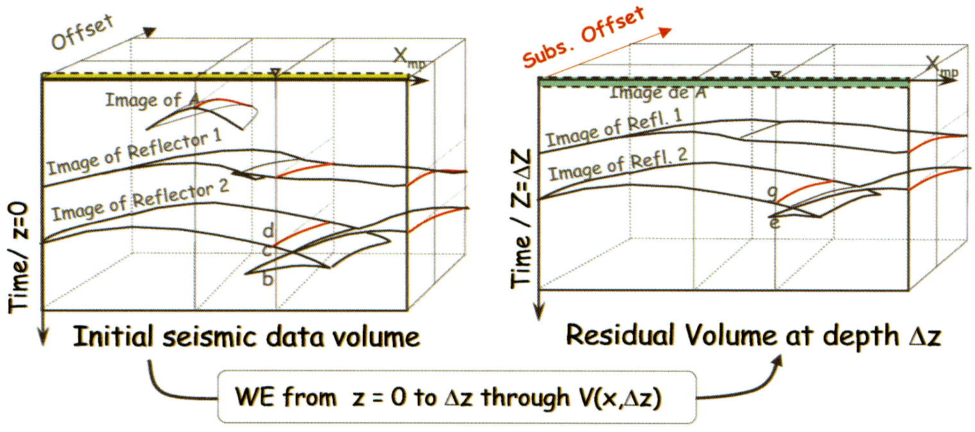

The Double Square Root (DSR) dispersion equation $$k_z = \sqrt{\frac{\omega^2}{V_s^2} - \frac{1}{4}\left(k_{Xmp} - k_h\right)^2} + \sqrt{\frac{\omega^2}{V_r^2} + \frac{1}{4}\left(k_{Xmp} + k_h\right)^2}$$

Figure 6.15: Survey Sinking using the frequency - wave-number 2D 'Double Square Root' equation in the mid-point - offset domain. Fourier transform assumes a constant velocity in the layer [0-Δz]. For more realistic 3D cases where velocity varies laterally, more complex schemes need to be applied.

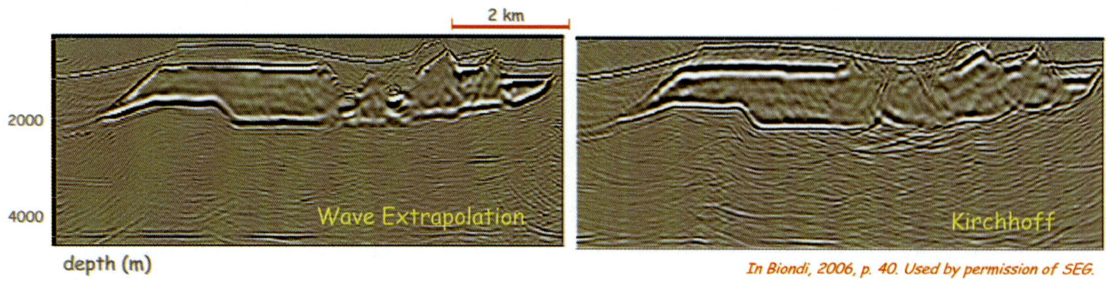

Figure 6.16: Comparison between WE migration and Kirchhoff migration. Same velocity model. Synthetic 3D SEG-EAGE salt model. Courtesy B. Biondi.

migration step, the downward propagation of the information departing from a shot and arriving at a receiver takes place in the same plane.

These methods have lost part of their interest with recent trends towards Wide Azimuth Towed Streamer acquisitions (often referred to as WATS or WAZ). They are still used mainly for marine surveys acquired with a limited number of streamers and for fast wave equation-based migration velocity analysis. This is because they are still faster than shot point migration and even more so than RTM methods, while benefiting from the advantages of WE migrations and particularly, proper handling of complex velocity models.

Principle and workflow of Reverse Time Migration

We move now from OW migration techniques to the 'Reverse Time Migration' or RTM, which is intrinsically a 'two-way' extrapolation method.

In this section we deal with the migration of one single shot only, knowing that encoded-shot techniques are implemented for RTM as they are for OW Shot Point migrations.

The principles of RTM were already given in figures 6.02 and 6.03: if we are able to reconstruct the whole 'history' of the down- and upgoing wavefields at a given location (x, y, z) in the subsurface, the imaging condition tells us that the reflection coefficient at that location is the scaled correlation coefficient of these two 'histories'. For a single shot, they make the two data volumes sketched at the top of fig. 6.03. Vertical constant-time slices through these volumes are, once again, 'snapshots' of the down- and upgoing wavefields, respectively.

Fig. 6.17 presents on the same cartoon a series of snapshots for our 2D model, firstly in red on the left for the downgoing field and secondly, in blue on the right for the reconstructed upgoing wavefield. Note the difference with the constant depth wavefields in figures 6.04 and 6.05. We show on top of these computed fields their respective 'boundary conditions': the source signature for the downgoing wavefield and the shot record for the upgoing one.

With these 'snapshots' and the imaging condition in hand, we can define a RTM theoretical workflow such as the one presented in fig. 6.18, that would go as follows:

- Starting from the source signature, a modelling exercise yields the snapshots of the downgoing wavefield at successive time steps dt, 2dt, 3dt, etc. We store these snapshots.
- The process is recursive, i.e., snapshot at time t + dt is computed from snapshot at time t, using the whole (possibly anisotropic) velocity field V(x, z). A typical value of dt is 1 or 2 ms. As an example, for a 6 s recording time, we have to compute 6000 or 3000 such data sets and store may be half of them (at least as many as required for an accurate correlation), each having in principle the size of the final migrated image.
- When this is completed, we start the second recursion, now in *reverse time*. It consists in the computation of the backward extrapolated snapshots of the upgoing wavefield.
- In application of the imaging principle, at every time step of this second recursion each sample in (x, z) of the downgoing field is multiplied by its corresponding sample of the upgoing field and added to the corresponding pixel in (x, z) of the migrated image. The migrated image is thus built progressively, each time step bringing its own contribution to the correlation in the imaging condition. Recall that we also need to compute the scaling term that appears at the denominator of the imaging condition equation in fig. 6.02.

The RTM strategy presented in fig. 6.18 necessitates computing first all snapshots of the downgoing field and *storing them* in order to have them ready when we recursively compute the back-extrapolated snapshots of the upgoing field. However, these one does not need to store, given the recursive character of the technique. Storing 6000 or even 3000 snapshots as suggested in the figure is a very

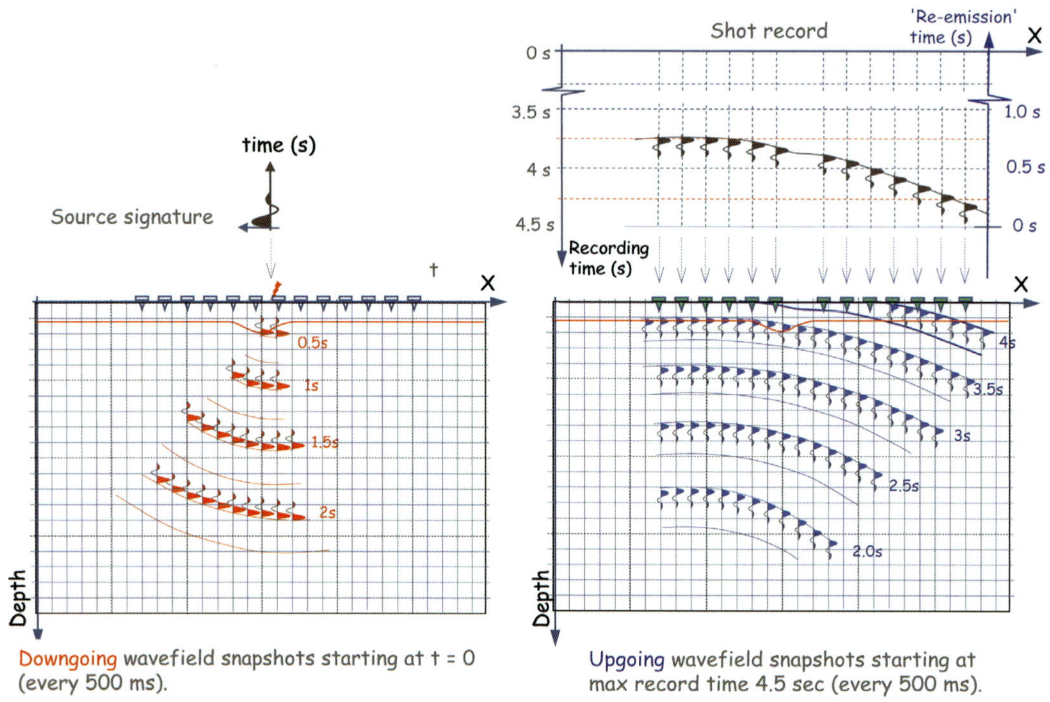

Figure 6.17: Reverse Time Migration. Snapshots of the down- and upgoing wavefields are computed typically every 1 or 2 ms (shown here every 500 ms). The 'time boundary conditions' are the source signature for the downgoing field and the 'time reversal' shot record for the upgoing field.

memory-intensive job and has prevented the development of RTM for a long time. We currently find three ways to address this issue:

- to adapt the computer architecture to actually store and handle the thousands of downgoing snapshots, as outlined in fig. 6.18;
- to recursively propagate the downgoing field and store only 'virtual recordings' at the edges of the model, as if we had laid out receivers there. These records are then back-propagated in time synchronously with the true shot record;
- to use 'random boundaries' as proposed by Clapp (2009).

We discuss the last two options later in the chapter. Once the recursion has been carried out until zero time, it suffices to scale each pixel by the computed scaling factor and the RTM migration of one shot is done.

Handling all shots in RTM

The way several shots are treated in RTM is in no way different from what we said for OW shot point migration. More explicitly:

- Each shot is processed separately with the same anisotropic velocity field and all shot contributions are added together to obtain the final RTM result. The cartoon of fig. 6.09 that we drew for OW Shot migration remains valid for RT Migrated shots.

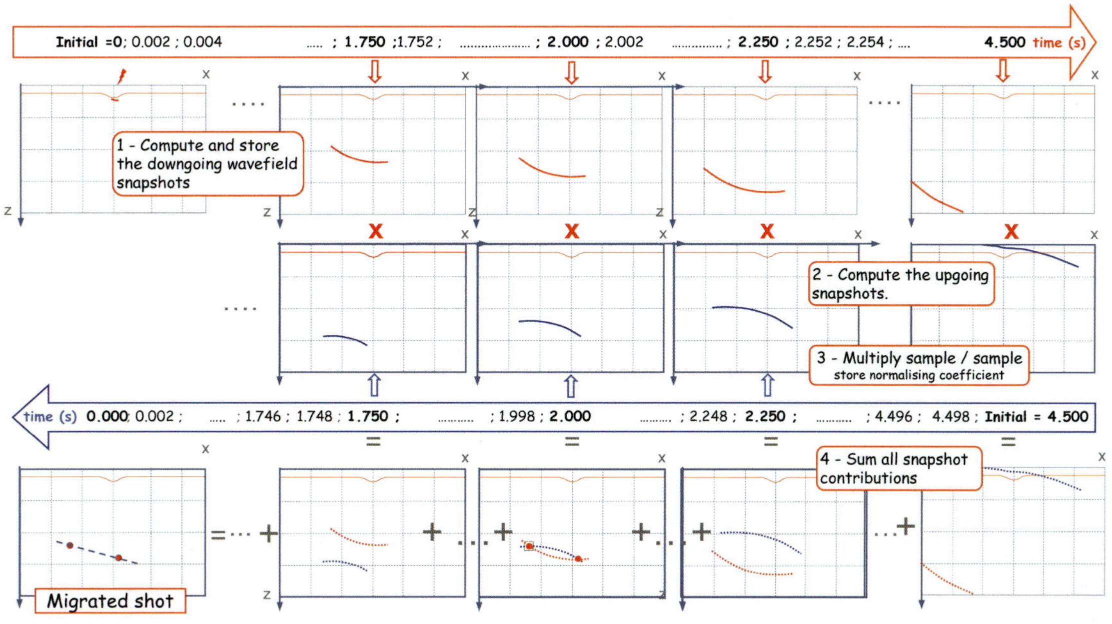

Figure 6.18: Reverse Time Migration workflow of one 2D shot. Snapshots of the down- and upgoing wavefields that have the same time are multiplied sample by sample and normalised according to equation in fig. 6.02. The result is summed to the image, which builds up as more snapshots are computed. One difficulty of this process is that snapshots of the two types of wavefields must be available simultaneously, while they are computed in inverse time order.

- The definition of the shot aperture (figures 6.08 and 6.09) is however more critical in RTM because of the memory challenge inherent to the process.
- Shot Index CIG are produced if requested, either for QC purposes or for migration noise reduction before summation. Note that the concept of 'shot index' is easy to handle in 2D, as shown in fig. 6.09; it is much trickier when dealing with real-life 3D, wide azimuth acquisition!
- Alternatively, shots may be migrated jointly in an encoded-shot.
- Migration velocity analysis after RTM may be approached in a very similar fashion to the OW Shot migration case. This is discussed in more detail in chapter 7.

Time extrapolation using a Finite Difference scheme

Time extrapolation is the process that allows us to compute a snapshot at time t from the snapshot at time t+dt (for the computation of the downgoing field) or t-dt (for the reconstruction of the upgoing field). As mentioned, this process is applied thousands and thousands of times per shot in a RTM project. Consequently, it must be as fast as possible and at the same time, very accurate. Unfortunately, the 'Gazdag's trick' to speed the process up is not applicable directly, since at a given step propagations takes place throughout the whole velocity model, which makes Fourier transform inapplicable.

Extrapolation is currently performed by applying numerical Finite Difference schemes 'in time'. The positive consequence is that one is no longer restricted to one way approximation: RTM uses '*two-way propagators*' based on the *acoustic wave* equation, at least for current industrial applications.

The first elastic results are starting to emerge from research labs but are not yet viable on an industrial scale.

It goes beyond the scope of this book to describe all the mathematical and numerical developments behind efficient schemes. Nevertheless, we present in fig. 6.19 the simplest scheme possible (even though the figure is busy!), called *'second-order stencil in space and time'*. This is

- firstly to give the reader a flavour of the principles of Finite Difference schemes for the integration of the two-way wave equation (we met it for OW in fig. 6.10);
- secondly, to explain a bit more how we 'back propagate in time' and how we 're-inject' the shot record into the wavefield;
- and thirdly, to introduce the discussion on intrinsic relationships between acquisition and processing parameters.

The wavefields in the computer are digitized functions, with time- and space sampling intervals: dt, dx, dy (in 3D), dz. The wavefield P_{up} that we are concerned with here becomes a multidimensional matrix with elements denoted $P_{i,j,k}$ for $P_{up}(i \cdot dx, j \cdot dz, k \cdot dt)$. As shown in the figure for the 2D case, the numerical representation of a snapshot is a 2D matrix and the whole field is a gridded 3D volume. Note that we assume regular grids (dx, dz , dt are constant).

Next, we need to transform the wave equation into a 'Finite Difference' scheme applicable to these digitized functions. This is done by *approximating derivatives by differences.*

The most straightforward formula for this is the first-order Taylor's expansion:

$$\left.\frac{dP}{dx}\right|_{x+dx/2} \approx \frac{P(x+dx)-P(x)}{dx}$$

When applied twice, this approximation yields the approximation of the second-order derivative given in fig. 6.19. The derivative in one point is approximated by a linear combination of the value of three points. Similar schemes can be applied for derivative in time.

More sophisticated schemes are used in practice to approximate derivatives in space, often based on Taylors series expansions up to the 8th and even 10th order. We talk then of 'nth order in space, mth order in time' schemes. We discuss later the benefits and drawbacks of high order schemes.

We stick to this simple second-order one in the framework of this presentation. Injecting approximations of the derivatives for our three variables (x, z, t) into the wave equation yields a polynomial equation. This equation, which we denote *'FD stencil'* in fig. 6.19, links a total of 7 elements $P_{i,j,k}$ of snapshot matrices at *three different times*. This comes from the use of a second-order derivative approximation in time.

Before moving forward, we note that only dt^2 appears in the equations: the wave equation is independent of the sign of dt. The same equation drives the wave propagation whether we want to predict what 'comes next' (dt>0) or what 'was before' (dt<0). The difference between the two is in the boundary condition that needs to be causal, i.e., know what was before if one wants to predict what comes afterwards, and *vice versa*!

In the case of the upgoing wavefield, we invert the sign of time in order to 'run the movie backwards', including for the shot record, which now becomes a 'source term' and we run it from the maximum recording time t_{max} until zero time, when the shot was physically triggered.

Coming back to the FD stencil, we recall that the elements of the snapshot matrix that we want to compute at time t-dt has a subscript k+1; we may thus isolate the element $P_{i,j,k+1}$ and note that it is a function of 5 elements of the snapshot matrix at time k·dt, but also one element of the previous matrix at time (k-1)·dt.

Assuming that the two previous snapshot matrices have been completed in preceding iterations, we have in hand the scheme to compute the new snapshot matrix. According to the FD equation, each of its elements is in fact a linear combination of elements in the two previous matrices and can thus be computed one after the other until the matrix is filled. When this is done for all elements, we have the snapshot at time t-dt and the process is repeated for the next time step t–2dt. The recursion is carried on until one reaches time zero.

It is worth realizing that the parameters of the linear combination are direct *functions of the sampling rates dx, dy, dt* and that the velocity used in each stencil position is averaged over a 3 × 3 samples area, highlighted in grey in fig. 6.19. Apart from the short wavelength smoothing effect, velocity is thus allowed to vary very rapidly both horizontally and vertically. This is one of the benefits of WE-based techniques.

Next, we need to address what happens at the limits of the model:

- Two earlier snapshots are necessary in order to compute a new one. This reflects the fact that the integration of differential equations of the second order requires two boundary conditions. In our case we solve it by assuming that the matrix at t_{max} + dt is identically nil: no more reflections arrive after the last recorded sample. This is intrinsically wrong and shows up in the final migrated result as 'migration bowls' at the bottom of sections. Practical solutions to mitigate this effect include recording data for longer (typically one or two seconds) than the time of the deepest target arrival time, to taper the data in order to 'smooth' the sharp truncation of the record and finally, to mute the noisy bottom part of the migrated image.
- Similarly, when we place the FD stencil in fig. 6.19 in a position to compute the first and last column of the new snapshot, we see that we need data outside the limits of our sections. To assume that the values outside the recording area are zero is equivalent to assuming that the limits of the sections *are total reflectors*: no elastic energy escapes. The result of such an assumption would be that any waves hitting this boundary would be reflected back into our zone of interest, introducing unacceptable artefacts in this way. Codes thus include 'absorbing boundaries' to reduce these artefacts as much as possible, for all angles of incidence. Absorbing boundaries methods are largely discussed in the literature. Current schemes imply padding the edges of the model with extra columns of data where the outgoing energy will 'die' without bouncing back. The size of such padding areas must be as high as 4 or 5 dominant wavelengths (i.e., several hundreds metres), adding to the memory challenge already mentioned.
- Things are different but not easier at the surface. A (over) simplified view is to consider that the energy transmitted in the air is negligible, in which case the integration of the wave equation generates the 'stress-free surface multiples'. Depending on whether these are desired or not, the 'absorbing condition' at the surface will be 'on' or 'off'. Current practice is to attenuate if not totally remove multiples from the data and to add absorbing boundaries at the surface. SRME (for Surface Related Multiples Elimination) has become the standard tool for that purpose (Verschuur, 2006).
- Concurrently, we know what was the wavefield at the surface at any time from t_{max} to t = 0: it is the shot record. Consequently, one needs to force the wavefield to be equal at each time step to what we have recorded at this time. This is sketched in the figure as the green dots in the upper-right panel. In this way, we see that everything happens as if the shot was reversed in time and acting as an areal source (as discussed in a previous section), injecting its information as time unrolls backwards. The shot record is the boundary condition of the backward propagation of the upgoing field!

The numerical integration of the wave equation is repeated at each time step. A very similar scheme applies to the downgoing wavefield. The only differences are the 'sign of time', that increases, and the initial condition that should be a faithful representation of the source signature. As mentioned for OW Shot migration, it is often a zero-phase band pass covering the 'necessary frequency' bandwidth, a concept we return to soon. This assumes that the data have been properly 'signature-deconvolved' to zero-phase.

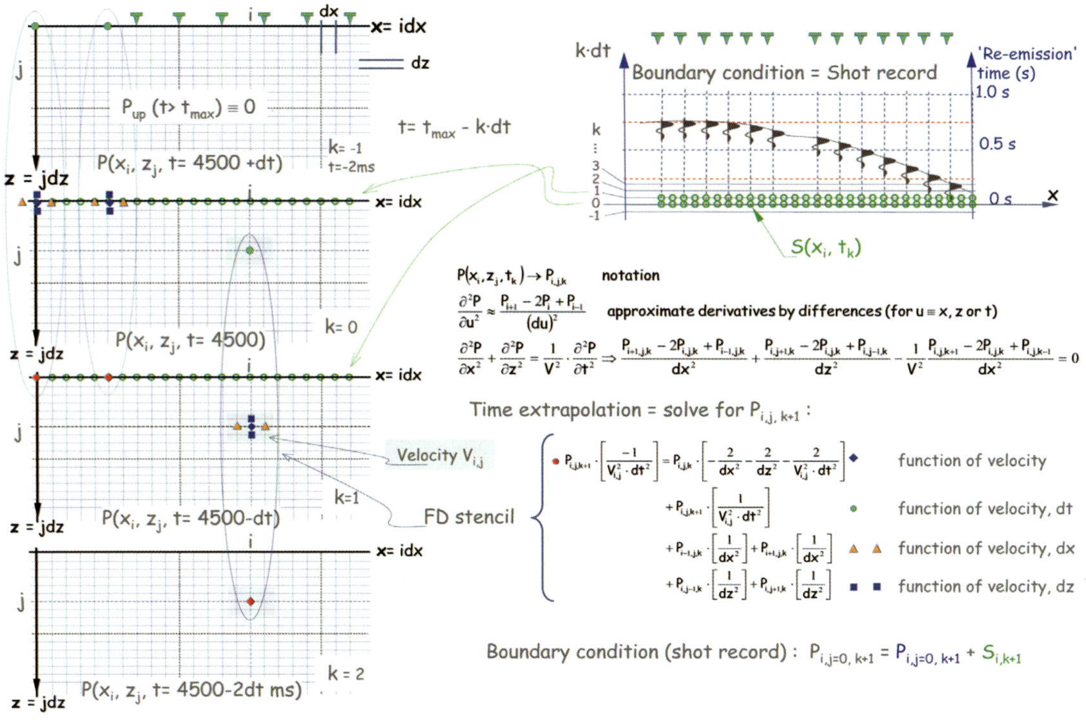

Figure 6.19: Reverse Time Migration. Extrapolation in time using Finite Difference. A sample in the P(x, z, t+dt) panel is a linear combination of samples in the P(x,z,t) and P(x,z, t-dt) panels. Weights depend on V(x,z), dt, dx, dz. Note the boundary conditions and the 'time reversed' shot record acting as a 'source term'.

The RTM 'noise'

As a preamble to this paragraph we note two important points.

- Firstly, there are no waves reflected during the two extrapolations in the over-simplified example that we use in the various figures mentioned above. This is what a very smooth velocity field would produce. For the more realistic models used in practice, some reflected events are generated at velocity contrasts of the model, with the consequences we discuss here. If we wanted to perfectly mimic the Earth's response, we should explicitly and accurately introduce each relevant velocity contrast into the model. In doing so, the energy of the multiple reflections would automatically contribute to the migrated image. Not only there would be no more need for multiple removal, but the image and the velocity model would in theory be improved! How to estimate and introduce these reflectors into the model remains, however, a topic for research very close to full waveform inversion that we discuss in chapter 7.
- Secondly, as already mentioned (fig. 6.02), the waves do not 'know' the existence of the reflectors that generated them and carry on to the edges of the model.

Fig. 6.20 explains the origin of the low-frequency noise present in RTM images. Imagine that the reflector in the model corresponds to a sharp velocity contrast and that we have explicitly placed it in the velocity migration field. Then the two-way character of the propagator used in RTM makes it such that part of the energy is properly transmitted and carries on downwards, while the remaining

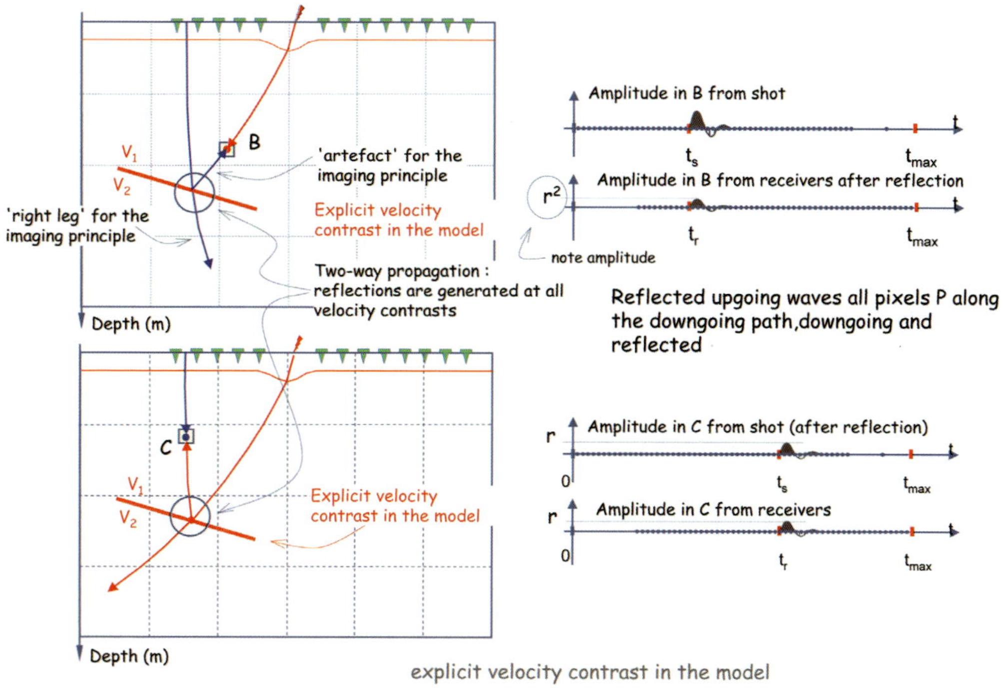

Figure 6.20: Low frequency 'noise' in RTM. At any contrast in the velocity model, a reflection occurs to the down- and upgoing wavefields. The phenomenon induces false reflection coefficients all along rays that show up as low frequency noise. The higher the velocity contrast (f.i. water bottom, top salt), the higher the noise. This artefact needs to be removed from the final RTM image.

energy is reflected back to the surface. In any point in the subsurface, such as the point B in the cartoon at the top of the figure, this reflected energy will coincidence in time with the downgoing wave computed from the source, producing a non-nil correlation at that time. The amplitude of this correlation is of the order of magnitude of r^2 (neglecting spherical divergence, etc.) but happens all along the ray path. A similar effect appears for the reflected downgoing wave, as shown for any point C on the bottom cartoon of fig. 6.20 and for diving waves even though these do not have actual reflection.

The overall result is high-level, low-frequency noise above the shallow contrasts in the velocity model that intrinsically have a high reflection coefficient. This is visible in the synthetic example of fig. 6.21, where we compare an OW Shot migration and RTM using the same velocity model. The benefit of RTM in terms of imaging steep dips is clear, but the low-frequency noise, already attenuated in the figure, is obvious too.

Several remedies have been devised and are used to remove this artefact:

- Smoothing out the velocity contrasts in the model. This reduces the undesired reflected waves and hence, the artefacts but bias the accuracy of the depth rendering.
- Simple low-cut frequency filtering. This is efficient but the image of very steeply dipping reflectors, which are precisely one motivation for RTM, appear as very low frequency objects and may be filtered too. This remains, however, the most popular technique.
- Modifying the imaging condition (Kaelin and Guitton, 2006; Guitton *et al.*, 2009).
- Directional damping: the down- and upgoing wavefields at all points such as B or C in fig. 6.20 are exactly parallel, since they follow the same ray. This criterion may be used to select the cor-

responding events in the imaging condition. This is, however, expensive (need to analyse the slope of the wavefield at each correlation point).

Why RTM is a computer-intensive process and possible remedies

Fig. 6.18 in itself shows why RTM is an expensive process. Each of the two sets of 3000 to 6000 snapshots sketched in the figure is in principle a 3D object of the size of a full migrated image, and each of this thousands of snapshots is derived from two others, following the heavy computing scheme of fig. 6.19 after having extended it to 3D.

Many 'tricks' or processes have been developed and are now applied in order to limit the cost of a project as much as possible. Amongst them we may cite:

- Optimizing the grid size;
- Selection of the aperture;
- Pre-computation of the downgoing wavefield at the model boundaries;
- Code parallelization; domain decomposition;
- And also encoded shots techniques, as discussed for OW Shot migrations;

The *grid size* directly influences the number of mathematical operations to perform to integrate the wave equation according to numerical schemes such as the one in fig. 6.19. It also impacts the memory management challenge related to the number and size of data volumes to manipulate. The optimization of grid size is constrained by the sampling theorem and the accuracy and stability of the numerical integration scheme. This is illustrated in fig. 6.22.

There are two kinds of parameters involved: the dominant or the shortest wavelength in the wavefield on the one hand and the number of points necessary to properly sample this wavelength on the other. The wavelength is in turn related to:

- The lowest migration velocity V_{min} in the model. This is imposed by the geology. For marine data, it is fixed to about 1480 m/s. For land data it may be lower in the weathered zone but will become a real issue when dealing with shear waves in the future.
- The dominant or the highest frequency f_{max} in the data. The lower the value, the faster the RTM. To go from a maximum frequency of 10 Hz to a maximum frequency of 100 Hz reduces the grid size in x, y, z but also the time step by a factor of 10, increasing the computing effort by a factor of 10^3 in 2D and 10^4 in 3D! One key reason for the success of the technique in sub-salt geological settings is that only frequencies not higher that 10 to 15 Hz are present below salt, making RTM affordable with the computing power available in the early 2000's.

The second constraint is the number of points necessary per wavelength for an accurate wavefield extrapolation. This point is related to the accuracy with which one approximates derivatives by differences. A second-order scheme, as the one sketched in fig. 6.19, would require at least 10 points per wavelength. This corresponds to a grid size in space equal to one tenth of the wavelength, i.e., 1.5 metres for a maximum frequency of 100 Hz and a lowest velocity of 1500 m/s!

Once the maximum frequency has been selected, it is preferable to apply a band-pass filter to the data with this frequency in order to avoid the numerical dispersion noise of any signal (or noise) that includes higher frequencies.

An *optimal selection of the shot aperture* is another key factor of efficiency. The concept was introduced in fig. 6.08 for 2D and presented again in the 3D case in fig. 6.23. The aperture is the space in which one allows any events to migrate, whatever its dip, azimuth and depth. A 'full aperture' would correspond to the whole target migrated image, i.e., the 3D survey area plus its classical migration margins.

This would be needlessly costly. We thus define for each shot its aperture in both x and y directions. As mentioned earlier, it can be either a distance away from the farthest receiver or a distance from the greatest mid-point distance. Cost constraints reduce the size of the aperture. The trade-off is however given by the physics of reflection. In complex geological settings, such as the ones where RTM is required, it is hard to foresee where the reflection in the subsurface may come from. Too limited an aperture leads to events not migrated, such as the area in orange on the schematic example in fig. 6.23, which may have been recorded by the shot but migrate outside the shot aperture.

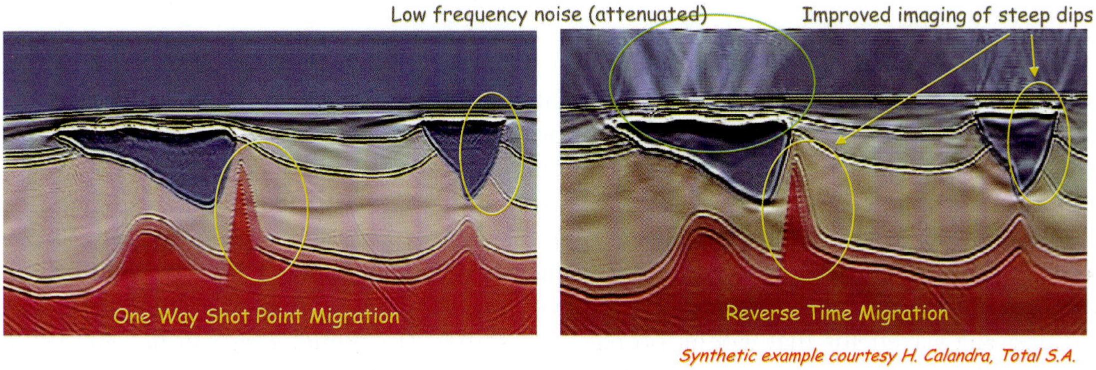

Figure 6.21: A comparison of OW-Shot Point versus RTM migrations on a synthetic data set showing the improvement in the image of steep dips for RTM, but the existence of low frequency noise, especially above 'hard' shallow reflectors.

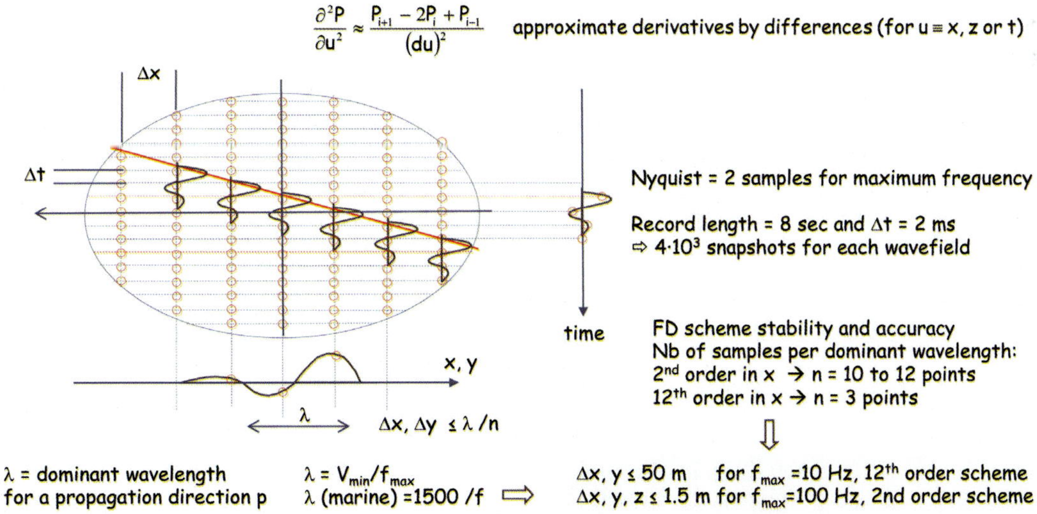

Figure 6.22: Impact of frequency in the performance of RTM. The size of the grid intervals impacts severely the computing cost of RTM. The optimum grid intervals are related to (1) the frequency content of data to migrate and (2) the choice of the FD scheme. Depending on implementations, the computing effort may increase as much as the cube of the inverse of the grid size.

The way the aperture is populated by receivers is also a key factor. The figure features an aperture for a given shot filled with regularly spaced geophones. Two comments are worth making at this stage:

- Firstly, a regular sampling of receivers is necessary to apply the numerical wave equation integration scheme sketched in fig. 6.19. The 'dx' term is supposed to be constant in the FD stencil. Data that are inherently irregularly sampled in the field (cable feathering, obstacles and limited access on land, positioning accuracy of OBC and nodes) need to be properly prepared to emulate the desired regular acquisition. Conversely, this 'interpolation step', necessary to account for the true location of receivers and shots, may be included in the RTM process itself, with limited 'pre-processing' of the data. Both approaches are possible and still debated, but it goes beyond the scope of the book to discuss this crucial step in data pre-processing.
- Secondly, the cartoon in fig. 6.23 shows a set of receivers regularly placed all around the shot. In this case the azimuth – offset distribution for this shot would be optimal. This kind of acquisition layout may be achieved directly in land data. It can also be achieved with OBC and nodes by inverting the role of shots and receivers, with one receiver on the sea bottom and many regularly spaced air-gun shots at the surface. It may result from Wide Azimuth Towed streamers after trace sorting and reconstruction of 'macro-shots', i.e.; the collection of traces belonging to various shots all in the same limited area. For 2D data or classical Narrow Azimuth Streamer marine acquisition, the receiver locations would be 'polarized' along a line. It should be clear that in terms of wavefield propagation, we have in this last case a poor sampling of the down- and upgoing wavefields that will be propagated in the 3D volume. This will be only partially compensated by the migration of the neighbouring sailing lines and will inevitably impact the quality of Subsurface Offset CIGs, yet another type of CIG used for migration velocity analysis after wave extrapolation migrations, which we introduce in the next chapter.

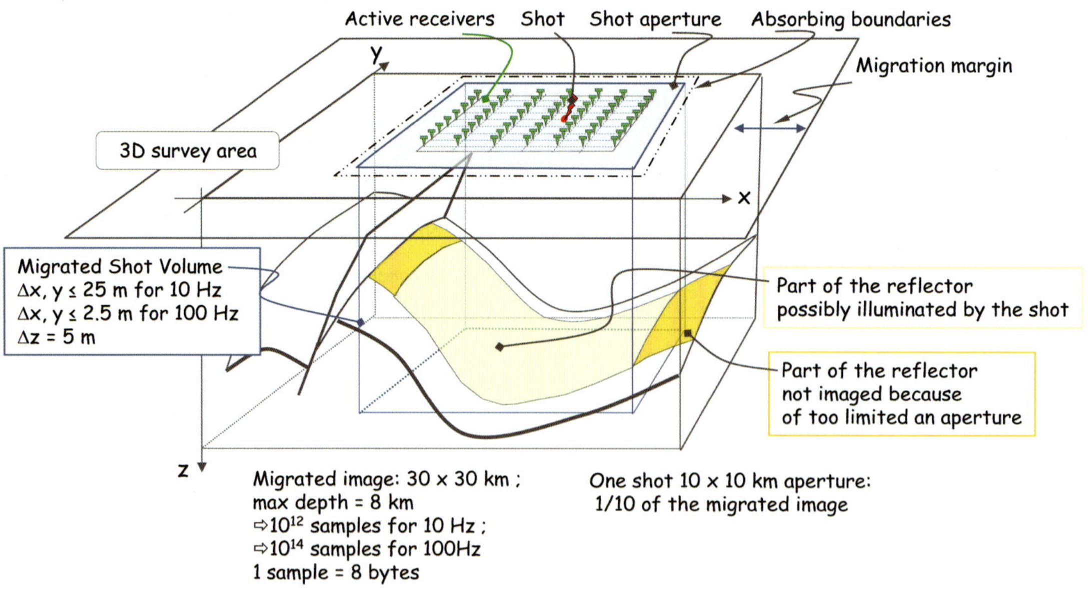

Figure 6.23: 3D lay-out geometry for OW and RTM shot migration. Shot migration is well adapted to wide azimuth acquisition (land, streamer, nodes) as all receivers are migrated together. The size of the data volume to manipulate is huge and data must be regularly sampled in all directions

- The important message here is that RTM, as well as OW Shot migrations, are well adapted to Wide Azimuth acquisition.

As mentioned earlier, a huge memory management issue in RTM is the storage of thousands of downgoing wave snapshots, as sketched in fig. 6.19. Dussaud *et al.* (2008) explored various strategies to overcome this issue. They included:

- Simply storing all snapshots. The simplest way if the computer architecture available allows it.
- Check-pointing schemes. Instead of storing all snapshots, one can store couples of consecutive snapshots at regularly time steps. These are used as intermediate 'starting conditions' to restore the necessary snapshots at each time step.
- The approach already suggested by Etgen (1986). It consists in letting the extrapolation carry on recursively without any storage until the maximum time t_{max} but recording the wavefield at the lateral and bottom boundaries of the model, as if we had placed virtual receivers there. When this is done, the downgoing and the upgoing wavefields are in similar situations. We have the wavefield histories from time 0 to t_{max} at the boundaries: the surface for the upgoing wavefield and all other limits of the model for the downgoing wavefield. We can then apply a very similar scheme to both wavefields by 're-injecting' from time t_{max} to time zero the real records and the computed virtual ones respectively, for the down- and upgoing wavefields. In this way, we have access to the two snapshots synchronously. The memory requirement on the downgoing wavefield side is the records of the virtual receivers, which is orders of magnitude smaller than the storage of all snapshots. We need to 'record' the downgoing wavefield over a number of 'virtual geophone layers' equal to half the size of the FD stencil to obtain the set of boundary conditions necessary to integrate the second-order differential equation. As a side comment, we may note that a similar idea is used in acquisition when we try to record a pair of wavefields at the surface to help generate the 'full wavefield'. This could be the pressure at two different depths with the 'Over-Under' acquisition technique or with the pressure and particle velocity in the case of the dual sensor Ocean Bottom Cables or now dual-sensor streamers.
- Smarter ways of applying a similar idea, including 'random boundaries' (Clapp, 2009), are currently emerging and will further reduce the computational effort. The random boundaries concept is illustrated in fig. 6.24. It is based on ideas similar to blended shooting or time reversal techniques (Fink, 2006). Instead of making the boundaries 'absorbing', we make them *'randomly reflecting'*, i.e., made of randomly distributed reflectors (for instance add random velocity disturbances over the velocity grid in the boundary areas). The reflected energy is incoherent and does not correlate with the extrapolated upgoing wavefield at the imaging condition step. However, when it is back-propagated through the exact random boundary, it reproduces exactly the initial downgoing field. We still have to compute the extrapolation twice, once forward in time, but we need no storage and once again backwards in time, as sketched in fig. 6.24. In this second part of the computation, the upgoing field is extrapolated simultaneously. However, the benefit is that no more snapshots than the number required by the numerical scheme in time (i.e., only two for a second order in time) are stored to be able to reconstruct the downgoing field almost perfectly. This is currently a hot topic of research.

To include anisotropy has a positive effect on the quality of the data but this is at a non-negligible computing cost in RTM. The dispersion equation in fig. 6.25 gives the reader an idea of the extra complexity it brings in the VTI case. In the case of STI, the cost and stability issues are even greater. The VTI case may require as many as three times the number of arithmetical operations as the isotropic case and the figures go to 15 times in the STI case. The overall run time may not be increased in the same proportion but remains a few times longer than the isotropic case. A rule of thumb may

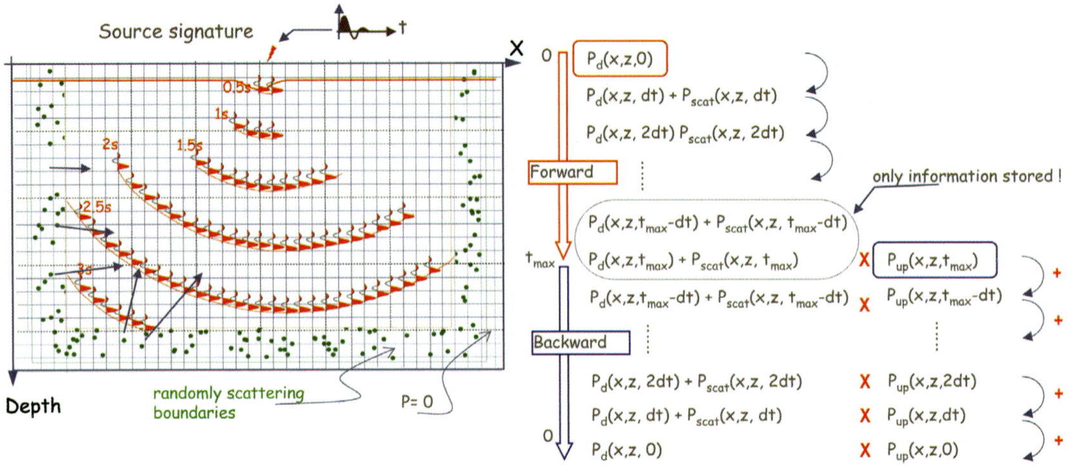

Figure 6.24: The 'random boundary trick' (Clapp, 2009). The source is propagated in time in a model including random velocities at the boundaries. This generates a scattered 'information', which is part of the downgoing field. When arriving at t_{max}, the extrapolation carries on, but now backwards in time and for both fields. The randomly scattered part of the downgoing wavefield correlates poorly with the upgoing wavefield, introducing limited noise in the correlation process. No storage of the downgoing snapshots is needed.

be a factor of 1 in computer's time for isotropic, 2 for VTI and 3 for TTI (E. Dussaud, personal communication) but again, this depends on how the code is structured.

Fig. 6.25 compares a synthetic result of OW Shot Point migrations with RTM in an anisotropic context. The result justifies the extra cost of running such anisotropic codes, provided that we are able to estimate the proper velocity and anisotropy parameters.

RTM and more generally two-way extrapolation propagators do not benefit from the possibility to parallelize codes along frequency planes. Other more complex tricks then need to be developed to fully benefit from the computer architecture available, or conversely to acquire a computer with a view of the size of the survey that will be processed and the way the RTM code will be developed! For very large projects 'tricks' may have to include 'domain decomposition', where the model is decomposed into various volumes of more manageable size. The cost is then the exchange of data between domains, without loss of efficiency.

This comment is made here simply to emphasize the imbrications between 'pure' geophysics, numerical analysis and computer sciences that are at the heart of the current success of seismic imaging.

Recapitulation: benefits and drawbacks in RTM

One may conclude this section on RTM by mentioning the benefits and drawbacks of this technique, mainly related to the fact that we extrapolate the wavefields using *two-way propagators*. They are schematically presented in fig. 6.26, and listed as:

- The possibility to automatically image very steeply dipping reflectors and even overhanging ones via 'turning waves'. This is true of course if the geological setting, and particularly the vertical velocity profile, allows for the development of such waves, if the recording time was sufficient (they have a long wavepath in low velocities) and if the migration aperture is sufficiently wide.

- The applicability to wide azimuth data sets.
- Other potential benefits rely on the possibility to introduce some relevant reflectors into the model. 'Prism waves' or 'multi-bouncing waves' are one example already exploited (Jones *et al.*, 2007 ; see chapter 8, fig. 8.08). Migration of multiples is a longer-term hope, shared with Full Waveform Inversion.
- The main drawbacks are cost and low-frequency noise that affects mainly the shallower parts of the 3D migrated volumes. Cost is decreasing steadily, thanks to computers' increasing power and clever tricks, such as 'random boundaries'.

VTI dispersion equation :

$$k_z = \frac{\omega}{V_{vert}} \cdot \sqrt{1 - \frac{q \cdot V_{hor}^2 \cdot (k_x^2 + k_y^2)}{\omega^2 + (q-1) \cdot V_{hor}^2 \cdot (k_x^2 + k_y^2)}} \quad \text{with } q = \frac{1+2\cdot\delta}{1+2\cdot\varepsilon} \quad \text{and } V_{hor} = V_{vert} \cdot \sqrt{1+2\cdot\delta}$$

V_{vert} and V_{hor} = phase velocity along and normal to the vertical axis of symmetry

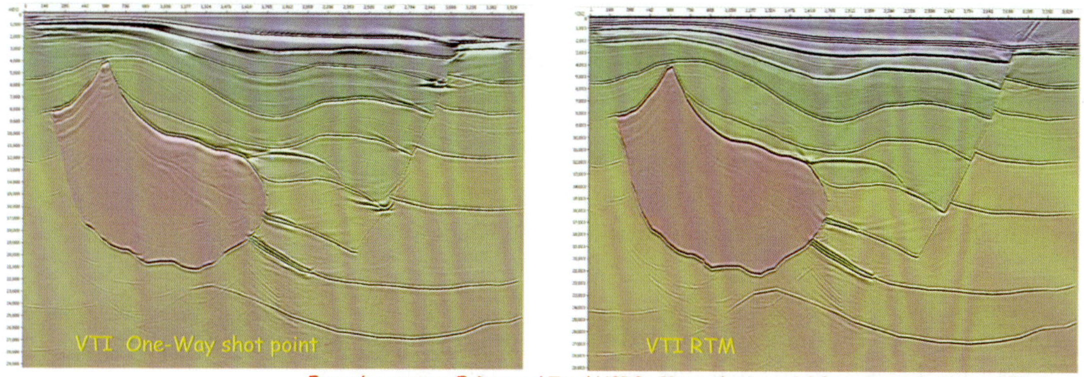

Example courtesy E. Dussaud Total HGRG. 2D synthetic model from Amerada-Hess - SEG.

Figure 6.25: A comparison between One Way-shot point and RTM migrations in the anisotropic case. The ratio of numerical operations between isotropic and anisotropic STI is about 15, resulting in elapse time increased by a factor 3.

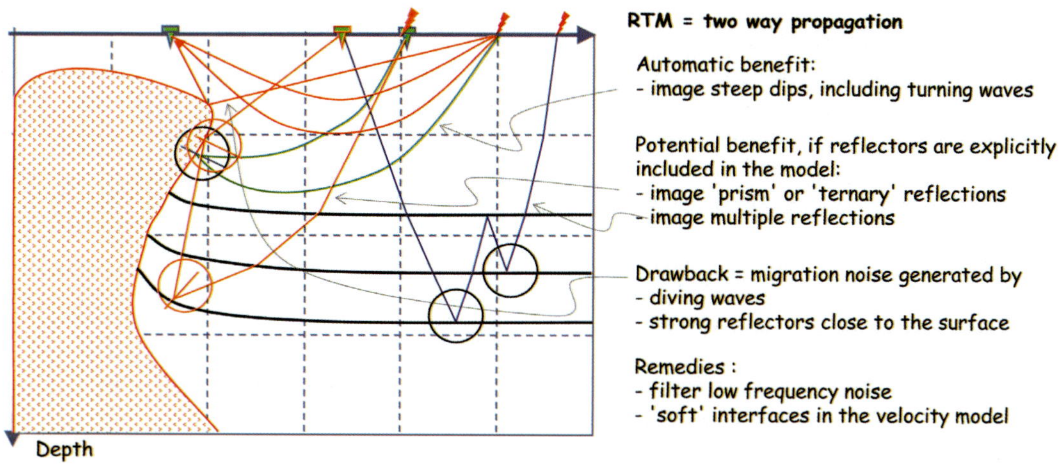

Figure 6.26: Benefits and drawbacks of RTM related to two-way extrapolation.

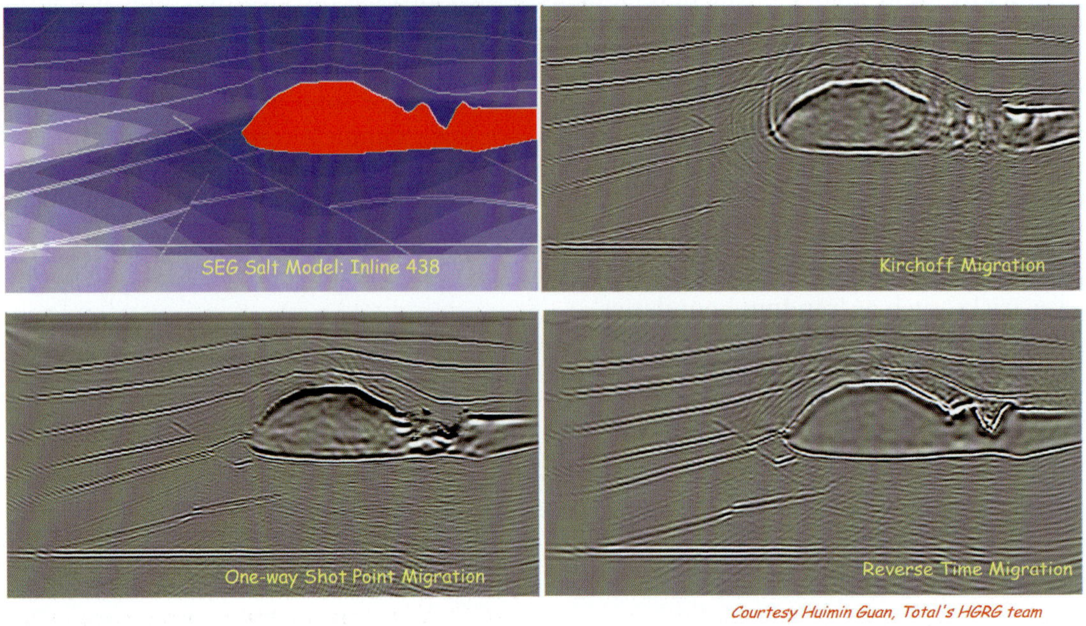

Figure 6.27: A comparison of three migrations techniques for the synthetic SEG-EAGE 3D Salt model. RTM produces the best image of the complicated top of salt, that includes steep dips. The lower resolution comes from the migration of a limited bandwidth.

To conclude, fig. 6.27 compares three migration methods, Kirchhoff, OW Shot Point and RTM on the well-known 'SEG-EAGE 3D Salt model' synthetic data set.

This result, confirmed by hundreds of other real data examples in salt tectonics and other environments, explains why RTM is considered as the 'ultimate' imaging tool of the coming years.

7 Wavefield Extrapolation Migration Velocity Analysis and Inversion-based Techniques

Common Image Gathers are classical tools for estimating the migration velocity field. Their computation is, however, not straightforward in all Wavefield Extrapolation methods. As a consequence, Migration Velocity Model Building is a topic *per se* in WE-based methods, still under development in many ways. In the first part of this chapter, we discuss in some detail the Subsurface-Offset image gathers resulting from One-Way Shot Point- and Reverse Time migrations. The discussion includes extended Imaging Condition and explanations of how to obtain Angle Common Image Gathers.

If Angle CIGs are used as inputs to ray-based tomography, Subsurface Offset CIGs are better suited to 'focusing analysis'. The second part of the chapter is consequently dedicated to 'Differential Semblance Optimization', a method that searches for the model that optimizes the 'focusing' of the image at zero-subsurface offset.

The third part of the chapter is devoted to Full Waveform Inversion, yet another method aimed at the estimation of high-resolution velocity fields. The principles of FWI are not new (Lailly, 1983; Tarantola, 1984) but this highly computer demanding method that shares points with RTM, has been revived in recent years with the publication of promising first results at industrial scale. Different implementations of the method and some synthetic and real examples will be discussed in the last part of the chapter.

The concept of Subsurface-Offset Common Image Gathers

With this paragraph, we enter the domain of velocity analysis using Wavefield Extrapolation methods. Common Image Gathers in the surface offset domain and consequently tomographic methods, are no longer available.

Migration Velocity Analysis will in fact be based on another object: the Subsurface-Offset CIG, or SO-CIG in short. There are currently two routes to perform migration velocity analysis with these gathers:

- Techniques based on 'focusing analysis', which look for the highest correlation at zero-time lag and/or zero-offset (Symes, 1993) of the multi dimensional correlation function of the down- and upgoing wavefields.
- Transformation of SO-CIG into Scattering Angle CIGs', an object that we met already in chapter 4 when we dealt with Common Reflection Angle Kirchhoff Migrations (figures 4.19 to 4.22), used as an input to tomographic inversion techniques such as the ones described in chapter 4.

Both routes necessitate the extension of the imaging condition presented in chapter 6 (figures 6.02 and 6.06). The extension in time was already shown in the middle panel on the right of fig. 6.06. instead of computing the correlation at zero-time lag, one computes the correlation as a function of time shift. This extension was widely used in 2D in the 1990s to estimate migration velocities via 'focusing analysis' (Jeannot *et al.*, 1986).

Alternatively, de Bruin *et al.* (1990) showed that we can estimate the reflectivity as a function of scattering angle in wavefield extrapolation methods by extending the imaging condition through the computation of the correlation function along the spatial horizontal dimension. Rickett and Sava (2002) established the notion of 'subsurface-offset CIGs' and proposed methods to derive from them Scattering Angle CIGs. By combining both time and depth, Sava and Fomel, (2006) and Sava (2009) recently introduced the concept of 'extended Image Condition' or (eIC).

Fig. 7.01, close to fig. 6.06, is a reminder of the imaging condition applied in One Way Shot point migration, again in the case of our simple model of fig. 1.13. The down- and upgoing wavefields estimated at depth 5000 m are correlated and the correlation coefficient taken as an estimate of the value of any reflection coefficient at that depth.

In the figure we show that instead of computing the correlation at zero time lag as requested by the classical imaging condition, we may as well consider the whole correlation function along time. This is the middle panel on the right of the figure. The correlation function has some organization that has an important physical meaning and we present it as a first hint as to how slopes of the wavefields and their correlation may be related to the various angles around the reflection point. For instance, the slope (double-arrow in red) of the downgoing field is a measure of the ray parameter of the source ray, which we can relate to its incidence angle if we know the local velocity and anisotropy parameters. Similarly, for the upgoing wavefield. One can show as a consequence, that the slope of the time correlation function is related to the scattering angle ψ.

The zoom indicates the geometrical relationships between rays, wavefronts at the time-coincidence and slopes on the snapshots. These relationships may be easy to derive in the isotropic case. As one might expect, they are much trickier in the anisotropic case (Alkhalifah, 2009).

With fig. 7.02, we move towards the spatial dimension. The upper right panel of the cartoon shows the two wavefields estimated at depth 5000 metres but now, each one *shifted laterally and symmetrically* with respect to the location x_{cig} of a quantity + and – h/2 respectively. x_{cig} is the location where we want to compute the Common Image Gather.

Next, we compute the correlation function along time of the two traces at location x_{cig} and place the result at the abscissa –h of the panel. By repeating these operations for a series of + and – h values, we create a new correlation panel, a function of the two dimensions h and t.

The *shift h is called 'subsurface offset'*. We now take the zero-time lag of this 2D function that composes the 5000-metre row of the new 'Subsurface-Offset CIG', sometimes called the '*extended image condition in the subsurface offset dimension*'.

This row displays a maximum at zero subsurface offset simply because we have chosen x_{cig} at the location where the dipping reflector of the model in fig. 1.13 is precisely at the current depth of 5000 m in the recursive migration process. Since we used the exact velocity, the correlation is maximum for zero subsurface offset!

The sequence of operations just described is performed at each depth step of the Shot Point migration recursion, from zero depth to z_{max}. Let us look with fig. 7.03 at what happens at a depth of 4800 metres. The top right panel superimposes the two wavefields captured at that depth by the extrapolation process. When applying the extended imaging conditions we note:

- Firstly, that the two fields are in phase at a location denoted x'_{cig} (in orange) in the figure. At that location, they will correlate at zero-time lag yielding the image of the dipping reflector at that depth.
- Secondly, if one shifts the two fields laterally by – and + $h_s/2$ respectively, as we did in fig. 7.02, the two shifted fields will be in phase again (wavefronts are synchronous) at the x_{cig} location, where we are currently computing the SO-CIG. The bottom right panel sketches the zero-time lag of the time correlation, i.e. the image of the reflection at depth 5000 m as it 'leaks' on the SO-CIG at depth 4800 m.

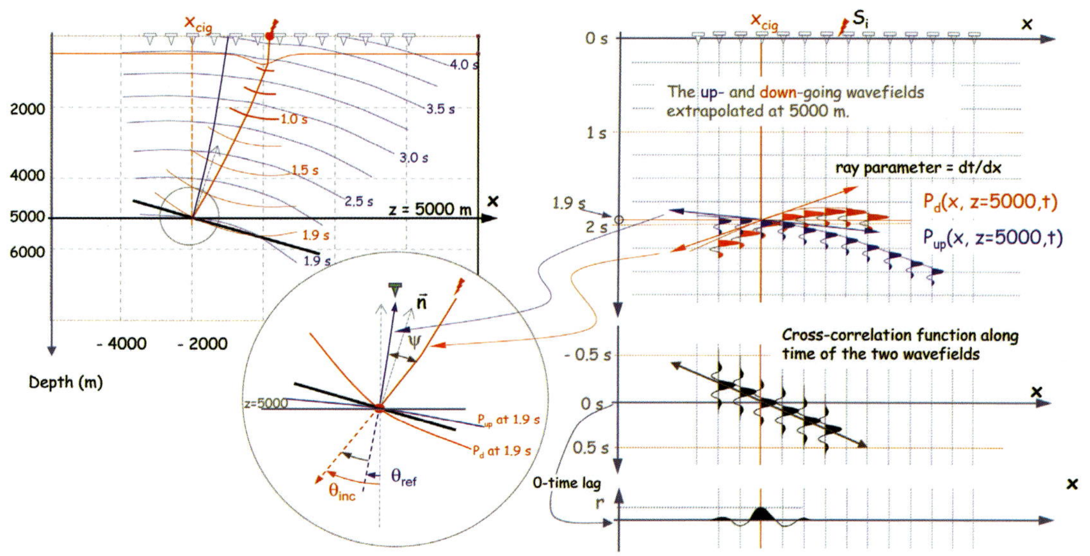

Figure 7.01: Imaging Condition in One Way Shot point migration (applied here at 5000 m depth; same as fig. 6.06). The slopes of the two wavefields at the correlation point are functions of the local velocity and anisotropy and of the incidence and reflected angles w/r to the vertical. The slope of the correlation function is a function of the scattering angle ψ.

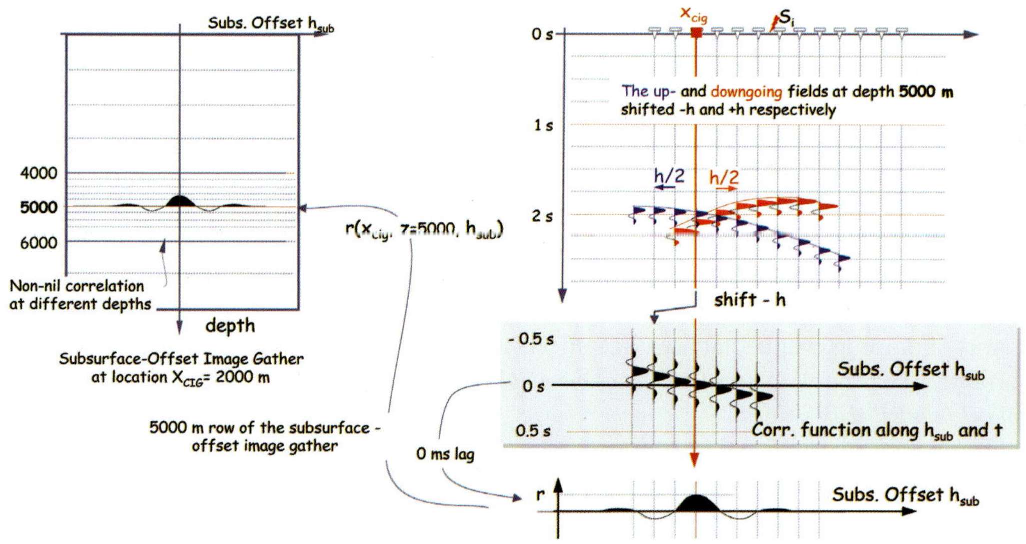

Figure 7.02: Principle of the computation of Subsurface-Offset Common Image Gather. A row of a gather at one location x_{cig} is obtained for one shot by shifting laterally the down- and upgoing fields and compute the correlation coefficient at the imaging location x_{cig}. This operation is done for all depth steps of the recursive imaging process, thus filling the SO-CIG for the specific shot.

We show what happens at 4800 metres for the sake of the demonstration but the same phenomenon happens at all depths of the recursion above and below the correct image depth. The result is the 'subsurface offset response' of a single reflection (such as the one at depth 5000 m), which appears not to be a single point but stretches over depth as a dipping event above, but also below, the reflector's depth. Energy appears below the reflector's depth because the wavefield extrapolation does not stop at reflectors, as already mentioned.

As one might expect for a dipping reflector, the farther from the right depth, the larger the subsurface offset at which the correlation at zero-time lag appears, hence the dip in the SO-CIG.

The positive counterpart of the spread of the event is that its dip is directly related to the scattering 'reflection', or 'opening' angle ψ, as Rickett and Sava (2002) showed. This should not be a surprise when considering fig. 7.01. The authors showed more precisely that if one assumes a homogeneous isotropic 2D medium in the vicinity of the reflector, the local velocity may be eliminated from the equations and the slope is simply the tangent of $\psi/2$. The relationship involves the measurement of the reflector's dip and azimuth in 3D and moreover, becomes much trickier in the presence of anisotropy (Alkhalifah, 2009).

Subsurface Offset- and Scattering Angle CIGs from several shots

With figures 7.01 to 7.03, we discussed the migration and the contribution to a SO-CIG of one shot only. Of course, all shots are processed in a similar way, each shot bringing its own part to the migrated image, as explained in the previous chapter (fig. 6.09) but also, to any SO-CIG located within the shot aperture. The stack of all shot contributions yields the final Subsurface Offset CIG at location x_{cig} and all others where it was selected to compute one.

Fig. 7.04 considers the contribution of two different shots at the same SO-CIG location x_{cig}. For each shot, the reflection at depth 5000 metres produces a dipping event similar to the one in fig. 7.03, but each one illuminates the reflector at x_{cig} *with a different scattering angle*. The dips of the response on the SO-CIG are consequently different.

The summation of the contribution from every shot has two different effects on the Subsurface Offset CIG.

- On the positive side, the stack of shot responses that presents different dips will sum up *constructively precisely at the right depth and at zero subsurface offset*. We say that the 'SO-CIG is focused' around that point. However, focusing will happen only if the migration was performed with the correct velocity model. We discuss next what happens when this is not the case. The central panel on the right of fig. 7.06 presents an example of a SO-CIG for a simple model made of one horizontal reflector at the bottom of a constant velocity layer, shown on the left of the figure. One notes on this SO-CIG the remnants of dipping events that are edge effects related to the finite number of shots used but more importantly, the otherwise *neat focusing of the amplitudes at zero subsurface offset and the right depth when the correct velocity model is used*.
- On the (apparently) negative side, we seem to lose the information on the scattering angle embedded in the contribution of each single shot. A linear tau-p transform (also called the Radon transform) is able to 'disentangle' each shot contribution and allows us to construct scattering angle gathers from the SO-CIGs.

The principle of tau-p transform was presented in a different context in fig. 5.08. The process to build a Scattering Angle CIG is schematically presented on the right of fig. 7.04. Once all shots have been processed, the SO-CIG undergoes the following steps:

- select the regularly spaced series of scattering angle $\{\psi_i\}$ for which one wants to compute the gather; according to Rickett-Sava's formula each angle value corresponds to a given slope $(dz/dh_{sub})_i$ in the SO-CIG;

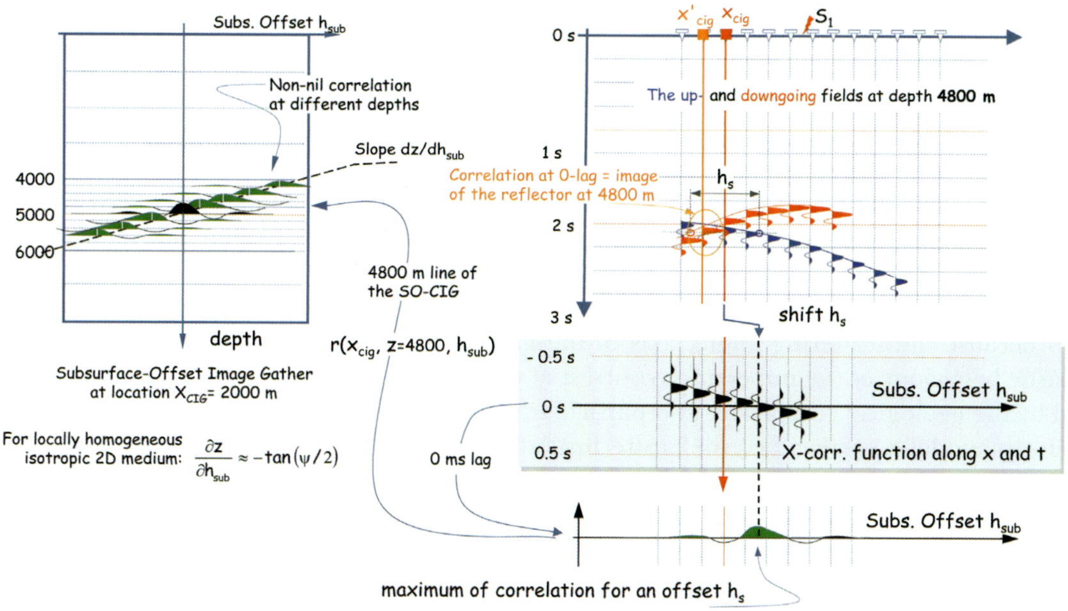

Figure 7.03: Response of the reflector around the image point. Correlation between the two fields is not nil above and below the depth of the imaged point. (Rickett and Sava, 2002) showed that the subsurface offset at which the correlation is maximum depends on the scattering angle ψ at the reflection point. They give the relationship between the slope of the event on the gather and the scattering angle ψ (fig. 7.01) in a locally homogeneous isotropic medium.

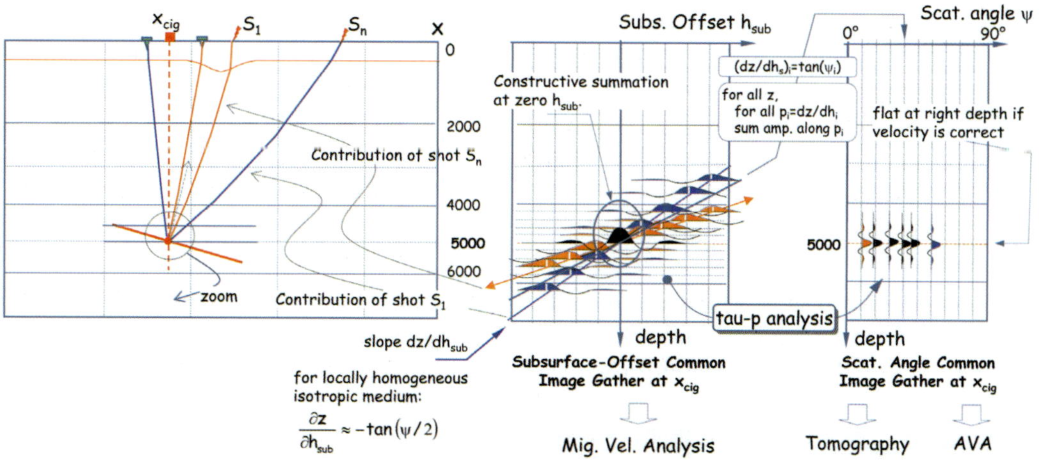

Figure 7.04: Contribution of two different shots to the Subsurface offset Common Image Gather. For a single shot, correlation of the two fields appears as a dipping event. Since the slope of the event depends on the actual scattering angle ψ, it varies from shot to shot. If the migration velocity model is correct the summation of all shot contributions will add up constructively at the right depth and for zero subsurface offset. Tau-p analysis allows to 'detangle' each contribution and to construct 'Reflection Angle CIGs'.

- for each slope $(dz/dh_{sub})_i$ repeat the following actions:
  - for each depth z from 0 to z_{max}, sum the amplitude of the SO-CIG along the slope centred at zero subsurface offset;
  - place the result at the appropriate depth and scattering angle column in the Scattering Angle CIG being constructed.

This process needs to be repeated at every selected CIG location. Examples of SO-CIGs and their corresponding SA-CIGs are given in fig. 7.09 for the synthetic 2D case of the 'Marmousi' model.

As mentioned, Scattering Angle CIGs are natural inputs to direct 'Amplitude versus Angle analysis'. This is because the scattering angle, axis of the gather, is the parameter that effectively controls the elastic behaviour of the impedance contrast at the reflector. In theory, the analysis is no longer biased by the geological complexity and particularly, the dip of reflectors, provided once again that the velocity model is correct, i.e., the Scattering Angle gathers are flat and amplitudes are properly managed in the migration process, which remains a challenge.

The second usage of these gathers is tomography, whose goal is precisely to flatten the gathers. As discussed in the chapter dedicated to Kirchhoff migrations, events on SA-CIGs as much as (surface) offset gathers should be flat. Non flatness is an indication of a poor velocity and/or anisotropy model. Picking of SA-CIGs are alternative inputs to tomographic inversion as described in fig. 4.04.

Despite their benefits, some issues prevent the generalized usage of SA-CIGs for the time being.

- Firstly, their cost. Remember that we have so far discussed a 2D process, but all modern data sets are 3D (and more and more of the wide-azimuth type). For 3D data, SO-CIGs are three-dimensional objects. We can easily generalize subsurface offset h to subsurface vector offset $\{h_x, h_y\}$, as we have done for surface offsets in fig. 4.19. This is discussed in a following section. Similarly, scattering in 3D depends on two angles denoted ψ and β in the same figure. The computation of one single SA-CIG amounts consequently to a two-parameters tau-p analysis in the three-dimensional SO-CIG resulting from 3D WE migration. The cost of the CIG may be considered prohibitive!
- Secondly, WE methods do not propagate waves exactly in the same way as rays, and use different velocity model parameterization. One likes the tools used on the one hand for imaging and on the other hand for velocity analysis, to be consistent.
- Furthermore, we generally choose to apply WE based methods precisely in areas where ray tracing technology fails.

SO-CIG and SA-CIG after RTM

Figures 4.01 to 4.04 were drawn for One Way Shot point migration. The concept of Subsurface CIG holds as well for Two Way Reverse Time migration techniques. In this case, the lateral shift necessary to introduce the subsurface offset dimension is performed at each time step. Looking at fig. 6.18, one can think of a process whereby at each time step, on top of the direct multiplication, one shifts the down- and upgoing snapshots laterally for a series of subsurface offset values. For each shift value, the multiplication is performed and the result stored in the appropriate CIG trace. This is done for all selected CIG locations. The process is repeated for all time steps and the results stacked as the recursion proceeds. The SO-CIGs are finally obtained at the end, when all times have been processed.

Scattering Angle CIGs are computed in exactly the same way as the ones coming from One Way Shot point migrations.

RTM being a computer-intensive process, the computation of SO-CIG adds an extra cost to. The synthetic examples in fig. 7.06 are actually results from 3D RTM.

SO- and SA-CIGS when the velocity model is not correct

When the velocity model is not correct, the two down- and upgoing wavefields are extrapolated but not at their correct location on snapshots. Equivalently, they exist but are distorted in the depth extrapolation methods.

Consequently, the extended imaging condition yields flawed answers as one might expect. For a given shot and a given reflector, a response such as the one in fig. 7.03 will exist, but will neither cross at the correct depth nor with the correct slope, as sketched in fig. 7.05. On the cartoon on the right of the figure we show the response of the reflector for the shot in blue in fig. 7.04. We spot two parameters on the figure: firstly, the depth at which the event appears at zero subsurface offset (we used to call it 'focusing depth' in focusing analysis) and secondly, the subsurface offset at which the event is seen at the true unknown depth.

Both parameters are shot-dependent, simply because each shot involves the propagation of wavefields at different angles and in different parts of the (wrong) velocity model. Consequently, 'focusing' on the SO-CIG will be poor and appear at the wrong location: lateral shift for dipping reflectors and wrong depth in any cases. The poor summation of the various shot contributions is equivalent to the poor summation of non-flat Offset CIGs in Kirchhoff migrations.

On the left-hand side of fig. 7.05 we also sketch the rays that would correspond to the parts of the wavefields that correlate and show up in the SO-CIG at x_{cig}. We draw those corresponding to the wrong depth z_f and the ones stopping at depth 5000 m, corresponding to the subsurface offset of maximum correlation at this depth.

It is interesting to realize that one had the same rays in fig. 5.07, when we considered 'ray migration quality' in the case of poor migration velocity. This cartoon intends to give more understanding

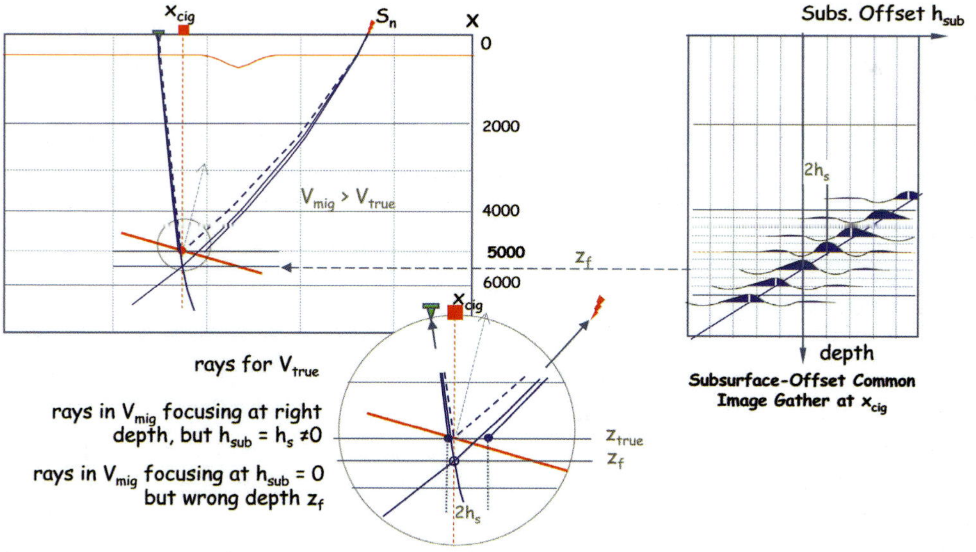

Figure 7.05: SO-CIGs when migrating with the wrong velocity (here V_{mig} is too high). The response of one reflector for one shot is mispositioned. We can define the 'focusing' depth z_f at which it appears at zero subsurface offset and the subsurface offset $2h_s$ (unknown !) at which it appears at the true depth. Note the similarity between h_s and 'migration quality' in fig. 5.07. Unfortunately, z_f and h_s change for each shot, even for the same V_{mig}. Consequently, the sum of the responses for a series of shots does not focus to a single location. This is exemplified in fig. 7.06.

of the 'physical' meaning of the concept of subsurface-offset, which is different from the one we introduced in fig. 6.14 for sinking surveys. It also shows that ideally, we would like to *compute the correlation along the reflector's dip*. Since this one is unknown, one can compute it instead along both x and z directions, at the cost of adding yet another dimension to the eIC. This is what is suggested by Sava and Vasconcelos (2009).

Fig. 7.06 shows SO-CIG for a simple model made of one horizontal reflector. Focusing is clear on the central CIG panel (true velocity used in RTM) and wrong migration depth and characteristic shapes of 'under-migration' and 'over-migration' show nicely on the two other panels. The symmetry of the panels comes from the acquisition system, which includes traces with interchanged shots and receivers (split-spread acquisition, over 670 shots used). The 'bowl-shaped' feature in the panel with too high a velocity may be seen as the *envelope of all events for all shots* - similar to the blue one on the right of fig. 7.05.

Migration Velocity Analysis using Subsurface Offset CIGs

The intention to obtain automatically, i.e., without picking, a proper velocity model of the Earth is not new. Techniques proposed in the late eighties were based on the optimization of the coherency of events either for un-migrated data as in Toldi (1989), or preferably using traveltimes from depth-migrated reflectors, as in Landa *et al.* (1988, 1989, etc.).

Recently, the idea was revived again by Zhou *et al.* (2009), suggesting a method this time in the Shot Index CIG domain constructed after RTM.

This is currently a hot topic of research and first applications on real data are just emerging. We choose to present here in more detail Automatic Migration Velocity Analysis, using Differential Semblance Optimization, which has the same ambition (Shen *et al.*, 2003; Shen and Symes, 2006; etc.).

The technique is by definition consistent with Wavefield Extrapolation imaging methods and is viewed as a promising (although not yet mature) alternative to the computation of Scattering Angle

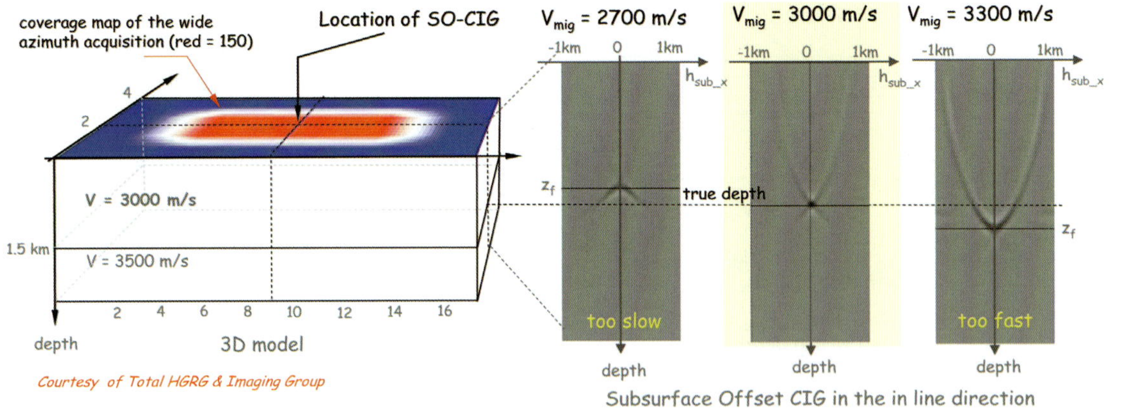

Figure 7.06: Subsurface offset CIG for a synthetic model. Wide Azimuth shooting is simulated over the 18 x 4 km model. 3D RTM is run with three velocities: 3000 (true velocity), 2700 and 3300 m/s and the SO CIGs along the x direction computed each time at the centre of the model. Note focusing at zero subsurface offset and correct depth when using the true velocity and the wrong depth and poor focusing in other cases. Symmetry comes from exchanging shot and receiver roles.

CIGs followed by tomography. It works in theory for both OW Shot point migration and RTM, although it is not feasible in the latter case at industrial scale in 3D with current computing capabilities.

It was cast into an inverse problem using the differential semblance principle proposed by Symes (1993).

The general structure of the iterative solution of non-linear inverse problems was already given in fig. 4.04 for tomographic inversion. We present a similar scheme in fig. 7.07, now specific to MVA using SO-CIGs.

The idea is to find, hopefully without direct human involvement, the (anisotropic) velocity model (**m**) such that Subsurface-Offset CIGs exhibit maximum amplitudes at zero subsurface offsets, and this at all depths and for all shots migrated via a Wavefield Extrapolation method.

Following the scheme of linearized tomography in chapter 4, the optimization problem consists in minimizing the cost function iteratively.

In the present case, the N unknowns are the model parameters updates $\{dm_i\}$ added to the model parameters $\{m_i\}$ at each iteration. The $\{m_i\}$ may be either grid points of the model or spline parameters at the nodes of a spline grid. The M 'measurements' are all sample values along z and h_{sub} of all selected SO-CIGs.

The '*forward problem*' is hence the computation of SO-CIGs after One Way Shot point migration or (in the future) RTM using the current velocity model (**m**). Key parameters at this stage include:

- The parameterization and the degree of smoothness of the model (**m**) and/or the model update (d**m**). The model should not include details much smaller than the dominant seismic wavelength (say 100 metres for the shallow part and up to several hundreds at the bottom of the model).

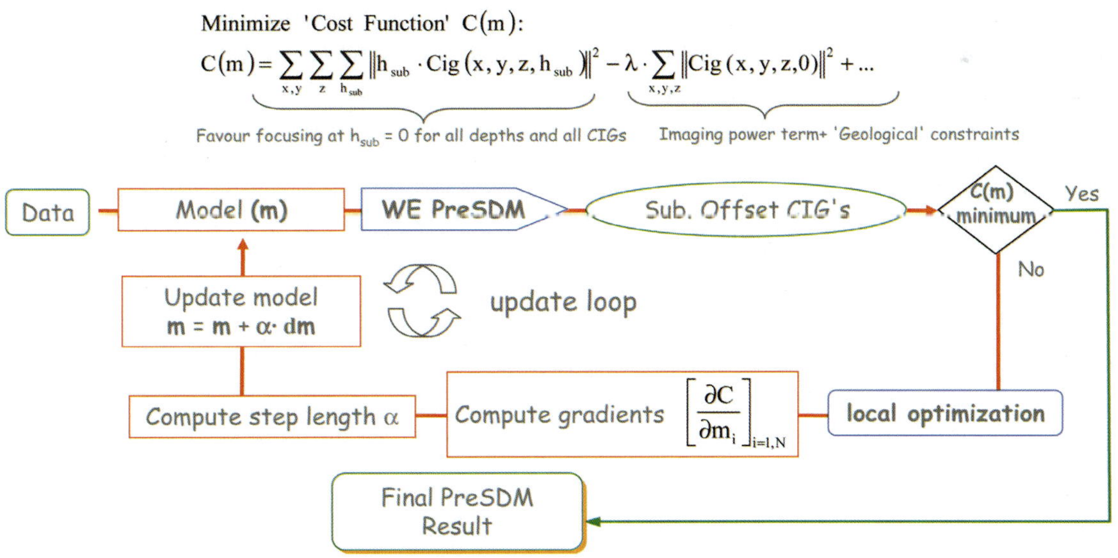

Figure 7.07: Principle of Migration Velocity Analysis using Differential Semblance Optimization. The 'forward problem' is the computation of SO-CIGs via OW or RTM migration. The cost function includes a 'Differential Semblance' term applied to CIG' to favour models that focus SO-CIGs at zero-subsurface offset and regularization terms. The gradients may be computed using 'adjoint state methods'.

- The discretization of SO-CIGs: Δh_{sub} (it should be the same as the bin interval of the migrated image); h_{sub_min} and h_{sub_max} (recall that these are vectors in 3D!).
- The location of the SO-CIGs: in the case of Shot Point migration, one may not wish to compute them at all bin locations for computation time constrains.

The *cost function* is based on 'Differential Semblance', a concept that is well-adapted to SO-CIGs. It says that finding a model (**m**) that produces a maximum amplitude at zero subsurface offset is equivalent to searching (**m**) that minimizes the sum over h_{sub} of the function $|| h_{sub} \cdot Cig(h_{sub}) ||^2$ where $Cig(h_{sub})$ represents the amplitude of a SO-CIG at location x_{cig}, depth z and subsurface offsets h_{sub}. $Cig(h_{sub})$ is also denoted $r(x_{cig}, z, h_{sub})$ in figures 7.02 and 7.03. This is the first term of the cost function presented at the top of fig. 7.07.

Shen and Symes (2008) observed that this term was not sufficient to reproduce the middle-wavelength information in the velocity model in synthetic cases and suggested that a second term, called 'image power', be added to the cost function. Image power is the square of the amplitude at each point of the current image itself, i.e., the value at zero subsurface offset, as per application of the 'basic' imaging condition, summed for all shots. A scaling factor $-\alpha$ (possibly in the form of a variance matrix) is there to balance the respective weighs of the two terms. It is negative to minimize the overall cost function.

The linearized problem is different from the tomographic case in chapter 4 but close to the Full Waveform Inversion scheme that we address later in the chapter.

In a tomography-like approach we would have to compute the matrix $[\partial Cig/\partial m_i]$ of partial derivatives of the 'measurements' (the value of the Cigs at all subsurface offsets, all depths and all shots) with respect to the model parameters m_i. This is a huge matrix of the size N x M that happily does not need to be computed, thanks to the technique called the '*adjoint-state method*' (Lailly, 1983; Tarantola, 1984), now a 'standard' when inversion based on Wavefield Extrapolation is concerned. The 'adjoint-state method' allows us to compute at 'limited' cost the vector $[\partial C(\mathbf{m})/\partial m_i]$ of the gradient of the cost function C with respect to the model parameters. We come back on this specific important technique in a next section dedicated to Full Waveform Inversion. Suffice it to say here that we use, for example, the '*steepest descent*' method that says that at each iteration (n) of the update loop, we assume that we are close to the final result (local optimization) and the iteration consists in a linear update of the model, using the relation (classical first-order Taylor expansion):

$$\{m_{i,j}\}^{(n+1)} = \{m_{i,j}\}^{(n)} - \alpha \cdot \left\{ \frac{\partial C}{\partial m_{i,j}} \right\}^{(n)} \tag{7.01}$$

where (n) indicates the iteration number.

Once the gradient is computed, and 'scaled', the so-called step length α is key to determine. This is often done by the '*line search method*'.

Once this is done, the new cost function $C(\mathbf{m})_{n+1}$ is computed. $C(\mathbf{m})_{n+1}$ should be smaller than $C(\mathbf{m})_n$ if the process converges. The update loop process is stopped when a minimum of $C(\mathbf{m})$ is thought to be achieved.

As the reader will realize, this type of inversion-based model estimation is particularly computer intensive. The cost of one single iteration amounts to approximately three times the cost of a single PreSDM: one to compute the current SO-CIG and two to compute the gradients matrix. The number of iterations of the update loop may be of several tens to a few hundreds. It is consequently not yet viable at industrial scale, especially in 3D and when RTM is the migration tool to compute SO-CIGs.

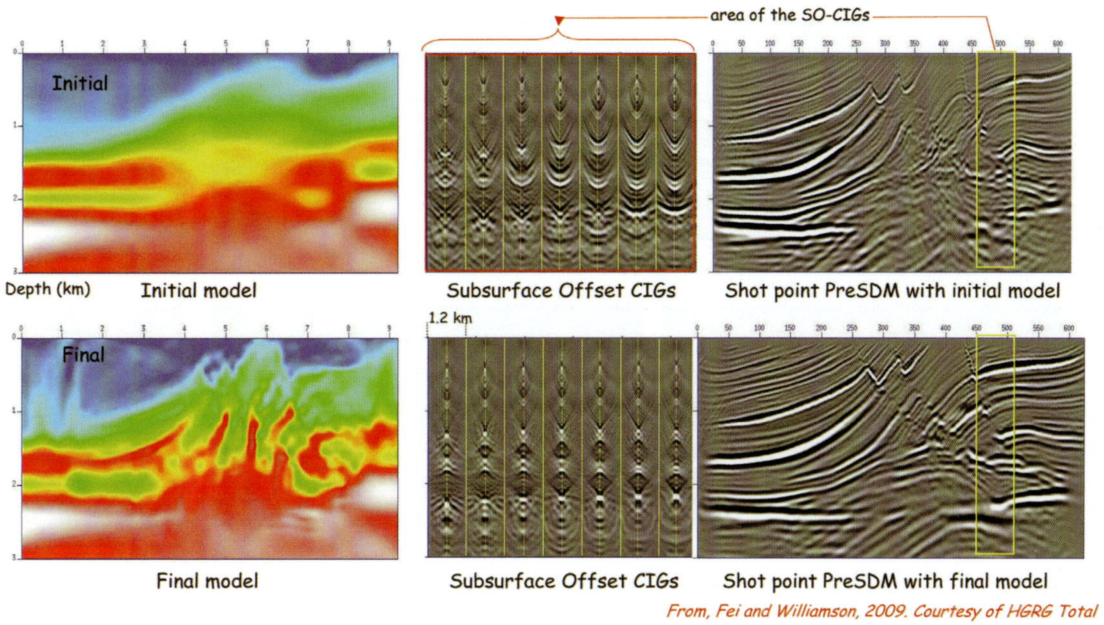

Figure 7.08: Example of MVA using Shot Point migration and Differential Semblance Optimization on the synthetic 'Marmousi' data set. The top row presents respectively the initial model, some SO-CIGs and the result of migration with the current model. The bottom row shows the same after 50 iterations of the inversion scheme of fig. 7.07.

It is however a route investigated as an alternative to Full Waveform Inversion. Fig. 7.08, courtesy of W. Fei, at Total HGRG, presents a synthetic, though realistic, example that illustrates the promise of such techniques. The example uses the famous '2D Marmousi' model but 3D examples are given as well in Fei *et al*, (2009). The model includes close to a thousand shots.

The top row panels in fig. 7.08 exhibit respectively from left to right, the *initial velocity model* used in the test, a series of eight SO-CIGs computed through OW Shot Profile migration and the corresponding PreSDM image, where the location of the CIGs is marked. The SO-Cigs are computed for h_s from –0.6 to 0.6 km with a dense sampling interval $dh_s = 10$ metres.

The bottom row panels show similar features but after 50 iterations of the inversion MVA, with the DSO scheme sketched in fig. 7.07. The process is fully automatic. One notes the details in the model recovered by the method, that yields a clearly improved migrated image, especially below the complex structure. The comparison between the 'before' and 'after' SO-CIGs illustrates how such panels 'focus' when one improves the velocity model.

The central column of images in fig. 7.09 are enlargements of these 'before' (top) and 'after' (bottom) SO-CIG panels. Images in the left-hand column are the corresponding Scattering Angle CIGs computed via the Tau-p (or Radon) transform illustrated in fig. 7.04. SA-CIGs are computed for angles ranging from – to + 60°. The equivalence between 'focused SO-CIGs' and 'flattened SA-CIGs' is clear.

The technique works but not all issues are solved. The building and the sensitivity to the initial model is one of them, but the computer cost is probably the more challenging one, at least for the time being.

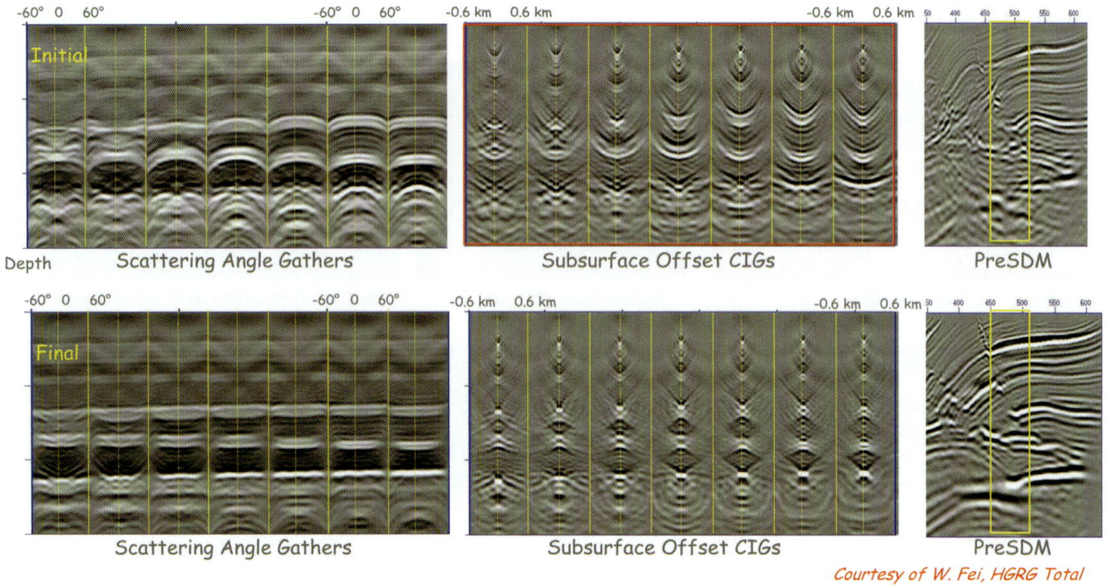

Figure 7.09: A zoom on the Subsurface Offset CIGs of fig. 7.08 and their corresponding Scattering Angle Gathers before (initial) and after (final) DSO inversion. Note how the Angle Gathers have been flattened in the MVA process.

Impact of 3D acquisition on Subsurface Offset CIGs

So far, all figures in this chapter have dealt with 2D acquisition. In this section, we address some of the issues related to acquisition and imaging in 3D.

Fig. 7.10, based on the simple model already presented in fig. 7.06, sketches the acquisition of a Wide Azimuth lay-out (top left cartoon): the shot is surrounded by receivers. This may be the result of a real acquisition (land data), or a 'macro-shot', reconstructed from a Wide Azimuth acquisition using towed streamers, or ocean bottom nodes acquisition, where we would transpose the role of shots and receivers, using a reciprocity principle.

We show in the model the area of the reflector at z = 1.5 km that is illuminated by this lay-out.

In a OW Shot Profile migration, one has to compute the two 3D down- and upgoing wavefields at each depth step. These are sketched at the bottom left of the figure at the precise depth of the reflector (1.5 km) that has generated the upgoing field. These fields are computed in practice over the whole shot aperture. They are limited in the cartoon for the sake of simplicity to the illuminated area only (and without edge and boundary effects) and to their wavefronts.

Applying the imaging condition at that depth yields a 'map' of the reflection coefficients. They will all be identical, since the reflector is flat. This also means that the downgoing and upgoing wavefields are only different by a scalar multiplier, the reflection coefficient r (laterally variable because of AVA effects). The 'map' of reflection coefficient r(x, y, z =1.5) is placed at depth z = 1.5 km. At that depth, the two wavefields are perfectly synchronous and we obtain a maximum value for r. When processing the other depths around z = 1.5 km, we obtain in fact the typical cross-correlation of the source wavelet and receiver field wavelet, shown in the zoom at the top right of the figure, which reflects once again the limited vertical resolution of seismics due to its limited bandwidth.

Fig. 7.11 outlines the many steps necessary to obtain the Subsurface Offset gathers in 3D.
As mentioned, subsurface offset is now two-dimensional: this means that we have to shift the two

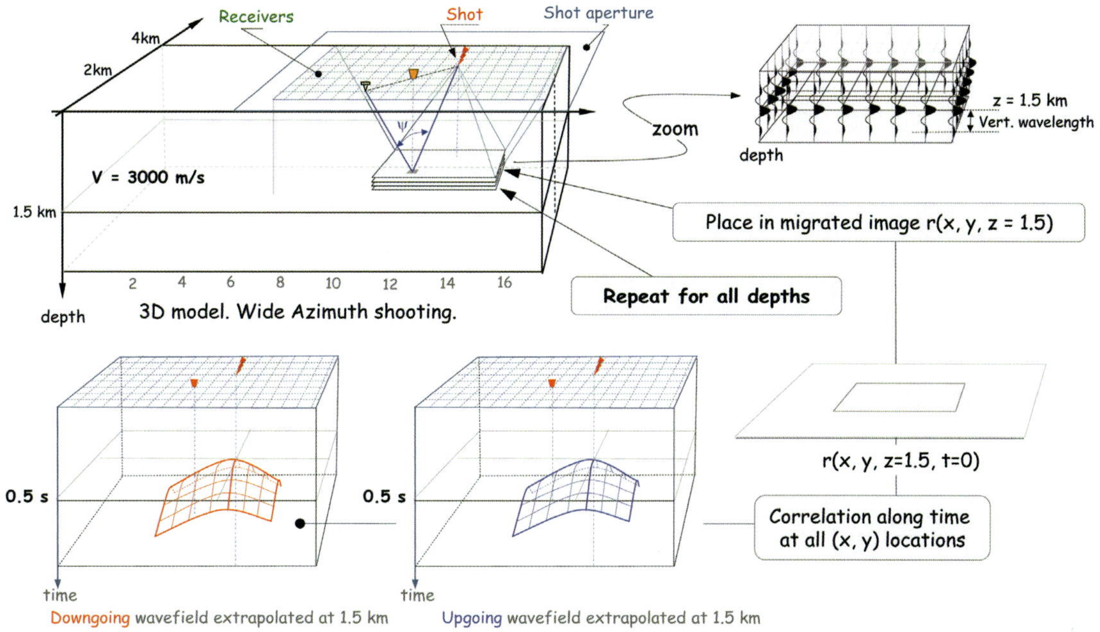

Figure 7.10: Wide Azimuth OW Shot Profile migration. The down- and upgoing wavefields at a given depth (here the depth of the horizontal reflector) are 3D objects. The reflection coefficient at that depth and for each (x, y) location is the zero time lag correlation coefficient. Contrary to fig. 7.01, the two fields are equal because the reflector is horizontal.

wavefields both in the x and y dimensions, as the cartoon on the left of the figure shows. This gives the surface of the three-dimensional SO-CIG at depth z. Note that we construct simultaneously all CIGs located in the shot aperture. Repeating this operation at each step of the recursive process, we end up at depth z_{max} with all SO-CIGs filled with the contribution of one shot. The whole process is repeated for every shot with its own aperture and the results are added to each CIG concerned. When all the shots have been processed, we have the final series of CIGs.

A SO-CIG sampled typically every 25 metres in subsurface offset h_x and h_y, with maximum offset being +/- 1000 metres represents 6400 traces. Computing a SO-CIG at each bin location of the migrated image would mean handling a data volume representing 6400 times the final migrated volume!

On the right of fig. 7.11, we draft two SO-CIGs after the migration of one shot and represent schematically the response of the reflector at 1.5 km. The equivalent of the dipping event in fig. 7.03 is now a planar event (at least close to zero subsurface offset). For this simple model, the maximum slope of the plane is in the vertical plane containing the shot and receiver pair of the trace that illuminates the reflector at that SO-CIG location. The corresponding rays are traced for the location $(x_{cig}, y_{cig})_1$ marked by an orange symbol in figures 7.10 and 7.11. This is the plane of reflection azimuth β in depth in which Rickett and Sava's, (2002) formula applies to derive the scattering angle ψ. Note that β is equal to the shot-receiver azimuth α at the surface for this simple model, but this is not the case for complex velocity fields (see the relationship between α and β in fig. 4.19).

Remember that a three-dimensional Radon transform needs to be run to transform the three dimensional SO-CIG in h_{sx}, h_{sy} and z into Scattering Angle CIGs along the dimensions ψ, β and z.

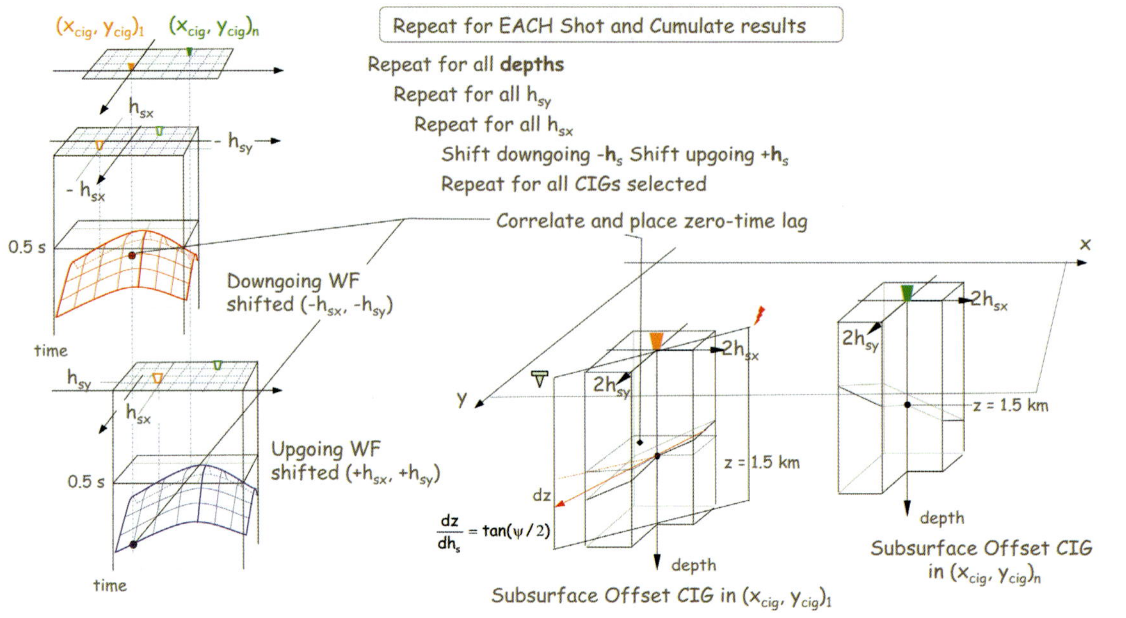

Figure 7.11: Computation of SO-CIGs in Wide Azimuth OW Shot Profile migration. At each depth of the migration of a shot, the down- and upgoing wavefields are shifted laterally by (h_{sx}, h_{sy}). The zero time lag correlation coefficient is computed and placed at the corresponding coordinates and depth in all CIGs. The process is repeated at each depth step of every migrated shots and results stacked. Note the 3D version of Rickett - Sava's formula. There is an extra loop over frequencies for FFD.

This gives an idea of the computing and data handling effort necessary to obtain dense SO-CIGs and even more so, SA-CIGs after WE migrations.

Figures 7.10 and 7.11 were drafted assuming a Wide Azimuth acquisition. Fig. 7.12 now investigates what happens if we deal with a Narrow Azimuth, or at the limit with a single streamer acquisition. Now the shot record exists only along the streamers' position. As shown in the figure, even for marine cases, one can simulate *split-spread shooting* by virtue of the '*reciprocity principle*' (receivers' and sources' roles may be exchanged). This is a necessary step to solve edge-effects along the in-line direction.

When we migrate a shot in a 3D sense we have no choice but to assume that the upgoing wavefield is nil at the surface away from the streamer. We nevertheless need to migrate the shot over the whole aperture to capture out of the plane reflections.

The extrapolation of the downgoing wavefield is the same for any type of acquisition. In contrast, the backward extrapolation of the upgoing fields is very dissimilar. Assuming that the reflected information at the surface is nil apart from a line along the streamer, would be equivalent in a constant velocity medium to assuming that the Earth is made of cylindrical reflectors having as a horizontal axis the streamer itself. A line migrating as a cylinder is the 3D extension of the point that migrates along a circle in the 2D case illustrated in figures 1.08 and 2.08. This explains why the extrapolated wavefield at depth 1.5 km, sketched in blue in the middle of fig. 7.12, has a saddle shape. It is synchronous with the downgoing field (in red) only in the vertical plane along the streamer, hence producing at this depth the image of the reflector just below the streamer (for the

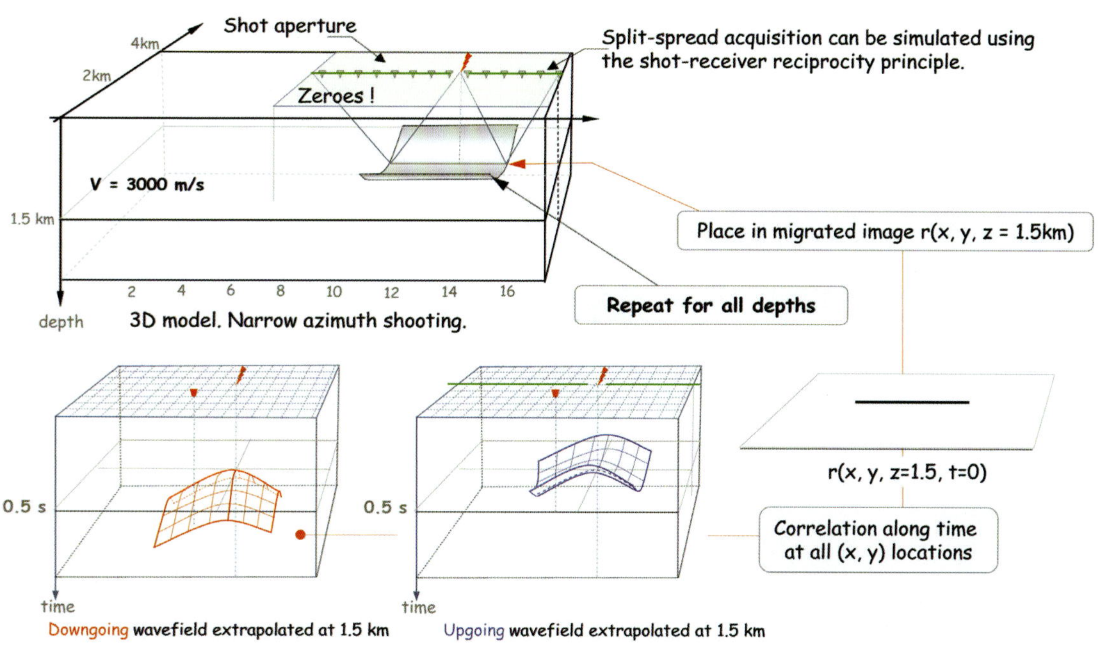

Figure 7.12: OW Shot Profile migration of a Narrow Azimuth acquisition. The downgoing wavefield is computed as in fig. 7.10, but the 3D extrapolation of the upgoing wavefield acts as if the recorded data at the surface was nil everywhere but along the streamer resulting in migration smiles in the cross line direction. These smiles are attenuated when summing the contribution of the shots along the parallel sailing lines only if the shot sampling is small in this direction.

constant velocity and horizontal reflector case). At shallower depth, we would find similar synchronism for lines parallel and symmetrical to it. The result of the migration has thus a cylindrical shape, as the one sketched in the cartoon at the top of the figure. Seen in successive constant y planes, the section of the cylinder may be called 'migration smiles'. These 'smiles' will stack out when processing and adding the contribution of the neighbouring sail lines, to leave eventually the image of the flat reflector only, but only if the cross-line shot sampling is smaller than a Fresnel zone.

A dense narrow azimuth acquisition survey may therefore produce a comparable result to the expensive 'Wide Azimuth' one, but only for the final results of a dense acquisition (as are the marine surveys today) in simple geological settings and for the true velocity.

However, even in this over simplified case, the impact of the acquisition on the velocity analysis is not negligible. It is clear that replacing the extrapolated upgoing wavefield in fig. 7.11 (Wide azimuth acquisition) by the one in fig. 7.12 will yield a different result. One can anticipate that we obtain a similar result when shifting and correlating the two fields along h_{sx} for $h_{sy} = 0$, but very different ones in any other cases.

This is confirmed when looking at fig. 7.13, where we display the SO-CIGs computed after 3D RTM at the centre of our model (they are the same anywhere out of the edge effects!) for a single streamer dense acquisition (top row of images) and a wide azimuth one (bottom row). The acquisition parameters are given in the figure. Note that the aperture of the wide azimuth is not square but still wide enough in the cross-line direction (2km), given the depth of the reflector (1.5 km).

The two images of the central column are obtained using the correct velocity and the left- and right-hand columns correspond respectively to a migration with 90% and 110% of the true velocity.

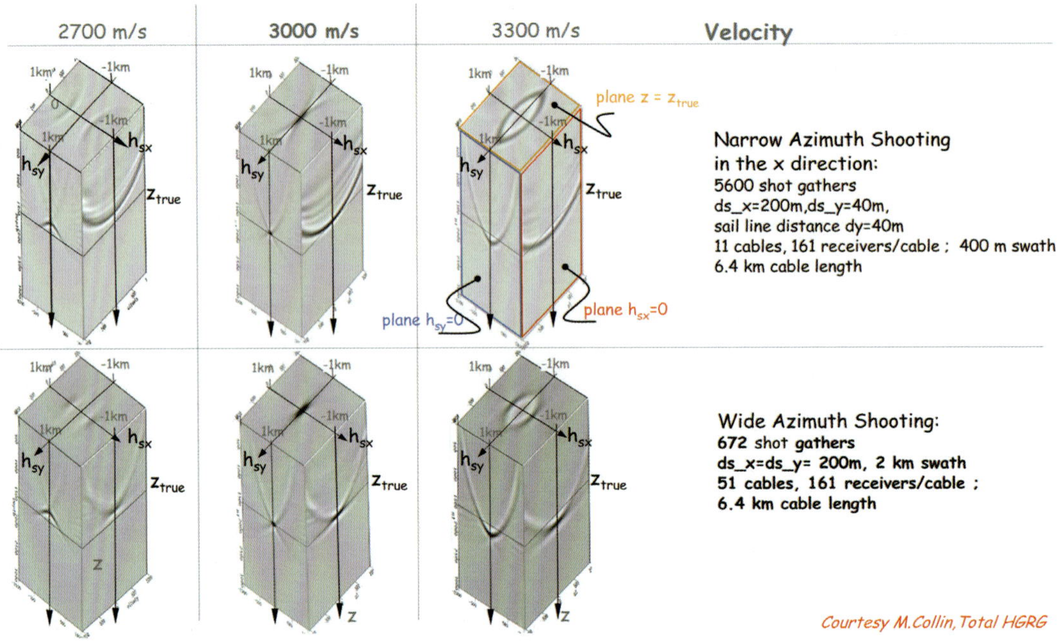

Figure 7.13: Comparison of SO-CIGs computed using RTM at the centre of the model in fig. 7.10 for three different migration velocities (V_{true} = 3000 m/s + and - 10%) and single cable (top row) and Wide Azimuth acquisition. Note the 'bowl-shaped' response in the cross-line direction for narrow azimuth.

In the in-line direction (subsurface offset h_{sy} = 0), the panels are the same as in fig. 7.06 for the wide azimuth experiment. As expected, we find similar results for the narrow azimuth case.

When looking along the cross-line direction (subsurface offsets h_{sx} = 0), we find a similar 'bow-shaped' response as for the image itself for the Narrow Azimuth case. For the Wide Azimuth case, focusing is not as good as in the in-line direction, simply because the cross-line aperture is 2 km instead of 4 km in the in-line direction.

Constant depth panels at the true reflector's depth give a more precise idea of the 3D 'structure' of the SO-CIGs. Recalling that the goal of DSO MVA is to find the model that focuses the SO-CIGs best in both h_{sx} and h_{sy} directions, one can infer that wide azimuth shooting facilitates the convergence of inversion.

Principle of Full Waveform Inversion

With MVA using DSO we were already in the domain of inversion methodology applied to the estimation of propagation velocity in the Earth and we have seen how computer intensive this may be.

With Full Waveform Inversion, or FWI in short, we are even more ambitious. The ultimate goal is to estimate directly the elastic property of the Earth, at discrete but as densely sampled locations as the seismic image of the subsurface itself. In migration techniques discussed so far, we use the recorded data and build a velocity model that helps us to construct an image of the elastic reflectivity of markers in the Earth with the resolution of the actual seismic bandwidth. In FWI, one searches for a model **m** of 'absolute value' elastic parameters that 'explains' the actual recorded data.

The ambition is for FWI to replace both velocity analysis and migration and to eventually obtain logs of calibrated P and S impedance (and even density) at each bin location at a vertical sampling of the order of a few metres. Not only do we need more powerful computers and sophisticated algorithms for this but also, breakthroughs in the way we acquire data in the fields (low to very low frequencies; reliable and faithful multi-component recording; etc.)

For the time being, the goal remains principally to obtain as detailed a model as possible of the propagation velocity field in the Earth, which will be used to produce the final 'best' image of P (and maybe soon S) reflectivity via a final RTM and we leave to 'stratigraphic inversion' the task to 'merge' the two pieces of information calibrated with borehole data to produce 'impedance' or 'pseudo-impedance' volumes.

The principle of FWI may look very similar to the other inversion applications that we have seen so far (tomography; MVA) but there are critical differences in the 'details', a major one being that the cost- (or objective) function is directly estimated in the data domain and no longer in the image domain (residual moveout of the Offset or Scattering Angle CIGs for tomography and focusing of the Subsurface Offset CIGs). FWI is of the *'data-fitting'* type.

Fig. 7.14 gives the general scheme of FWI, as pioneered in the eighties by Lailly (1983) and Tarantola (1984). We find again the same 'ingredients' as in fig. 4.04 for tomography and fig. 7.07 for DSO MVA, namely:

- a *parameterized model* of elastic properties in the Earth, which is the unknown of the problem;
- the resolution of a *forward problem*, here a seismic modelling exercise;
- the estimate of a *cost function* that may include 'regularization terms' and needs to be minimized to obtain the optimal parameterized model;
- the *initialization* and some *linearization* of the problem at each step of an *iterative update loop*;
- the computation of *gradients* necessary to estimate the update of the model at each iteration;
- a stopping criterion to limit the number of (costly) iterations.

We will discuss successively these various 'ingredients of the Full Waveform Inversion recipe'.

The model is at the heart of the process. It is the unknown of the inverse problem. The way it is *parameterized* varies from one implementation to the other. The most current one described on the left panel in fig. 7.15 consists in a regular grid of elastic parameters. In current implementation at industrial scale, this parameter is typically P-wave propagation velocity V_p or equivalently P-wave slowness (1/velocity). Other ways to parameterize the model include spline nodes as in tomography, which allow us to reduce the number of unknown parameters for 'smooth' velocity fields.

The same grid can be used to compute the wavefields, i.e., to propagate waves as required by the solution of the forward problem. It is consequently linked to the dominant seismic wavelength and the FD scheme used, as it is for RTM (fig. 6.22). We come back to this point later on.

For the more ambitious objectives that we mentioned above, the set of parameters to invert is vectors of parameters at each grid point, including P-wave and S-wave velocity and density or Lamé's coefficients or compressibility and rigidity.

In our notation, the ensemble of parameters describing the model is denoted $\mathbf{m} = \{m_{ij}\}$, with $m_{ij}=m(x_i,z_j)$, a matrix of a total of N elements per type of elastic parameters.

Regular grids are not adapted to complex geometry and more importantly, to rough topography for land data. The current research trend is hence to investigate spectral elements (Luo et al., 2009) or finite elements as for instance, discontinuous Galerkin (DG) methods. Fig. 7.16 from Brossier (2009) and the Seiscope consortium, presents an example of such a velocity parameterization for a 2D synthetic test in a simulated foothills environment. One of the difficulties in finite elements

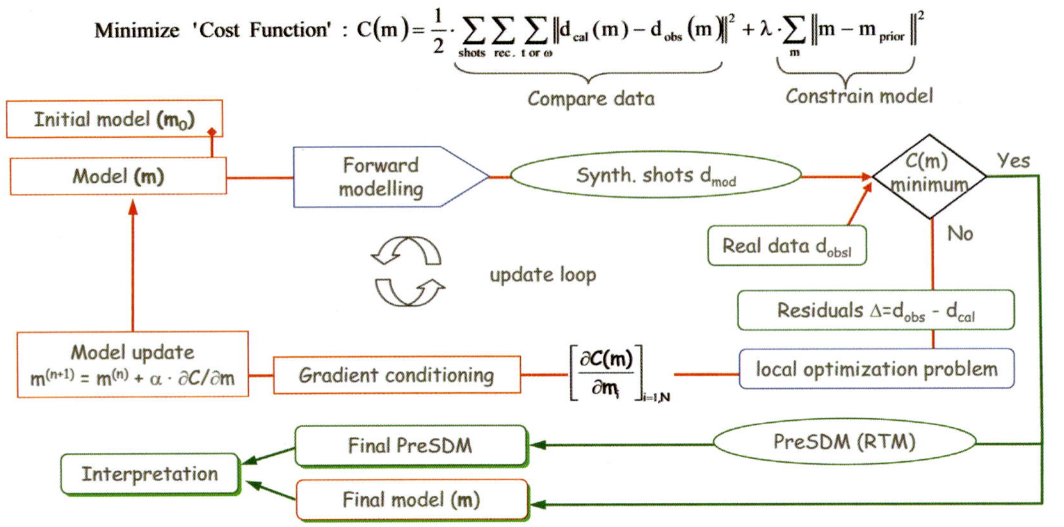

Figure 7.14: The Full Waveform Inversion workflow. The 'forward problem' is the computation of synthetic shot gathers from a discretized model of velocities for the acoustic case. The cost function quantifies the discrepancy between the synthetic shot records and the real ones. The high resolution final model may become part of the interpretation.

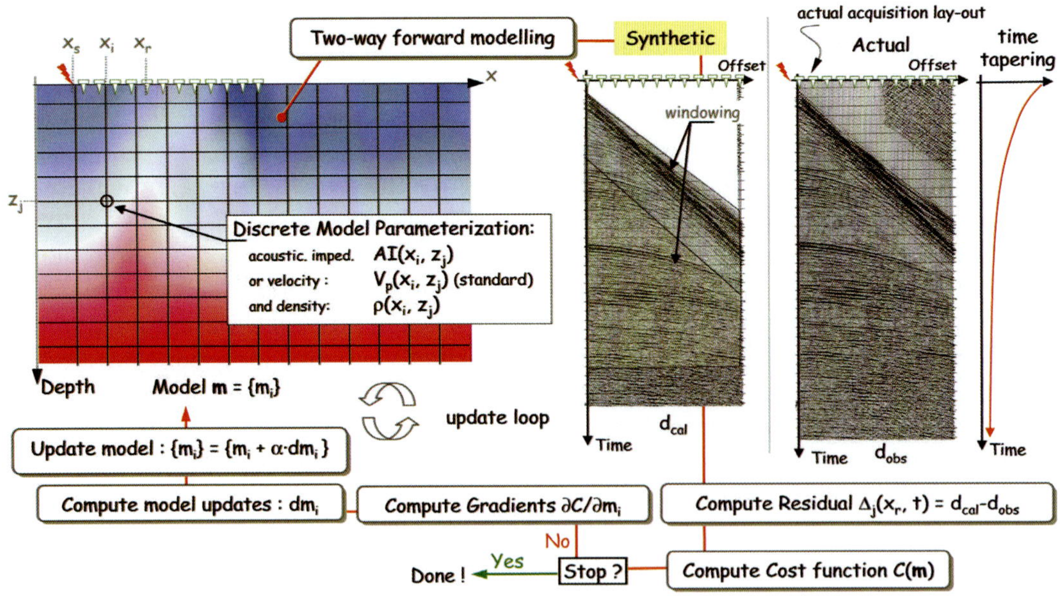

Figure 7.15: The goal of Full Waveforl Inversion is to find a discrete parameterised model of the subsurface that explains the recorded data through forward modelling. The process is solved iteratively. The gradients of the cost function are obtained directly using the 'adjoint state method'.

methods lies in the intricacy between the model parameterization and the discretization of the wave equation, their mathematical formulation and numerical resolution schemes.

The forward problem in FWI

The *forward problem* consists in the computation of synthetic shot records with the objective to compare them with the actual ones. The ensemble of simulated shots is classically denoted in the literature 'd_{cal}' for calculated data. The data recorded in the field and that we want to fit with our model are denoted 'd_{obs}' for observed data.

The comparison between d_{cal} and d_{obs} is performed by subtraction. It is therefore crucial for the method to obtain both the time and amplitude of every reflected element in the simulated shots as exactly and realistically as possible. This implies some conditions:

- The '*propagator*', i.e., the scheme that simulates wave propagation, must be accurate. It should work 'two-way' to generate diving waves and more importantly, the reflected waves that are recorded at the surface. Propagators in FWI are in fact the same as the ones used for RTM. Fig. 1.03 gives examples of snapshots of such a modelling exercise where the primary reflections, but also the *multiple reflections,* are generated at reflectivity contrasts.
- To emulate the actual acquisition lay-out: the synthetic shots must be 'triggered' at their right locations and with their own signature and traces should also be 'recorded' at their proper coordinates. It is advisable that shot records undergo basic signature deconvolution, for instance, being zero-phased and homogenized over the survey. Signatures are not always recorded in the field (they are not on land!), probably not for every shot and would anyway be known at best to a scalar multiplying factor. Either the whole signature or the scaling factor only are therefore *considered as unknowns* of the inverse problem, which are updated along the inversion iterations. Calculated traces d_{cal} are part of the computed wavefield on the numerical propagation grid and 'captured' or interpolated at the actual receiver locations (together with elevation for land data and water depth for nodes).
- The propagator in principle should account for all desired and undesired propagation phenomena in the Earth that impact the amplitudes and traveltimes of P waves, including surface waves (ground roll), refraction, diffraction and conversion. The latter would imply that one propagates elastic waves in an elastic medium. The model should consequently be defined by elastic

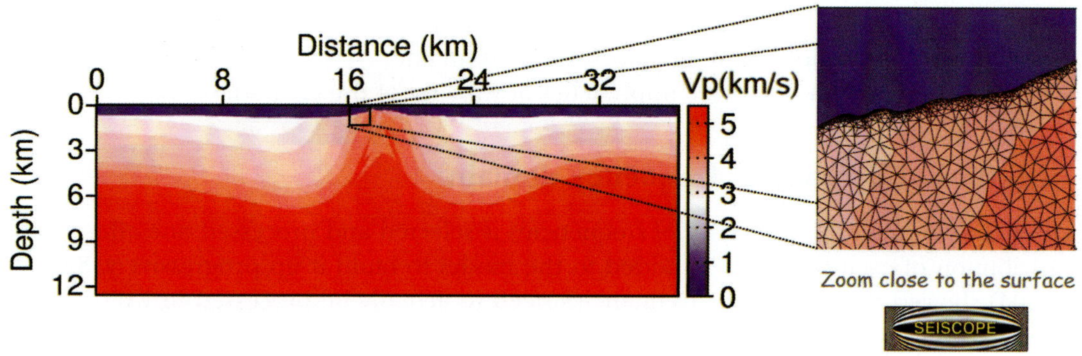

Figure 7.16: Example of a meshed model to account for topography and weathered zone for the application of FWI to land data in complex geological settings. Courtesy of R. Brossier and the Seiscope consortium.

parameters at each grid point. These could be V_p and V_s, or Lamé's coefficients. Moreover, as mentioned several times, anisotropy is of importance for a proper image and the related parameters should also be added to the model. In an elastic medium, the wavefield is vectorial (particle velocity in three directions and pressure or stress, as discussed in chapter 1). Elastic and anisotropy parameters have to be inverted too. In current practice *one inverts P-wave velocity* V_p *only*, using sophisticated P-wave propagators but neglecting the impact of conversion on amplitudes. Anisotropy may be added to the model and can be inverted, but recall the extra cost of RTM in STI media, discussed in chapter 6!). Elastic FWI that goes together with multi-component recordings on land, at the sea floor or in borehole seismics, is a hot topic of research (Brossier *et al.*, 2009).

Forward modelling is currently performed in four domains and there is still an ongoing debate to decide which is the best one.

- Historically, Tarantola (1984, 1987) and others (Bunks *et al.*, 1995) developed FWI in the *time domain*. In this case, shots are computed as a function of recording time and the comparison between modelled and real shots is simply the difference between both, as sketched in fig. 7.15. We have already discussed modelling in the time domain for RTM in chapter 6 and particularly, with fig. 6.19. Time domain FWI uses exactly the same Finite Difference schemes of high order in x, y and z. There are actually lots of similarities between RTM and FWI, as we shall see. We compute d_{cal} in FWI as we extrapolate the downgoing wavefield in RTM. Snapshots are computed recursively and the trace samples are captured at the surface, at the actual receivers' location as time unfolds.
- Pratt and Worthington (1990) and Pratt (1990) introduced FWI in the frequency domain. As mentioned for Gazdag and the FFD methods applied to OW Shot Profile migrations discussed in chapter 6 and as a virtue of the superposition principle and the assumed stationarity of the Earth (i.e., its elastic properties do not change significantly during the seismic survey!), the source signature may be decomposed in its frequency components via Fourier transform and propagation of each frequency can be considered and processed independently. The result in the time domain is reconstructed, if needed, by inverse Fourier transform of the processed frequencies. We detail FWI in the Fourier domain in a following section (fig. 7.17).
- Sirgue *et al.* (2008), Vigh and Starr (2008) show the issues and merits of using a 'hybrid' form, where the propagation is performed in the time domain but the iterative inversion in the frequency domain.
- Shin and Ha (2008) and Shin and Cha (2008) proposed to use Laplace transform. Shin and Cha (2009) extended it recently to the Laplace-Fourier domain.

We present the results of the last three approaches at the end of the chapter.

The workflow of FWI in the frequency domain

In the frequency domain FWI, a series of frequencies is selected and processed one after the other. The contribution of each frequency is added to the model to build it up progressively. Sirgue and Pratt, (2004) showed that only a few frequencies need to actually be processed in FWI in order to obtain detailed velocity models, provided large offsets are used at the acquisition stage. The reason is linked to the intrinsic phenomenon of 'stretch', briefly mentioned in chapter 3 (fig. 3.06), related to the presence of long spatial wavenumbers for high scattering angles. This reduces the cost of FWI and makes the frequency domain approach very popular to date.

Another important message is that one must *start with the lowest frequency possible*. This frequency may be as low as 3 Hz if recorded data allow (for instance for OBC acquisition). The first low frequency controls and optimizes the initial velocity model. Subsequent higher and higher frequencies will add details of shorter and shorter wavelength in the model. The set of frequencies to use is a function of the maximum offset recorded, as Sirgue and Pratt (2004) pointed out. This *'multi-scale approach'*, which was proposed by Bunks *et al.* (1995) in the time domain can be applied straightforwardly in the frequency domain.

Each selected frequency is processed independently. Fig. 7.17 presents the key steps of the iterative FWI for *one frequency* ω_1. For the sake of simplicity of the figure otherwise already very busy, we show the process in 2D and for one single shot only, knowing that all shots are processed simultaneously and that the cost function and gradients are the sum of all single shot contributions. Recall as well that each shot illuminates only part of the reflectors and contributes to a limited volume of the model but all shots together hopefully illuminate the whole model volume and hence, contribute with redundancy to its construction.

Under these conditions, the $\mathbf{d}_{obs}$ is the vector of the complex valued component for frequency ω_1 of the Fourier transform of the traces at the receivers' locations, expressed, for instance, as their Real and Imaginary part in fig. 7.17.

The *initial model is a smooth first guess*. It could be a multi 1D model or even a constant velocity one. The choice of this initial guess is unfortunately not insensitive to the final inversion result.

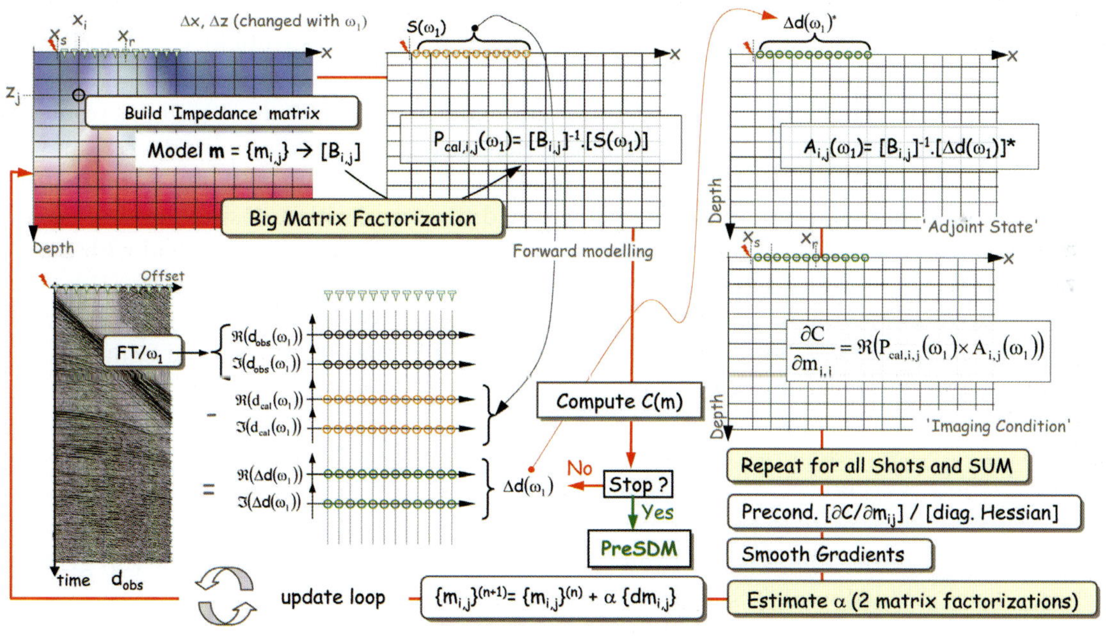

Figure 7.17: Workflow of the Full Waveform Inversion of one shot and one frequency ω_1. $\mathbf{d}_{obs}$ is the vector of the ω_1 frequency component of shot traces. Forward modelling is performed via the resolution of a linear system. The gradient is computed according to the 'adjoint-state method' as the product of the forward model and the propagated time reversed residual Δd^* acting as a 'virtual source'. The optimization can be a 'steepest descent' (α is a scalar) or other techniques. The process is iterated until C($\mathbf{m}$) is considered to be minimum. It is then repeated for higher frequencies, each of them adding details to the model ($\mathbf{m}$).

The model is discretized as a regular grid in x, y and z. The waves will be propagated in this model, which thus needs to be sampled accordingly. The frequency domain methods do not escape this constraint, discussed in fig. 6.22. As an example, selecting a 4th order scheme and a criteria of 4 grid points per wavelength, a minimum velocity of 1500 m/s for a marine case yields a sampling rate of dx = dz = 75 m for 5 Hz and 25 m for 15 Hz. Between the two frequencies, the number N of unknown parameters $\{m_{ij}\}$ is multiplied by 9 in 2D and by 27 in 3D. The *grid sampling has a huge impact on the cost* of the process. It is critical to adapt it to the frequency being processed. It is changed when moving to a higher frequency.

Next, we need to '*solve the forward problem*'. It consists in the propagation of the shot with its signature into the model and capturing the wavefield at the surface to emulate acquisition. One calls on the wave equation again for that purpose. In the space-temporal frequency domain, the scalar (constant density) wave equation takes the following form in 2D, which is often referred to as the 'Helmholtz equation' in seismics literature:

$$\frac{\partial^2 P_\omega}{\partial x^2}+\frac{\partial^2 P_\omega}{\partial z^2}=-\frac{\omega^2}{V^2}P_\omega(x,z) \tag{7.02}$$

In this equation P_ω (x, z) represents the components for the radial frequency ω of the wavefield at the location (x, z) in the subsurface. This component is complex-valued and obtained by Fourier transform along the time direction of a full wavefield P(x,y,t).

Equation (7.02) is used to compute the ω component of the full wavefield $P_{i,j}$ in all locations (x_i, z_j) of the propagation grid, which results from the source triggered at a location x_s. Only the ω component is 'sent'. This corresponds to a thought experiment where a source in x_s would emit indefinitely in all directions a sinusoidal signal of radial frequency ω and amplitude $||S(\omega)||$.

Similar to what we did in fig. 6.19, this differential equation is transformed into a '*finite difference equation*' using formulas to approximate derivatives by differences, with a desired level of accuracy. The difference with the time domain however, is that for each frequency ω, we deal with a system that has only two dimensions (in 2D) of discretization: x_i and z_j.

Using approximations of second order derivatives similar to the ones in fig. 6.19, and the notation $P_\omega(x_i, y_j)$ and $V(x_i, y_j)$ ➔ P_{ij} and $V_{i,j}$ respectively, the Helmoltz differential equation (7.02) becomes the 'finite difference equation':

$$\frac{1}{dx^2}\cdot P_{i-1,j}+\frac{1}{dz^2}\cdot P_{i,j-1}+\left(-\frac{2}{dx^2}-\frac{2}{dz^2}+\frac{\omega^2}{V_{i,j}^2}\right)P_{i,j}+\frac{1}{dz^2}\cdot P_{i,j+1}+\frac{1}{dx^2}\cdot P_{i+1,j}=S_{ij} \tag{7.03}$$

The $P_{i,j}$ and the $V_{i,j}$ are the values respectively of the wavefield and the velocity at the (i,j) grid point of the discretized gridded model. S_{ij} is the 'source term'

Equation (7.03) is valid for each grid point. It represents consequently *a set of equations*. Note how the discretization intervals dx and dz and the velocity at each grid point V_{ij} enter into this set of equations and that the $P_{i,j}$ are complex valued.

Equations (7.03) are valid everywhere where waves propagate. The source imposes 'its initial condition' as the source term $S_{ij}(\omega)$ on the right hand side of equation (7.03) that is non nil for $x_i = x_s$.

The equation needs to be modified as well along the lateral and bottom boundaries of the model, where '*absorbing boundary conditions*' need to be applied in order to avoid waves bouncing undesirably back into the model. Discarding this last point, the set of linear equations (7.03) can be cast in a compact matrix product notation:

$$\lfloor B \rfloor \cdot \lfloor P_{ij} \rfloor = \lfloor S_{ij} \rfloor \qquad (7.04)$$

[B] is often referred to as 'Impedance matrix': in the case of variable density and elastic models, the Wave Equation is different from (7.02) and [B] includes compressibility and density terms related to acoustic impedance.

[B] is of size N x N, N being the number of unknowns $\{P_{ij}\}$. It is sparse i.e. it contains many nil elements. The non-nil ones are the coefficients of the $P_{i,j}$ in equations (7.03); they contain the velocity information at every location of the digitized model $V_{i,j}$.

$[P_{i,j}]$ is the vector of the unknowns of the forward modelling exercise, i.e. the ω component of the digitized wavefield $P(x_i,y_j,\omega)$ and [S] is the 'source vector' assumed here to be nil everywhere but at abscissa x_s at the surface.

We get $[P_{i,j}]$ by solving equation (7.04). A 'brute force' solution to the forward problem would be to invert matrix [B] and left-multiply the matrix [S] by its inverse $[B]^{-1}$ (assuming it exists). This is simply not achievable with today's most powerful computers. Other techniques are better suited to this type of matrix. They include 'LU factorization' and iterative solvers.

Since the wave equation (7.02) used corresponds to a *'two-way propagator'*, this field includes all energy reflected at any V_{ij} contrasts. These upgoing reflections are derived at the surface as a real experiment would do: recording of the wavefield at the surface (z_j = 0) at the actual location x_r of the receivers in the field. Since the computation grid does not match the acquisition lay out ($x_r \neq x_i$), some interpolation is needed here. This yields the modelled shot record $d_{cal} \equiv P_{i=r,j=0}$ for the current velocity model (**m**) = $\{V_{i,j}\}$. Because we are in the frequency domain, d_{cal} are complex valued. If one would repeat the operation for all frequencies in the seismic bandwidth and perform an inverse Fourier transform of the result, one would get the whole shot record x_s as a function of space and time, as sketched in fig. 7.15 but as mentioned, we do not need all frequencies: the forward problem is solved for ω_1 in the frequency domain!

The modelling may seem simpler than in the time domain. The reader should however realize that we are facing here a serious numerical issue. [B] is a sparse but huge matrix to factorize even in 2D, not to mention 3D. A frequency ω = 3 Hz for a criterion of 4 points per wavelength corresponds to a grid of Δx = Δy = Δz = 100 m, which means, as we saw earlier, 9 million unknowns for a standard-sized acoustic survey of 30 × 30 km with a maximum depth of 10 km. This is already at the limits of current computing capacity for direct matrix inversion. For the 15 Hz frequency, the spatial sampling rate should be 25 metres, corresponding to 4^3 = 64 times this value. For the elastic case, the situation is even more dramatic. Firstly, the size of the sampling rate is further reduced by the minimum velocity at the surface (500 m/s is possible, i.e., another factor 3^3 = 9) and the number of parameters at each point of the grid is multiplied by 6 (3 components of the particle velocity, plus 2 elastic parameters plus density!

Factorization or inversion of matrix [B] is a numerical challenge by itself and appropriate solvers are requested for that. It is not in the scope of this book to discuss these methods but this numerical and computational issue is one of the key elements of FWI.

The reader can refer to Etgen and O'Brien (2007), Operto *et al.* (2007) or Virieux *et al.* (2009) for reviews on such modelling challenges in time or frequency domains, especially in 3D.

The good news however is that $[B_{ij}]$ is the same for all shots and we need to factorize it only once per frequency and iteration. The solution for one shot still requires some computation to insert the source term at the right place in the matrix equation (7.04). For iterative solvers in frequency or in time-domain FWI in contrast, the full propagation computing effort must be repeated for every single shot.

At this stage one can compute the *cost function*. Its formulation is given at the top of fig. 7.14. Its first term quantifies the 'distance' or the 'misfit' between the data d_{obs} recorded in the field and the data $d_{cal}(\mathbf{m})$ modelled with the current model **m**.
The d_{obs} are the samples of the shots, with limited pre-processing:

- It is advisable to remove incoherent noise to avoid it 'leaking' into the velocity model.
- *Multiple reflections should not be removed.* They are in principle generated by the two-way propagator and ought to contribute to the estimation of the model **m**. In theory at least, FWI is a tool that uses explicitly multiple energy to update the model, contrary to migration techniques, which necessitate removing them. This will work however only if the model is sufficiently detailed and updated to generate free-surface and internal multiples and if amplitudes are sufficiently reliable, particularly in handling mode conversion (elastic models and propagators). Vigh *et al.* (2009) suggests multiple removal before FWI but Sirgue *et al.* (2009) advocate the benefit of multiples in improving the image over the Valhall field. An open, important debate ...

In the case of FWI performed in the time domain, the misfit between the modelled and the actual data is estimated by simply taking the sample by sample difference between the recorded and the modelled traces of each shot record over the selected time-window, as sketched in fig. 7.15. For each shot, one computes the difference $\Delta d(h, t) = d_{obs}(h, t) - d_{cal}(h,t)$. The misfit is quantified as the sum of the square of all these differences, hence the symbols of summation over time, offset and shots in the equation in fig. 7.14. The sum of the squares of the differences is known as the 'L2 norm'.

In the case of frequency domain FWI as in fig. 7.17, for one trace and one frequency ω, d_{obs} is limited to one complex value, as is d_{cal} (see above). The $\Delta d(\omega)$ is hence the difference between these two complex samples, of which one takes the real-valued norm $|| \Delta d(\omega) \times \Delta d(\omega)^* ||$, where * denotes the complex conjugate operator. The data term of the cost function $C(\mathbf{m})$ for this frequency is the sum of all these scalars for all receiver locations (traces) of all shots. Hence, the summation symbols in the formula in fig. 7.14.

On top of the data term, we write in the cost function equation a 'model term' that can be seen as a regularization or preconditioning of inversion. This term constrains the solution (**m**) not to depart too much from a first guess ($\mathbf{m}_{prior}$) of the model. Playing with the factor λ in the equation, we may either 'let the data dictate the result' ($\lambda = 0$) or conversely, if the data are considered of poor quality, favour the *a priori* knowledge of the answer embedded in the prior model ($\mathbf{m}_{prior}$). In current implementations, this term is not used. Regularization is applied in contrast as a conditioning process, such as a smoothing of the model updates, for instance.

Other weightings are also used in practice. Typical ones include:

- Simple 'time windowing' of the data. For instance, we want to use only first arrivals, including near-surface refracted arrivals.
- Less radically, data can be tapered with time (sketched in fig. 7.15) to again favour first arrivals and correlatively the velocity background and possible shallow anomalies that are detrimental to the overall quality of the image. This relates also to the discussion of tomographic inversion in chapter 4. The taper may or not be hanging to the first arrival time. Data tapering is not a straightforward process in the frequency domain but is possible with the use of 'complex frequencies'.
- The Laplace-transform is a way to apply exponential decay tapering. We come back to this method in a later section.
- Similarly and in parallel, tapering can be applied in the offset dimension to favour long offsets that are more sensitive to the background velocity model.
- Data could also be weighted according to their quality. For instance, shots signalled as having a poor signal / noise ratio or including 'dead traces' (broken down receivers, poorly planted geo-

phones, etc.) should have less or zero weight in the cost function. For that purpose, weighting is expressed in mathematical terms as a 'covariance matrix' W_d acting on the 'vector matrix' [Δd] of the data in the following form:

$$C(m) = \frac{1}{2}[\Delta d]^{\dagger} \cdot W_d \cdot [\Delta d] \quad (\dagger \text{ denotes matrix transpose conjugate})$$

- A similar covariance matrix can be applied to the model term to smooth it (preconditioning) or to weight up or down for instance, the 'non illuminated' zones of the model. The whole 'trick' or preferably 'know-how' or difficulty is in the way one fills these matrices, which are not really used in practice!

Coming back to fig. 7.17, we now have the cost function C(**m**) for frequency ω_1 that corresponds to the initial model. There is no reason why it should be a minimum and we must go into the iterative process to converge towards the model (**m**) that minimizes it. Given the multidimensionality of the problem (possibly millions of unknowns), only iterative methods will work and we initiate the update loop that should lead, after several iterations (typically 30 to 60), to the searched minimum.

In chapter 4, for tomographic inversion we showed that minimization is achieved through the solution of the '*normal equations*' that necessitates the explicit computation of the $N_{mod} \times M_m$ matrix of the derivatives of the M_m measurements with respect to the N model parameters and comes eventually to the inversion of the big $N_{mod} \times N_{mod}$ matrix.

In FWI, minimization is often addressed slightly differently, although by means of the same 'local optimization', which assumes that the current model is already close to the optimal one and there is no false local minimum of the cost function in its vicinity.

For example, the '*steepest descent*' method already mentioned for DSO-based MVA is often used; it would write as (classical first-order Taylor expansion, same as equation (7.01)):

$$\{m_{i,j}\}^{(n+1)} = \{m_{i,j}\}^{(n)} - \alpha \cdot \left\{\frac{\partial C}{\partial m_{i,j}}\right\}^{(n)} \tag{7.05}$$

where (n) indicates the iteration number.

To solve (7.05), i.e., to compute the updated model $\{m_{i,j}\}^{(n+1)}$, we need to compute the 'ascent slope' or *gradient* $\partial C/\partial m_{i,j}$ of the cost function with respect to the model parameters, and estimate the optimal '*step length*' α, such that $\{m_{i,j}\}^{(n+1)}$ takes us as close as possible to the optimal solution. The gradient has the dimensionality of the model, since it has one value at each model grid point.

Lailly (1983) and Tarantola (1984) showed an important result: the gradient of the cost function $\partial C/\partial m_{i,j}$ can be computed without the necessity to compute the derivative of the 'measurements' $\partial d_{obs}/\partial m_{i,j}$. It is in fact a scaled version of a Reverse Time Migration using the current velocity model, but where the shot record is replaced by the residuals $\Delta d = d_{cal} - d_{obs}$.

One way to explain this result in simple terms is to imagine a thought experiment whereby the Earth would be made of the current (smooth) velocity background with the searched updates $\{dm_{i,j}\} = \{m_{i,j}\}^{(n+1)} - \{m_{i,j}\}^{(n)}$ being added as local 'spikes' at the grid points of the model. These 'anomalies' in the smooth background would act as *diffractors*, such as the one we saw in chapter 2, fig. 2.05 (Born approximation).

The result of a seismic shot in this hypothetical Earth would be the sum of the response of all these 'diffractors'. Conversely, the migration of this shot with the background velocity field would focus these diffractions back to their locations, hopefully with their proper amplitude $\{dm_{i,j}\}$. If we accept the vision that the residuals are precisely the propagation effects of what has not be included in the current model (i.e., the searched updates), we may understand why migrating the residuals (in the migration sense seen in the previous chapters) yields information on the updates! The simplified explanation of the approach above hides in fact many rigorous mathematical developments and proofs, which show that what we compute is actually a scaled version of the gradient of the cost function that should be 'pre-conditioned' or normalized. This way of computing the gradient of the cost function is known as the '*adjoint-state*' or '*adjoint-wavefield*' method (see, for instance, Plessix (2006) or Operto and Virieux (2009) for a review).

The computation of the gradient matrix amounts consequently to a RTM exercise, as the one illustrated in figures 6.03 and 6.18 for the 'classical' migration case.

In the frequency domain, as in fig. 7.17, we have already the 'downgoing wavefield' $[P_{ij}]$, i.e., the shot $P_{cal}(x,y,\omega_1)$. The extra cost for the computation of a shot contribution to the gradient is equivalent to another modelling exercise, now with the time reversed residuals as 'virtual source' $\Delta d(\omega_1)$ *, since the $[B_{ij}]$ matrix is already factorized. Time-reversal corresponds to complex conjugate in the frequency domain, hence the * operator applied to the vector in fig. 7.17. This exercise is equivalent to the upgoing wavefield extrapolation from the shot record in conventional migration and yields the 'adjoint wavefield' $[A_{i,j}]$ (the upgoing residuals). The application of the imaging principle is a simple term-by-term matrix multiplication (the equivalent in the frequency domain of the correlation in time) of the two matrices $[P_{i,j}]$ and $[A_{i,j}]$, the real part of which being the matrix of the gradient. The same process must be applied for all shots and the gradients accumulated.

The preconditioning term mentioned earlier can in turn be approximated by 'approximate Hessian terms' (Shin *et al.*, 2001) that cost approximately as much as the gradient computation.

Following this idea, let us consider the very first iteration in fig. 7.17. The initial model is smooth, as shown at the top left of the figure. The forward modelling of a shot into this model will produce no reflection. The calculated shot record (our 'upgoing field' in migration) contains only the direct arrival and the residual, the actual recorded shot minus this non-reflected wavefield. The gradient $\partial C/\partial m_{i,j}$ being the RTM migration of the residual, as mentioned, is in fact a scaled version of the RTM migration of the shot, plus some 'cross-talk noise'. This fact explains the classical statement that '*migration is the first iteration of full waveform inversion*'.

As soon as the second iteration, the 'full' information is included in the residual data and migrated with the current velocity model and it starts to contribute to the model update. More precisely, we use only a version of it taken over the coarse model grid and possibly smoothed, but the first 'depth-migrated reflectors' appear in the model and start to produce reflected energy at the next forward modelling step. The rich information in the recorded data will subsequently contribute more and more to the model as iterations progress, the frequencies used are higher and the model grid size is refined.

The next issue is to estimate the '*step length*' α. The so-called '*line-search*' consists in computing the cost function for three values of α (note that we have $C(\mathbf{m})$ already for $\alpha = 0$) and to assume that C is a parabolic function of α from which one can determine analytically the α_{opt} that minimizes it.

Other techniques, including the popular *conjugate gradient* one, are used (Virieux and Operto, 2009), but in any case the optimization step requires the computation of at least two other cost functions. In time domain FWI, this amounts to two extra forward shot modelling. In the frequency domain, it necessitates the building and factorization of two new impedance matrices $[B_{i,j}]$, since the model is (slightly) changed, followed by the modelling of all the shots to compute the cost function as a function of the two tested α values. There is a high cost to pay to converge in an optimal number of iterations!

Once the model is updated using the gradient and the step length, we can start the next iteration: modelling of the shot, computation of the cost function, decision to stop or not if the cost function has reached an asymptote. The amplitude of the last cost function gives an idea of the 'unexplained energy', which should hopefully be pure random noise.

To recapitulate fig. 7.17, one iteration of the frequency domain FWI requires three factorizations of large matrices [B] per frequency. It is, however, less sensitive to the number of shots and number of traces per shot. The reported number of iterations is between 30 and 60 and the number of inverted frequencies is between 5 and 10, always starting from the lower one. The $N \times N$ size of the 'Impedance matrix' [B] increases exponentially with the frequency: doubling the frequency means dividing the spatial sampling by 2 and multiplying the size of N by 8. In practice, a few frequencies (2 or 3) may be processed jointly to stabilize the process and improve the signal-to-noise ratio at each 'frequency group' iteration.

Laplace-Fourier domain and the initial model

The definition of the initial model in inversion remains a controversial issue, since FWI is a long way from being a mature technique. Authors suggest to start from a reasonably elaborated smooth model. It may come from first-arrival traveltime tomography (Operto *et al.*, 2004) or a previous PreSDM and tomography (Vigh *et al.*, 2009; Sirgue *et al.*, 2009).

Fig. 7.18, courtesy of the Seiscope Consortium, compares the results of Full Waveform Inverted velocity fields in the synthetic well-known 'foothills data set'. The frequency domain FWI result (panel d) when starting from a smooth version (panel c) of the true model (panel a) is almost perfect. When starting from a 'multi-1D' initial model, where the *a priori* information is some average of the background vertical slowness heterogeneity, the result is of a much lesser quality (panel b).

As mentioned, Shin and Cha (2009) suggest to invert seismic data in the Laplace-Fourier domain (complex valued damping constant), of which the Laplace domain is a subset (real Laplace damping constant).

The Laplace-Fourier transform of the wavefield P(x,z,t) is defined as an extension of the Laplace transform where the Laplace damping parameter s is complex valued and of the form $s = \sigma + j\omega$. The mathematical expression of the Laplace-Fourier transform of a wavefield P(x,z,t) writes as:

$$\tilde{P}(x,y,s) = \int_0^\infty P(x,y,t) \cdot e^{-\sigma t} \cdot e^{-j\omega t} \cdot dt \qquad (7.07)$$

For $\omega = 0$, the Fourier-Laplace transform reduces to a Laplace transform that is the integral over time of the traces of the wavefield, tapered with a decreasing exponential function controlled by the damping parameter σ.

An interesting feature with Laplace transforms is that the wave equation can be expressed in the Laplace domain, as it can be in the Fourier domain. FWI schemes can thus be developed in a way similar to the frequency domain, inverting one parameter σ after the other, as we invert one frequency after the other.

The higher σ, the more weight is placed on the first arrival. Inversion is then performed for decreasing values of σ and the model is updated iteratively. The expected benefit of Laplace domain

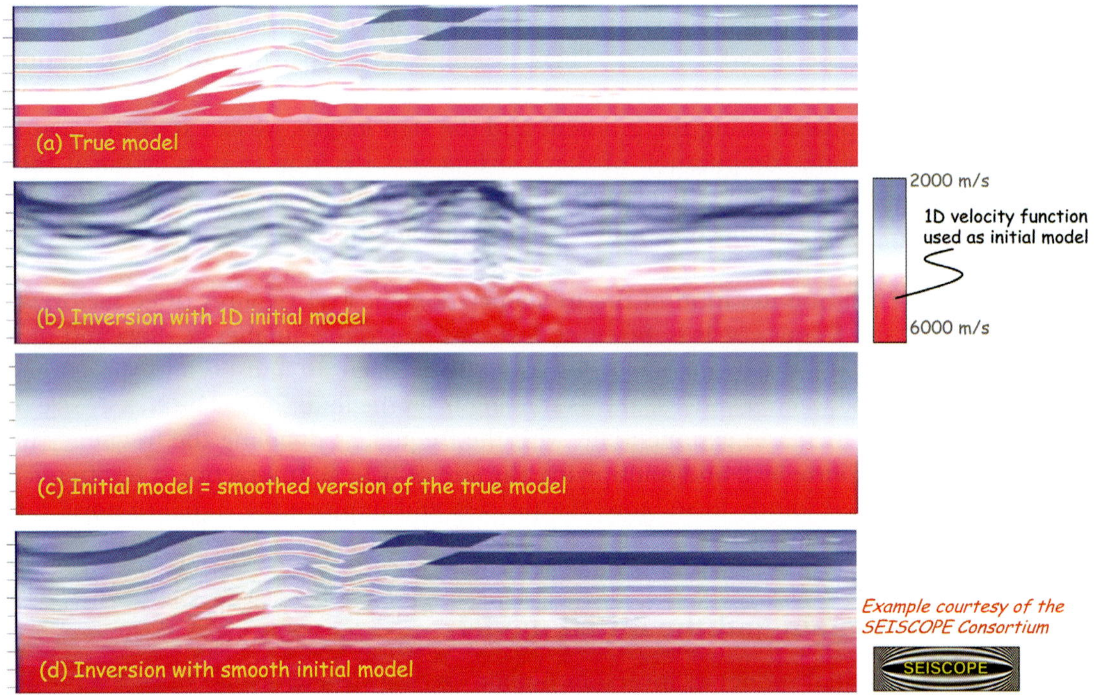

Figure 7.18: Example of frequency FWI using the overthrust model. With a simple 1D initial velocity field, some features are properly recovered, but not the very long wavelength velocity trend. With a smoothed version of the model, an almost perfect result is obtained. Four groups of 2 frequencies ranging from 3.5 to 20 Hz and 15 iterations per frequency were used in this case.

FWI is that it favours in the first iterations over σ the (very) low frequencies and first arrivals of the data. This is in a way equivalent to ray-based refraction tomography but included seamlessly in the general inversion workflow. Combination of the Laplace and Laplace-Fourier brings flexibility into the inversion strategy (Shin and Cha, 2009). Shin and Ha (2008) also investigated other expressions for the cost function that are better suited to the Laplace-Fourier domain.

2D and 3D results show that it may help to address the critical issue of building the initial model. Such an example in 2D, courtesy of C. Shin, Seoul National University, is given in fig. 7.19. In this example, Laplace domain inversion, starting from a very simple model (constant velocity) was able to recover almost perfectly the complex velocity model that emulates a salt tectonics environment. The two salt bodies were well delineated with limited prior knowledge of the velocity (average background velocity) and no picking of top or base salt, as is common practice (and time consuming!) in these geological settings.

Plane-wave and encoded shot approaches

The cost of FWI, especially for the time domain, is related to the fact that each shot, being a specific experiment, must be processed individually. We met the issue in chapter 6 for OW SP migration and RTM. The method is applied in migration, whether by plane wave migration or encoded shots.

The idea then comes naturally in FWI to try and process shots jointly, while trying to minimize the 'cross-talk' between the various experiments. Some approaches have been proposed recently,

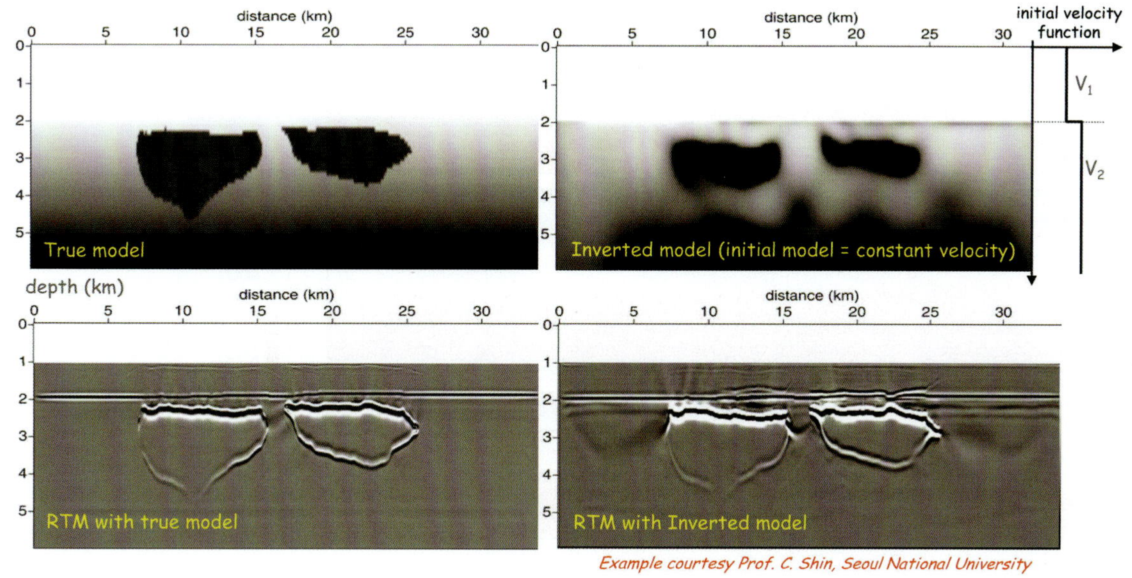

Figure 7.19: Synthetic 2D example of Laplace domain FWI. With a simple 1D initial velocity field made of two constant velocities, most of the features of the model are recovered, yielding a good final RTM image.

including Ben-Hadj-Ali *et al.* (2009), with promising results. We present these published results in fig. 7.20 as an example of current research directions.

The author proposes to combine all shots to create a single 'super-shot'. Instead of running N inversions of one shot, one has to run one inversion of this 'super-shot', with a potential benefit in the computational effort. The 'cross-talk' appears in the computation of the gradient, when one correlates implicitly the contribution to the 'super-shot' of one shot with the contribution of all other shots. The proposed solution is to encode each shot in the frequency domain by a random phase shift (equivalent to a time delay for that frequency). Phase shift does not change the contribution of one shot to the computation of the gradient. It would be equivalent to modifying the 'absolute time at which the shot was triggered'. However, the hope is that the fact that phase shifts change randomly from one shot to the other, from one iteration to the other and from one frequency to the other, will eventually make the cross-talk noise in the gradient reduce to an acceptable level, if not vanish.

Fig. 7.20 from Ben-Hadj-Ali *et al.* (2009) compares the multi-shot approach whose result is shown in fig. 7.18-b, with a result using the shot-encoding technique described above. A similar inverted model (although slightly noisier) is obtained with many more iterations (200 per frequency instead of 15), but the overall computational effort is still reduced by a factor of 7.

The Valhall 3D example

Full Waveform Inversion is not a fully mature technology as yet but its potential as a tool to estimate detailed velocity fields does not need to be proven any more. Industrial applications at full scale in 3D have been already published. The Valhall case (Sirgue *et al.*, 2009) is one of those. It addresses the delineation of a field under a gas cloud that blurs the seismic image. Multiple reflections is another seismic challenge here. The data were recorded on permanent Ocean Bottom Cables as

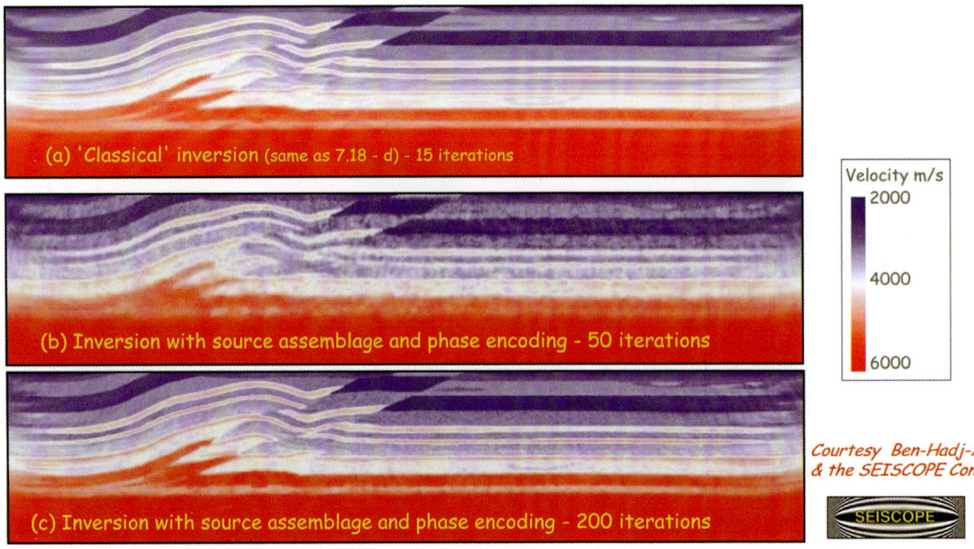

Figure 7.20: Comparison of (a) multi-shot frequency domain FWI and (b) source assemblage plus phase encoding (Ben-Hadj-Ali, 2009) after 50 iterations, and (c) after 200 iterations. Despite the number of iterations necessary to get rid of 'cross-talk' noise, the overall computation effort is seven times lesser.

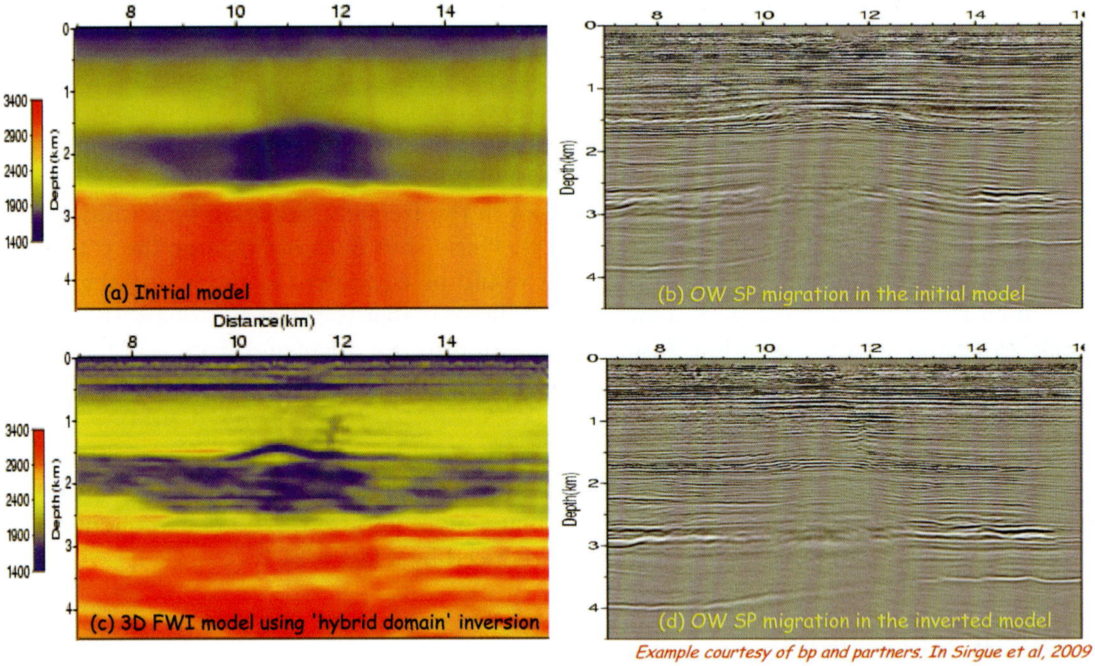

Figure 7.21: Example of 3D inversion of wide azimuth OBC data over the Valhall field using 'hybrid-domain' methods. The initial model is obtained using ray-based tomography. Inversion is performed using frequencies from 3.5 to 7 Hz. Note the high resolution character of the inverted velocity model that (1) improves the image quality and (2) is a new attribute for interpreters.

part of a 'Life of Field seismic survey'. Fig. 7.21, courtesy of bp and partners, presents on panel (a) a cross-section through the 'initial' velocity field, resulting from conventional approaches (OW SP Migration + tomography). The corresponding seismic image (OW SP migration) is presented on the panel (b). The result of a hybrid domain FWI inversion is shown on panel (c), with its corresponding seismic OW SP migration on panel (d). The increase in the resolution of the velocity model is clear. The validity of the added information as compared to the initial model is confirmed by the improvement in the resolution of the seismic image obtained by the PreSDM run with the inverted model.

The level of detail obtained in this example is comparable with the level of detail obtained in a different context by high-resolution tomographic inversion, as the one presented in fig. 4.14 in chapter 4. Both examples illustrate the trend of the industry towards higher-resolution velocity models.

In the case of tomography, the improvements come from high-quality data (dense acquisition; longer offsets; good signal/ratio; efficient 3D multiple removal), the development of fast and reliable high-density volumic picking tools and increasing computer capacity (inversion or factorization of large matrices).

In the case of FWI, the use of long offsets and very low frequencies are considered as key elements of the success of this project: the first frequency used in this case was 3.5 Hz, which is achievable in today's OBC multi-component surveys. Recording very low-frequency (below 5 Hz) reliable data with a towed streamer remains a challenge for the industry, although the dual sensor technology may help in this domain.

Both cases are good illustrations that seismic imaging progresses from the simultaneous improvement of acquisition, data processing and computer sciences!

As a conclusion to this section we may list the current limitations of FWI as:

- The computational cost. FWI is starting to be achievable today for the acoustic case but is not yet viable for the 3D elastic case.
- Acquisition requirements: the need for very low frequencies and long offset.
- The accuracy with which we can model the subsurface (impact of conversion and elastic effects; reliable generation of internal multiples; modelling in complicated models with topography, etc.).
- Accounting for or preferably inverting anisotropic parameters.
- Still no 'convergence' towards an optimized workflow (which domain or combination of domains; what QC; ...).

The benefits are:

- To obtain a detailed velocity model of the subsurface, in theory (but not yet in practice) in any kind of complex environment. It is still too early to assess in which manner it competes with or complements high-density high-resolution tomography ...
- A Wave Equation-based technology, hence going beyond ray-based methods in complex geological settings;
- Still a lot of room for improvement, noting that the ultimate goal is to obtain reliable 'Elastic parameters logs' at every trace.
- An automatic process.

This last point being a benefit or not may be debated. Experience shows that interactions with interpreters are paramount to obtain a decent seismic image in complex cases. Whether FWI may be able to directly obtain 'the' best image is doubtful. If 'interactive beam migration' is here, 'interactive FWI' is still a long way away and the interaction of FWI with interpreters will have to be developed in the future.

8 Examples, Discussion and Conclusion

This chapter concludes the course with discussions on various topics of general concern, such as resolution issues related to the acquisition layout. Least Squares migrations and non-conventional imaging methods are addressed in this context.

Next, without being exhaustive, we present simple examples of geophysical points to bear in mind while running an imaging project. Based on basic ray-tracing we again draw attention to illumination below fast velocity layers, converted waves, fault shadowing or imaging with 'prism waves'. We also come back to anisotropy and dedicate a section to the estimation of anisotropy parameters before briefly addressing azimuthal anisotropy.

In a concluding section, after examining a last comparison of results obtained with four current imaging tools, we stress again the need to select the 'right algorithm', knowing that there is no 'universal' imaging tool that can solve all imaging problems.

The last word is no surprise: team work, including processors, computer specialists, interpreters, geologists and even geomechanics, is the key to successful imaging projects.

Some considerations on resolution of Migration and Full Waveform Inversion

In this first paragraph, we discuss the information content of the various results of imaging.
Let us recapitulate a few points made in previous chapters.

- Migration uses a smooth version of the *background anisotropic velocity field*, discretized on a fairly coarse grid with a sampling rate in all dimensions going from one hundred to a few hundred metres. This wave propagation velocity field allows us either to compute 'Green functions' (i.e., traveltimes and divergence-related factors) in ray-based methods or to numerically propagate waves in WE methods.
- Using this field and the recorded data, we deduce by the various migration methods a band-limited version of the *reflectivity* field in the subsurface computed on the finely sampled image grid (typically 10 to 20 metres in x and y and 5 metres in depth). The resolution of reflectivity is intrinsically limited by the seismic bandwidth (see fig. 7.02 and 7.10).

The upper left panel of fig. 8.01 (taken from fig. 4.14) shows the superposition of these two elements together with some rays used in ray-based techniques.

We can 'integrate' reflectivity to obtain 'pseudo-acoustic impedance' (recall that reflectivity is the derivative of the logarithmic of impedance) but the result is still in the initial seismic bandwidth.

Unfortunately, combining migration velocity and integrated reflectivity does not yield the impedance logs that would be measured in virtual wells, which we dream of drilled at every migrated trace location! This is for the following reasons:

- Migration provides us with the velocity background and not the acoustic impedance one: we have no direct access to density.
- The frequency band or equivalently the wavelength range of the integrated reflectivity field on the one hand and of the velocity field on the other, do not overlap. There is still a 'gap' between both, already noted by Jannane *et al.* (1989).

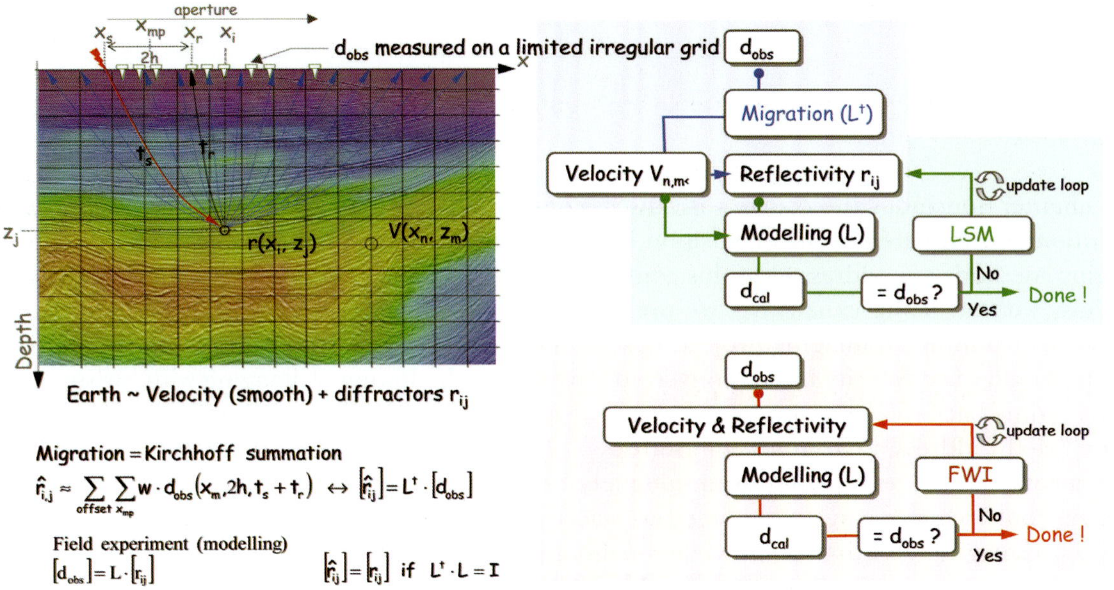

Figure 8.01: Schematic comparison between the workflows of migration, Least Squares Migration and FWI. In migration, we build the velocity model on a 'coarse' grid and get the reflectivity on the fine image grid. LSM complement migration by refining the reflectivity field to 'explain' through modelling the observed data in a least squares sense, partially compensating for irregular acquisition of data in the migration process. In FWI, we search directly a detailed field of elastic parameters that 'explains' the observed data.

As mentioned, in practice one tries to 'bridge this gap' in a post-processing stage using *stratigraphic inversion*, but for this we need extra information that comes mainly from borehole measurements, particularly calibrated sonic and density logs.

The frequency bandwidth of the reflectivity is limited on the high-frequency side mainly by non-elastic effects in the ground that reduce the signal below the noise and our ability to preserve and extract them from the noise in processing and imaging. It is constrained on the low-frequency side by our ability to send low-frequency energy into the ground and record it. All efforts are made to progress on both sides of the spectrum with constant and regular improvements.

The resolution of the velocity fields on the other hand, depends on several parameters:

- The method used to estimate it. As we have seen, techniques are improving regularly. For instance, high-density gridded tomography discussed in chapter 4 that uses volumic 'facets' picks yields velocity models that are considered as 'high-resolution'.
- The maximum offset acquired in the field. This is a well-known fact. However, long offset measurements often necessitate taking account of anisotropy and have by nature a lower frequency content (Sirgue and Pratt, 2004).
- The seismic bandwidth again: low-resolution data (lack of high frequency) do not help to accurately measure residual move-outs and the lack of low frequencies limit FWI in its capacity to estimate long wavelength components of the (initial) model (chapter 7).

Nevertheless, the ambition of Full Waveform Inversion remains to try and 'bridge the gap' directly. The parameters that we want to invert for are the elastic logs with their full bandwidth. We mentioned the difficulties in obtaining the very low trend. The short wavelengths in the elastic model are limited by the highest frequency present in the seismic data (at a given depth). FWI does not escape this limitation but it is hoped that the ultimate high frequencies effectively contribute to the result. Remember though, the computing costs of modelling high frequencies by numerical integration of the wave equation.

The amount of 'middle wavelengths' (the 'gap') that can be recovered through FWI depends on both sides of the seismic frequency bandwidth content and acquisition lay-out. We need as much *low frequencies* as high frequencies in the useful bandwidth and we need large offsets and possibly, a wide range of azimuths. How much each parameter brings to the 'gap-bridging' is still a matter of debate.

Migration, Least Squares Migration and Full Waveform Inversion

Other effects of the acquisition on the resolution of the imaging result are firstly, the limited aperture and secondly, irregularity in the acquisition system (obstacles, streamer feathering, etc.).

In fig. 8.01, we recall that the reflectivity in any point (x_i, z_j) of the image is computed as a *weighted summation* of the samples lying on the appropriate diffraction curve (fig. 4.02). This is explicitly the case in Kirchhoff migration but in effect holds for other migration techniques too. Solving complex linear systems in OW SP migration or RTM eventually comes down to obtaining a sample in the migrated image, being the result of a weighted sum of the appropriate samples in the data. The traveltimes and weights are implicit results of the various propagation exercises through the velocity model and scaled correlations.

Coming back to Kirchhoff summation, the shape of the diffraction surface, sum of the traveltimes from the shot t_s to the image-point and t_r to the receiver are computed by ray tracing. The weights take account of the 'divergence' of the energy, which is a long way from being 'spherical' in complex models. Theory, however, says that the summation is valid only for an infinite aperture and sufficiently dense sampling of the wavefield that is recorded at the surface. If these conditions are not honoured and they are not in real life, the result is perturbed by 'migration noise' in the shape of 'smiles' and limited lateral and vertical resolution. A synthetic example can be found in Wang and Sacchi (2009). Moreover, we do not obtain the correct reflection coefficient (poor amplitude preservation).

The computation of the weights {w} that go into the summation should include terms to try and compensate for it, but this is not an easy task. Other approaches, including '*Migration-Inversion*' (Beydoun W. and Mendes M., 1989) or 'True amplitude Migration' (Hanitzsch *et al.*, 1994) or '*Least Squares Migration*' (Nemeth *et al.*, 1999), were developed to address this issue.

Least squares migration or LSM is an expensive process. However, it has found new interest in recent years as it could be well adapted to RTM and to the processing of the 'simultaneous shooting' technology, yet another hot topic relating to acquisition and imaging.

LSM is again cast as an inverse problem. Its schematic workflow is presented on the right of fig. 8.01 in the light-green panel and compared with the schematic workflow of PreSDM and FWI.

The process starts after migration, which means that we have at our disposal a velocity field on the one hand and the corresponding reflectivity field resulting from migration on the other. The idea is thus to model the response of the reflectivity field through the model and to compare the result with the actual recorded data that were input to migration. Ideally, the result of modelling should be the same as the input data. Differences are symptomatic of problems in the migration that LSM will try and correct for.

With the notations used in chapter 7 for FWI, the recorded data are denoted d_{obs}, and the modelled ones d_{cal}. There are at least three reasons for the two data sets to be dissimilar:

- Firstly and to a limited extent, the velocity model. In LSM the velocity background model is not modified in the inversion process, contrary to tomography, DSO-MVA and FWI.
- Secondly, the irregular and limited acquisition lay out: to compensate for this is the main goal of LSM.
- Noise and any other propagation phenomenon (conversion and S waves, for instance) not accounted for in migration and modelling.

The forward problem consists in the computation of the modelled traces that would have been recorded given the velocity background V(x,z) and the current reflectivity $\{r_{ij}\}$. The problem is linearized in the sense that we again assume that each point in the image acts as a diffractor and that the modelled trace corresponding to a one source-receiver pair is the sum of the contribution of each individual diffracting r_{ij}. This linear approach to modelling can be written in a simple matrix form with the following usual notation:

$$[d_{obs}] = L \cdot [r_{ij}] \tag{8.01}$$

where $[d_{obs}]$ is the vector of all recorded data, $[r_{ij}]$ the vector of all reflectivity values at the grid of the migrated image and L is a linear operator in the form of a matrix including all 'propagation terms' (the Green's functions through the velocity model) and expressing that each element of $[d_{obs}]$ results from a weighted linear contribution of each r_{ij}. Note that we saw an example of numerical forward modelling expressed as a linear process with the solution of equation (7.04).

Kirchhoff summation that expresses the fact that the computed reflectivity in a point is the sum of the samples of the diffracted energy (see Kirchhoff summation equation in fig. 8.01), can in fact be written in a similar matrix form. Claerbout (1992) noted that the elements of the matrix operator are the same as the ones of the modelling operator (the Green's functions again computed through the migration model) and the migration operators can be written:

$$[r_{ij}] = L^t \cdot [d_{obs}] \tag{8.02}$$

where L^t is the matrix transpose of L.

As mentioned before, we would like modelling of the reflectivity to be equal to the input data d_{obs}. Substituting (8.02) into (8.01), this would be equivalent to saying that the matrix $L^t \cdot L$ should be equal to the unity matrix **I**. Unfortunately, this is not the case for the various reasons mentioned above and mainly because the acquisition lay-out is limited. We thus obtain an approximate reflectivity image $\hat{r}_{ij}$ out of migration instead of the 'true' reflectivity r_{ij}. The two are linked by the following expression :

$$[\hat{r}_{ij}] = L^t \cdot L \cdot [r_{ij}] \Rightarrow [r_{ij}] = (L^t \cdot L)^{-1} \cdot [\hat{r}_{ij}] \tag{8.03}$$

Applying the inverse of $(L^t \cdot L)$ to the migration result $\hat{r}_{ij}$ would transform it into the 'ideal one' r_{ij}. Nemeth *et al.* (1999) showed that the diagonal terms of this inverse bears the information of the irregularity of the acquisition system. This is to be related to the 'Beylkin determinant' often referred to in Kirchhoff migration literature for the compensation of irregular or sparse acquisition lay-outs. We also speak of an 'acquisition footprint' on the migration result.

L^t is known: it is the matrix expression of the migration process with the actual data acquisition lay-out. However, inverting $L^t \cdot L$ is not an easy task. The goal of LSM is precisely to do so in a least squares sense, with the objective of improving the quality of the estimated reflectivity field $\hat{r}_{ij}$ by compensating for acquisition scarcity.

The cost function of the LSM is the square of the differences between d_{obs} and d_{cal} for all samples of modelled traces at their actual acquisition location, plus possibly regularization terms. The reflectivity field $\{r_{ij}\}$ is updated at each iteration of the 'update loop'. The necessary gradient to do so can be computed with the appropriate version of the 'adjoint state method', discussed in chapter 7 for FWI. Different approaches to the optimization process, including steepest descent, conjugate gradients, etc., are reported in the literature. The convergence of the inversion is controlled by various ways of preconditioning (often smoothing the updates) the unknown reflectivity or adding terms in the cost function.

Because of its cost, LSM has not been extensively used so far. However, the demand for 'true amplitude' high-resolution images, particularly for reservoir delineation and monitoring, is steadily increasing. LSM is one way to address this issue. LSM is seen also as an appropriate approach to further reduce the cross-talk in encoded multi-shot imaging or even in imaging data resulting from simultaneous or 'blended' shooting.

Beyond conventional seismic imaging

This section borrows its title from Landa (2007) and the reader should refer to Landa's book for comprehensive discussions of the methods.

This paragraph is presented with two objectives in mind. Firstly, to recall that prestack migrations and full waveform inversion technology, although the main routes followed by the

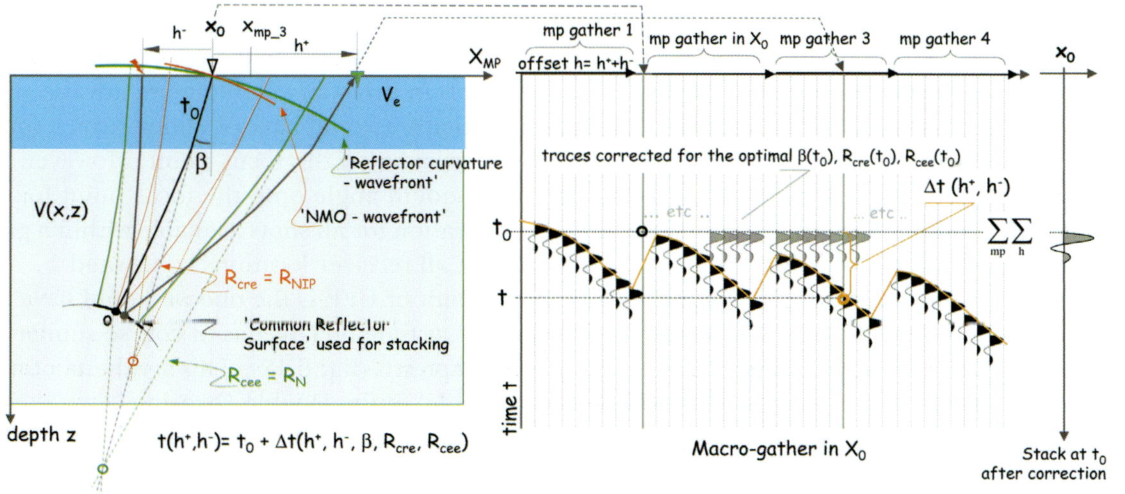

Figure 8.02: Normal Move Out travel times in CRS and multi-focusing analysis. We want to find the zero-offset trace in X_0, resulting from a stack of neighbouring traces presented here as a series of mid-point gathers. If we assume that the 'NMO-wavefront' (in green) and 'reflector's curvature wavefront' (in red) are locally circular, the moveout time t for a trace with a mid-point in the vicinity of X_0, has an analytical expression that depends on the velocity Ve at the surface, the shot and receiver offsets (h+, h-) with respect to X_0, the zero-offset emergence angle β in X_0 and the radius R_{NIP} and R_N of the two wavefronts. A three-parameter search based on this formula allows us to find the set of (β, R_{NIP} and R_N) that maximizes for each t_0 the coherency of events along the move-out curve (in orange) in all neighbouring traces. The end products are the field of (β, R_{NIP} and R_N) as a function of t_0 and the CRS or multi-focusing zero-offset stack trace in x_0. This is repeated of course for all X_0 locations.

industry, are not the only imaging methods offered by vendors. And secondly to recall, with a specific example, that data pre-processing or pre-conditioning is part of the success of an imaging project.

Alternative approaches to migration are often cast under the name of *'velocity independent imaging'* because they do not require building an explicit velocity model but trying to directly estimate operators' traveltimes.
Among these methods, we can list:

- Double Focusing and Common Focus Point or CFP (Berkhout, 1997).
- Diffraction Imaging was mentioned and briefly discussed at the end of chapter 4.
- Common Reflection Surface or CRS stack (Müller *et al.*, 1998).
- Multi-focusing or MF (Berkovitch *et al.*, 1994).

Double Focusing

In Kirchhoff migration, we 'focus' a diffraction by jointly removing receiver and shot propagation effects. Indeed again, for a given source-receiver pair, we search the sample that is at a diffraction time equal to the sum of the traveltime t_s from the shot to the image point, and the traveltime t_r back to the receiver. Weights take care of the effects of propagation on amplitudes and possibly, on irregular acquisition systems.

In *Double Focusing* (Berkhout, 1997), we separate the two effects. The image point is now called the 'focus point'. Take a shot record and imagine that we 'remove' from this record the shot-to-focus point propagation effects only. Putting amplitude issues aside, we note that in terms of traveltime, this means that we subtract from the recorded time the shot-to-focus point time $t_{s,}$ computed for instance by ray tracing through the current velocity model. In this way we obtain an 'advanced' (as opposed to 'delayed') record that emulates the recording that we would have obtained from a source buried at the focus (image) point. This new record is particular to the focus point. However, this buried source has a 'directivity response' related to the incident angle onto the focus point for this specific shot. Nothing prevents us from repeating this operation for all shots available within a given aperture and summing the 'advanced' records together at all receiver locations concerned.

This new 'macro-shot record', called Common Focus Point or CFP, is the one we would have had by triggering a multi-directional source at the focus point at time 0. This has of course similarities with the areal delayed sources that we encountered in chapter 6 (fig. 6.12). This synthesis of areal sources is the first focusing step, performed on the shot side, of the Double Focusing.

The point is that now the (time) response to this emulated buried source must be exactly equal to the operator that we used to synthesize it: we are dealing with the same traveltimes (now upwards) from the focus point to the receivers at the surface. Comparing, for instance in the tau-p domain as suggested by the author, the macro shot record with the operator, we can judge whether the underlying macro model is correct. If this is not the case, the idea is to analyse the time difference between them and change the operator and reiterate until convergence. In doing so, we update the migration operators without having to jointly change the underlying velocity model. Hence, the classification of Double Focusing in the 'velocity independent methods'.

Once we have the right operator, we can exchange the roles of shots and receivers and use this operator with exactly the same process as for shots to focus the macro-record back to the focus point and in this way obtain the reflectivity (averaged over angle) at the focusing (image) point. As pointed out by Berkhout (2007), the series of two focusing stages is equivalent to the Kirchhoff summation explained in chapter 4, apart from the intermediate step of operator's update, which occurs without gather flattening and explicit velocity model building. The price to pay, however, is that the image is at an approximate location in the 'initial' velocity model. Some kind of tomographic inversion is necessary afterwards to transform all operators, which are incidentally one-way Green's functions and exist for every focus point, into a depth model in which we want to accurately relocate reflectivities in the subsurface.

Another point worth noting is that it makes sense to focus the macro-gather in the second focusing step at focusing points in the vicinity of the 'true' one. The author shows how this gives access to reflectivity as a function of scattering angle, in a way very much related to the way SO-CIGs yield the same information (fig. 7.04).

The technique has not been offered as a service by commercial vendors. However, we mention it since it has been recently proposed again (Liu *et al.*, 2006).

CRS stack and Multi-focusing

We discuss now the last two methods listed above, i.e., CRS stack and Multi-Focusing. Both make use in fact of similar concepts that extend the Normal Move-Out velocity analysis, which was based firstly on a horizontally layered model and secondly, on estimation of traveltimes by 'shifted hyperbolas' (fig. 3.09).

CRS stack and Multi-Focusing deal with the more general case of a curved reflector and traces taken from a few midpoint gathers around a 'central point' at a location denoted x_0, where we perform the analysis. This 'extended' model is presented on the left of fig. 8.02.

The goal is to find an equation controlled by a limited number of parameters that relates the zero-offset two-way traveltime t_0 to the time $t = t_0 + \Delta t$ of any non-zero-offset reflection on the same reflector (but not necessarily at the same position) recorded by a neighbouring trace. The reflector is supposed to be a curved surface to be more general and realistic.

The neighbouring traces that will be concerned by one 'central point' can be organized as 'macro-gathers', which are the juxtaposition of a series of mid-point gathers, as the one sketched on the right of fig. 8.02.

If we have this equation, then we can imagine a process to estimate this set of parameters from the collection of traces in the 'macro-gather', using coherency or semblance along the selected traces in a scanning way, very similar to the nmo velocity analysis discussed in chapter 3. Once the set of parameters is estimated, we can correct all traveltimes accordingly (grey reflections in the figure) and stack them together. As in 'conventional' nmo stack, the result is the simulated 'zero-offset' trace at location x_0. Fig. 8.02 considers only one reflector for the sake of clarity but of course, the analysis is run for all t_0, hence treating any reflectors in the subsurface.

The moveout equation to be used must express the time delay Δt of the reflection between a trace located by its source-receiver coordinates with respect to x_0 (offsets h^+ and h^- in the figure) and the zero-offset traveltime t_0 in x_0: $t(h^+, h^-) = t_0 + \Delta t(h^+, h^-)$.

The expression of Δt differs between CRS and MF. The reader will find a discussion of the merits and limitations of each approach in Landa (2007). As for shifted hyperbolas, both methods consider time delays Δt measured in the near-surface layer of velocity V_e. They also use the same key parameters specific to the method, namely:

- The emergence angle β of the zero-offset ray in x_0.
- The radius of the fictitious wavefront that would be produced by a source at the reflection point marked as "O" in the figure and as it arrives at the surface in x_0. This radius is generally denoted R_{NIP} or R_{cre} according to the authors. We label this wavefront 'V_{nmo}' in the figure, since it is related to the 'old' nmo velocity around the normal ray (Robein, 2003).

- The radius of the fictitious wavefront that would be produced by a series of sources shot simultaneously on the reflector itself. This is Claerbout's exploding reflector concept. This radius is denoted R_N or R_{cee}. We call the corresponding wavefront 'reflector's curvature wavefront', as it bears information on the reflectors' shape (curvature). R_N would be infinite for a flat reflector in a constant velocity medium.

The moveout equations are of the 'double square root' type, i.e., with a Δ_s term for the source ray and a Δ_r term for the receiver ray. This opens the possibility to vary V_e in the land data case, if needed.

The equations are fairly complicated; interested readers can find them, for instance, in Müller *et al.* (1998) for CRS and Landa (2007) or Berkovitch *et al.* (2008) for MF. They make the assumption that the two fictitious wave fronts are circular and the same for all traces processed in a macro gather. This is the theoretical limit to the maximum offset to be used.

As mentioned, there are similarities between CRS and MF on the one hand and the 'old' nmo on the other, but major differences as well. We can list some of them:

- In 2D, we need to estimate at each time t_0 for each macro-gather three parameters instead of one for nmo or two in the case of 'bispectral search' (fig. 3.17). The theoretical number of parameters is even eight in 3D. The multi-parameter search is consequently much more tedious and crucial for the success of the method. Shortcuts are necessary to limit the dimensionality of the search in 3D.
- The search must be automatic and dense. The 3D volume of the sets of parameters thus obtained can be quality controlled and edited if needed.
- Compared to the conventional nmo stack, the stack trace is now the sum of a much higher number of traces (an order of magnitude). This results in a possibly dramatic improvement in the signal-to-noise ratio. This key characteristic is illustrated with the example in fig. 8.03.
- The counterpart is that each stack trace 'averages' the reflection over a reflection surface instead of (theoretically) a point of the reflector. Hence the name Common Reflection Surface! This can be viewed as a limitation (loss of lateral resolution) but it is mitigated if we ensure that the 'mixing' does not occur over more than a Fresnel zone. The size of the reflection area in the subsurface for one given trace is directly related to the size of the macro-gather (or macro-bin in 3D).

The improvement in the signal-to-noise ratio obtained by this optimal *reflector-consistent coherency enhancement* designates these methods for application in difficult areas, where the signal is poor or the acquisition is sparse, such as vintage data or data acquired in difficult access environments.

CRS and MF produce unmigrated zero-offset data images. They then need to be migrated. Post-stack migration, whether in time or in depth, may be sufficient if the image has been already optimized. We need a migration velocity model however, and this may be an issue. Two routes are available:

- The dense parameter estimates bear information on velocity in the subsurface. For instance, we already mentioned that the nmo wavefront bears information on the velocity along the normal ray of angle of emergence β:

  $$V_{rms}^2 = 2 \cdot R_{cre} \cdot V_e / t_0 \quad (V_{rms} = V_{nmo} \text{ for limited apertures}).$$

  This information can be used as input to either poststack time migration velocity model building (Robein, 2003) or ray-based tomographic depth model building.
- More recently, vendors offer the reconstruction of non-zero offset pre-stack data that benefit from the data 'cleaning' inherent to the method. Fig. 8.04 gives an example of this kind of MF

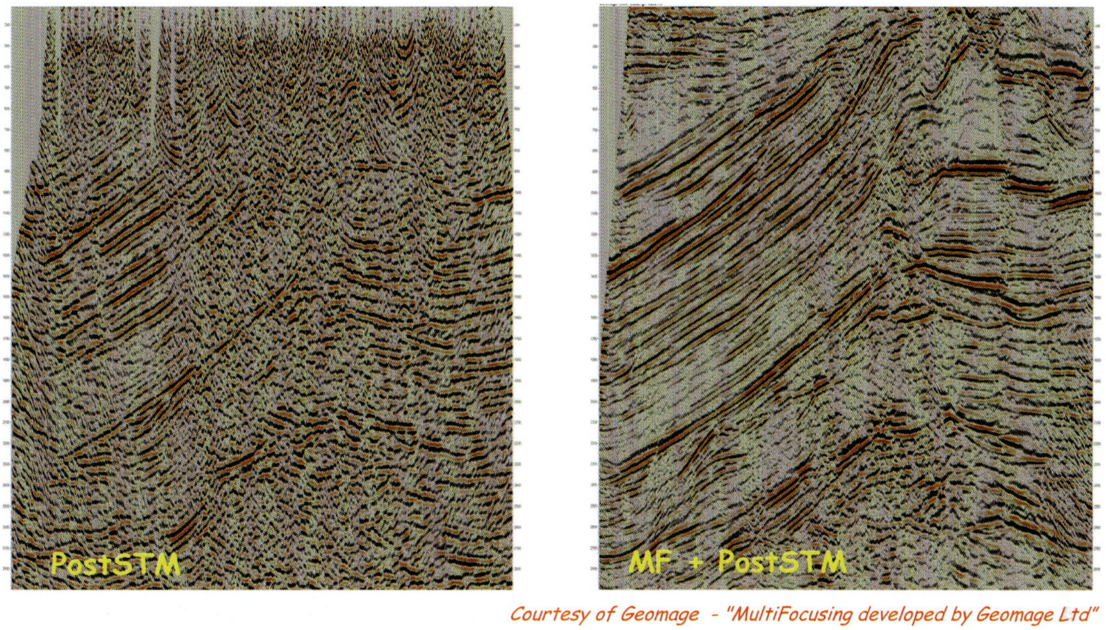

Figure 8.03: Example of Multi-Focusing processing on low fold land data. The fold is variable between 3 and 24. Both sections underwent post-stack time migration. The enhancement in Signal to Noise is dramatic in the Multi-Focusing processing.

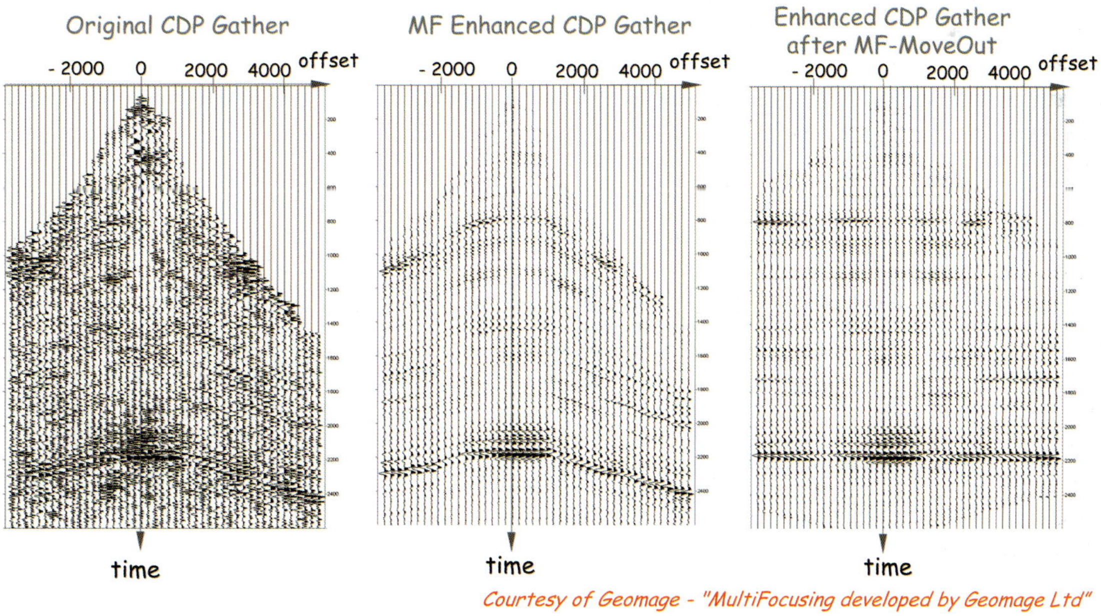

Figure 8.04: Example of reconstruction of a gather after Multi-Focusing. MF techniques allow us to construct not only zero-offset traces but also offset ones, which benefit also from the improvement in signal-to-noise ratio. These 'cleaned-up' gathers can be input to usual Pre-stack migration methods.

output. The idea is then to use these 'pre-conditioned' data as input to 'conventional' prestack migration imaging and velocity analysis routes, as described in this book. This approach remains to be evaluated.

Ray tracing as a tool for 'geophysical interpretation' while imaging

In this section, we give a few examples to illustrate how '*thinking propagation*', sometimes with the help of ray tracing, can assist the processor and the seismic interpreter with their imaging projects, specifically in complex areas.

The first example deals with the presence of *strong velocity contrasts*, such as between salt (4500 m/s) and clastic sediments (2000 to 3500 m/s). The point we make here is that there are cases where a reflector can simply not be illuminated by any acquisition system. This is illustrated on the right of the cartoon in fig. 8.05 (see also Biondi, 2006, p.138). Imagine a dipping reflector in sediments with a velocity of 2500 m/s below a salt layer with a velocity of 4500 m/s. The acoustic waves sent from the surface will pass through the salt, reflect on the reflector and hit the base of salt on their way up to the surface with an angle that is equal to or greater than the dip of the reflector. If this angle is greater than the critical angle at the sediment-salt boundary, then no acoustic energy can travel back to the surface: the reflector is simply not visible! A table on the right of the figure gives values of this critical angle as a function of the velocities above and below the refracting interface. In our cartoon case and with the velocities given above, the maximum dip angle that we could image below the flat salt is about 35°.

Note that for this specific simple example, wide azimuth shooting would not solve the problem. However, in more realistic cases, with complex salt bodies for instance, there might be a greater chance that a given acquisition lay out makes it possible to obtain 'windows' of azimuth-offset configurations that illuminate all expected reflectors. This has been the success of the WATS (Wide Azimuth Towed Streamer) technique in recent years.

However, in complex geological settings such as salt basins, a simple ray tracing exercise may not help. On the contrary, the survey must be properly designed to find the best compromise between cost and probability of an optimal illumination of targets. This is done more and more through

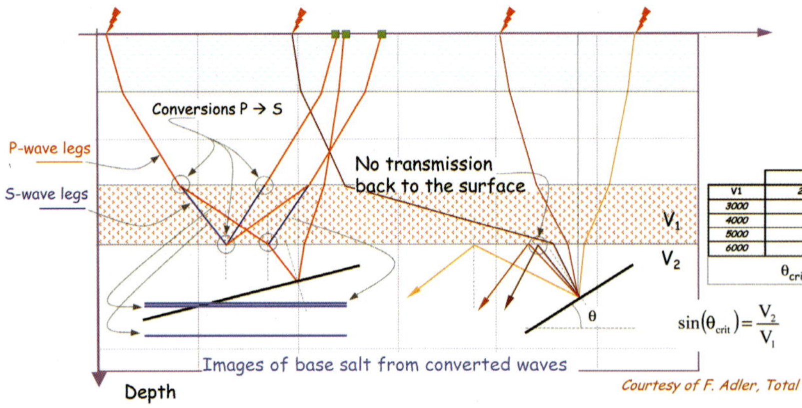

V1	V2 2000	2500	3000	3500	4000
3000	42°	56°	90°		
4000	30°	39°	49°	61°	90°
5000	24°	30°	37°	44°	53°
6000	19°	25°	30°	36°	42°

θ_{crit} as a function of V_1 and V_2

$$\sin(\theta_{crit}) = \frac{V_2}{V_1}$$

Courtesy of F. Adler, Total

Figure 8.05: Some issues related to strong velocity contrasts. Converted waves may have a high amplitude. However, in acoustic migration, they are migrated with the P-wave velocity and appears as artefacts often visible at the base of salt bodies. On the right, steep dips below high velocity layers may simply not be visible due to total refraction of the reflected waves at the base of the fast layer.

intensive 3D acoustic Wave Equation-based modelling. Survey design is inherently linked with imaging. It is not treated in the framework of this course but its importance is stressed here with the simple modelling exercise in fig. 8.05.

The second example is sketched on the left of the figure. Here, we take a reflector's dip smaller than the critical angle. The reflector can hence be 'seen' (we should say 'heard'!) from the surface. However, note the effect of refraction at the salt bottom interface: the *maximum scattering angle is very limited* even for a very large acquisition offset at the surface. The fact that the maximum scattering angle decreases due to refraction as velocity increases with depth was already noted on the right-hand side panels in fig. 4.22. In the salt case, the 'jump' towards limited angles of the curve representing the maximum scattering angle may be dramatic at the salt bottom.

Winbow *et al.* (2007) and Winbow (2009) illustrate with examples the potential benefit of Common-Reflection Angle Kirchhoff migration (see chapter 4) to address the issue of limited or variable scattering angles (together with multipathing issues) below or in the vicinity of complex salt bodies.

Another consequence worth noting of a limited scattering angle below fast-to-slow velocity contrast is that Amplitude versus Angle analysis becomes illusive: recall that it is the angle that is the physical parameter of importance and not the offset.

In fig. 4.22, the *angle-offset relationship* is ray-based but it is not the case in the WE-based computation of SA-CIGs from SO-CIGs and this presents a difficulty. The phenomenon is clearly visible on the left-hand panel of fig. 7.09: there seems to be information for large scattering angles at depth, although we know from simple ray tracing that this cannot be the case. It can only be tau-p artefacts or noise in Scattering Angle gathers. We would like to get rid of it by the application of some kind of 'mute'. Unfortunately, there is no evidence on where to place the limit between genuine information and noise ... unless we go back to rays. This is definitely an issue for WE-derived Scattering Angle CIGs.

A third point is made in fig. 8.05. Large P-wave velocity contrasts are prone to the generation of converted S-waves. These converted waves are treated as P-waves by all conventional prestack migration tools. S-wave velocities are slower, hence the converted waves travel paths are longer. Depending on the circumstances and the geometrical configuration of the fast layer, their poorly migrated but still visible image may interfere with or simply replace weak or non illuminated dipping reflectors, hence misleading the uninformed interpreter. It is classical to see two types of artefacts:

- The migrated image of the base of the fast layer corresponding to a conversion P to S at its top, reflection S to S at its bottom and conversion again S to P at the top on the way back to the surface. Both back and forth travels in the salt layer is done at the slow S-wave velocity: its image appears below the true base salt location.
- The migrated image of the base of the salt corresponds to only one way in the salt as a converted wave. There are two solutions for it corresponding to conversion at the top of salt either on the way down from the source or on the way up to the surface, as people processing converted waves are aware. The image is then doubled in theory and lying between the true base salt and the previously mentioned artefacts.

Artefacts induced by converted waves of course also appear in situations such as the one sketched on the right of fig. 8.05, in which case the interpreter may simply interpret erroneously converted waves instead of the true reflectors that are not visible!

Again, to have a workstation equipped with a ray tracing facility can be of great help to 'think in terms of propagation', including conversion! Ray-based migration tools have by definition ray tracing tools but unfortunately, not necessarily with S- or converted-waves options.

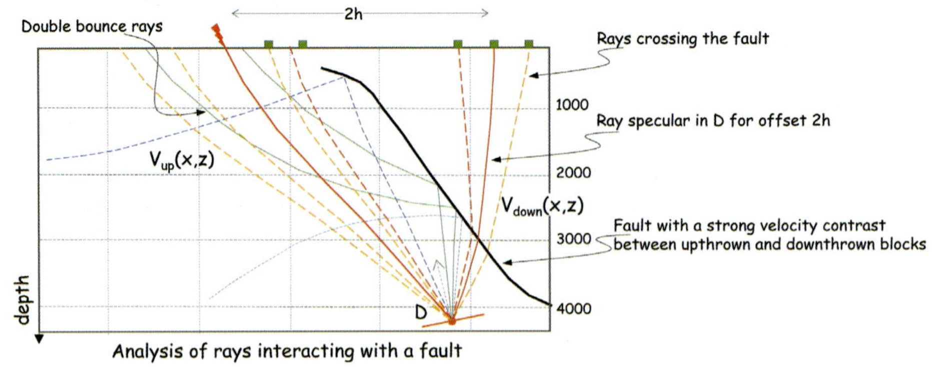

Figure 8.06: Ray path discrimination. Beam migration allows the analysis and discrimination of events according to their ray paths. Here a reflection in the 'shadow zone' of the fault . Rays that interact with it may or may not be selected and included in the image and in the CIGs depending on tomography strategy.

In fig. 8.06, we address the case of 'fault shadowing', which happens when faults separate upthrown and downthrown compartments that may have very contrasting velocities. In the cartoon in fig. 8.06, the model is inspired from a real example where the fault may put clastics layers having a velocity of around 2500 in contact with carbonate layers with a velocity up to 4000 m/s. Ray tracing is performed assuming such a high-velocity contrast.

Recall that in Beam migrations, each event picked in the tau-p or diplet domain has its corresponding pair of rays and 'properties vector' (fig. 5.15), which includes the location of the migrated event and the local dip. We can add to it a criterion that flags traces that have one propagation leg crossing the fault. As an example, a point such as 'D' in fig. 8.06 would be flagged. This allows us to build two sets of images and CIGs: one including all events for which one propagation leg crossed the fault and the one with events that have not crossed it.

The amplitude treatments or noise removal can now be different for each set. We can also decide to include or not the respective events in the CIGs in the tomographic process. To include the ones that have crossed the fault necessitates accurately picking it and introducing it explicitly into the velocity model. Inverting velocity with events that have not gone through the fault will be more stable but the velocity model must be extrapolated into the 'blind' fault shadow, since there is no more direct estimate of the velocity there.

Imaging with prism waves

The last example addresses the possibility offered by 'two-way propagators' to image prism waves provided two conditions are met: some careful data pre-processing and the inclusion of the reflector that generates the prism wave in the model.

A schematic example of prism wave (also called 'ternary' or 'double bounce' reflections) was already presented in fig. 6.26 as a potential benefit of RTM. The possibility to use them for imaging salt walls was first reported in the literature by Bernitsas *et al.* (1997) and more examples shown subsequently (Cavalca and Lailly, 2005; Farmer *et al.*, 2006). As mentioned, two-way propagators are needed because the second bounce happens in one of the two legs of the reflection path. This is illustrated in fig. 8.07, where we show, as in fig. 6.20, the situation where the downgoing field in two-way RTM is not only transmitted but also reflected at a sharp velocity contrast. When the reflected part of the downward wavefield correlates with the backward propagated upgoing wavefield, it may produce either RTM noise as discussed in chapter 6 (fig. 6.20 reproduced on left panel in fig. 8.07) or signal from a prism reflection, if there happens to be a reflector at the image location (as in 'C' in the right-hand panel).

It is important to note that the reflection's generator needs to be included in the model in order to emulate this reflected part of the downgoing wavefield. This will happen only if there is a sharp contrast in velocity at the reflector's location (limited smoothing of the model). The reflection could also be generated artificially as a density contrast, provided the propagator in RTM uses the appropriate wave equation and the model description includes this facility.

In contrast, the 'prism imaged' reflector does not need to be in the model. The information is included in the recorded upgoing wavefield provided it has been preserved at the pre-processing stage and the reflection is generated 'automatically' by the imaging condition in 'C'.

This point is well illustrated with the example in the right-hand panel of fig. 8.08 taken from Jones (2008). The RTM was run with no salt body in the velocity model. The result nevertheless exhibits the image of an overhanging salt wall, which actually results from the prism waves. On the other hand, the strong velocity contrast at the 'top Balder' is sufficient to generate the necessary reflected downgoing wavefield that would correspond to the red ray-leg on the cartoon in fig. 8.07.

The main point in Jones (2008) is the impact of pre-processing. The author shows how a ray tracing exercise through the current velocity model (left-hand panel in fig. 8.08) allows the processor to detect evidence of various propagation modes in the raw data set, including prism waves. This analysis demonstrates that these reflections have an unusual behaviour that can unfortunately be severely attenuated during traditional pre-processing and particularly by classical multiple removal. If this process is applied, the costly RTM cannot image the salt flank via prism waves of course and loses part of its benefits.

A salutary warning again that any link in the continuous imaging chain from survey design to interpretation has its importance!

As a last comment on this topic, we note that ray-based migrations are in theory able to image prism waves. However, this is at the cost of introducing all reflectors in the model and adapting ray-tracing in order for the processor to be able to 'tell the rays' which interfaces they have to interact with (ray-signature).

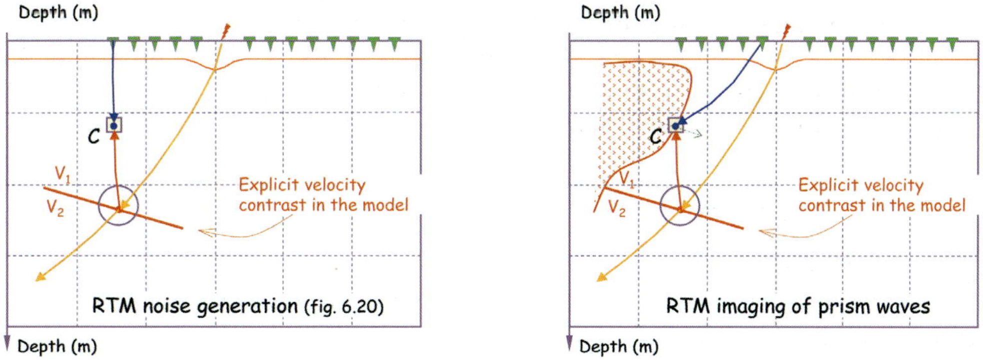

Figure 8.07: Migration of 'prism waves' by RTM. The two-way propagator used in RTM leads to time coincidence between the reflected downgoing wavefield (red) and the backward-propagated upgoing wavefield (blue). It is RTM noise when it does not correspond to a real reflection, as in fig. 6.20. It is useful signal otherwise and specially for prism waves, provided the reflection generator is included in the velocity model.

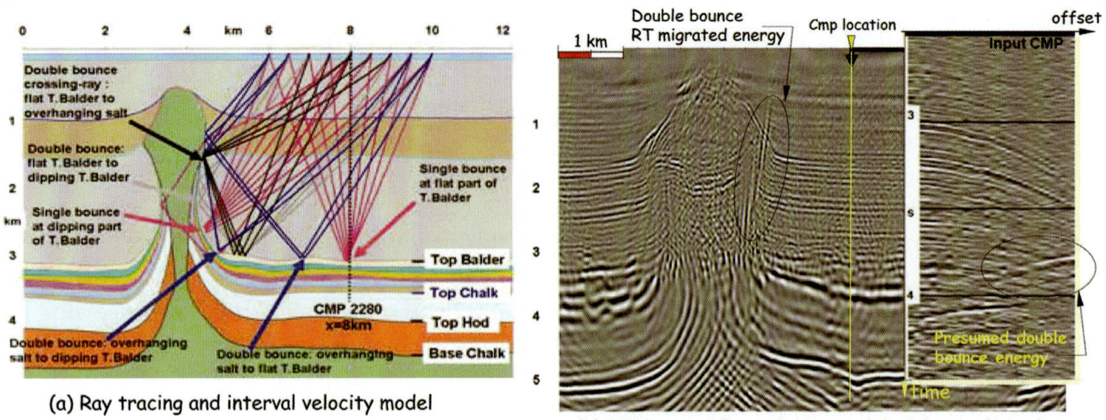

(a) Ray tracing and interval velocity model

(b) RTM run with a velocity model with no salt body

Example courtesy of I. Jones, ION GX Technology

Figure 8.08: Example of double bounce (prism wave) RT migrated energy (in Jones, I., 2008, with authorization). The ray tracing exercise facilitates the detection of various propagation modes in the data, and particularly of the double bounce. It is paramount that this energy is not attenuated by conventional multiple removal tools. The top Balder is a sharp velocity contrasts and generates reflected wavefields, but the velocity model yielding the migration result on the right does NOT include the salt body.

Anisotropy model: VTI or STI?

We discussed anisotropic Pre-stack time migrations in chapter 3. In the present section, we come back to the use of velocity anisotropy on seismic imaging but more specifically, in depth imaging.

The importance of accounting for anisotropy in depth imaging, including anisotropy with a tilted axis of symmetry or TTI, was documented already by Kitchenside (1992) and confirmed on real data by Ball (1995). It has since been widely recognized that inclusion of anisotropy in the migration velocity model was the only way in many cases to reconcile depths of reflectors measured in wells and depths of the same reflectors appearing on depth migrated images (sometimes called 'seismic depth'). The impact of anisotropy on accurate lateral positioning of dipping events and particularly faults, has been rapidly considered as another key argument for it.

Early papers on anisotropic PreSDM dealt with Vertical Transverse Isotropy or VTI. Fig. 8.09 presents such an example taken from Hawkins *et al.* (2001). Besides more sensible depth ties with wells, the Anisotropic PreSDM (or APSDM) seismics show considerable lateral shifting of bounding faults. In this comparison, the isotropic image had been produced from what was standard isotropic velocity updating, while the anisotropic image resulted primarily from the use of a layer consistent, well based $V_0 + k \cdot Z$ (chapter 1; fig. 1.09) velocity model and anisotropy parameters derived at other well locations.

Thanks to anisotropy, mis-ties up to a few hundreds metres in the North Sea could be explained and corrected. A better well tie also gives more confidence in the accuracy of depths away from the well(s). As an example, Suiter *et al.* (2004) illustrate how including anisotropy in the Kirchhoff migration velocity field not only improved the image, but also significantly modified the estimated rock-volume and hence reserves of a gas field, while reducing the range of uncertainty of these estimates.

We do not discuss uncertainty analysis in this book but the reader should realize that '*stochastic imaging*' is a topic of great importance for the future. Not only do we want to have the clearest

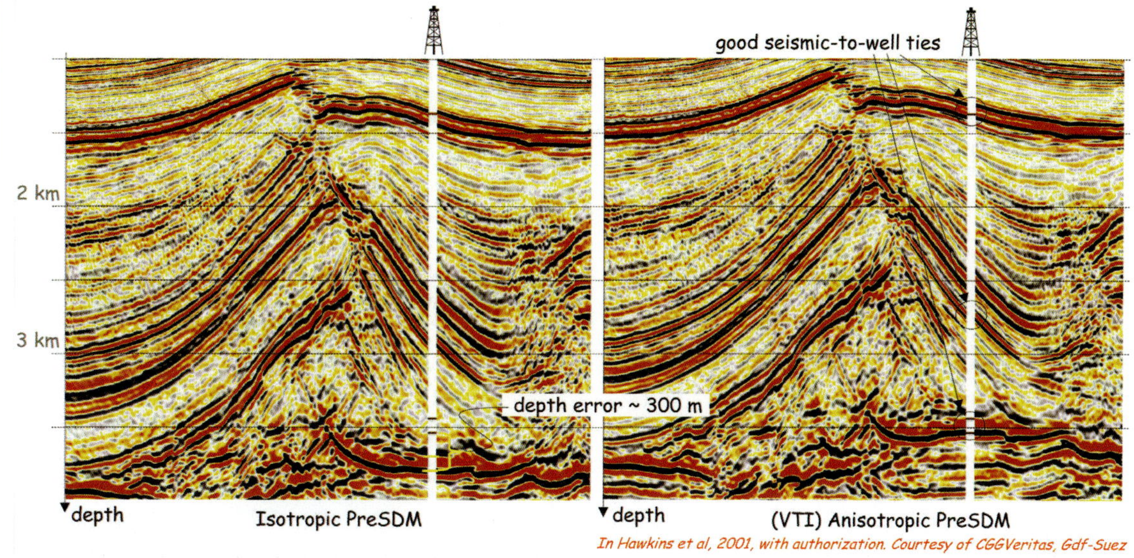

Figure 8.09: A comparison between 3D isotropic (left) and anisotropic (right) Kirchhoff migration in the North Sea. In this early example, VTI Anisotropy is assumed and the anisotropic parameters are estimated using borehole data and well control of depths. In this kind of environment, further improvement is reached when assuming TTI (Hawkins *et al*, 2002).

image possible, but also the most accurate one in terms of localizing reflections in the subsurface. Moreover, being able to place some kind of error-bars on these images would be a step forward and anisotropy will play a key role in this uncertainty analysis.

If VTI was a first step, it was recognized rapidly (Vestrum *et al.*, 1999; Hawkins *et al.*, 2002) that it did not make much geological sense in complex tectonics environments, as already illustrated in a caricatured way with fig. 1.10. In practice, we must consider tilted axis of symmetry or TTI. This brings, however, a further parameter that we must define. Geological intuition pushes us to assume that the axis of symmetry of anisotropy should be more or less normal to the bedding, as suggested in fig.1.10. The corresponding model is often referred to as 'Structural Transverse Isotropy' or STI. It has since become a standard in anisotropic PreSDM.

In practice, STI necessitates the picking of dips. It is interesting to note that in tomography based on dense picks (such as the one presented in fig. 4.11, 4.12), the dip information is readily available, facilitating the implementation of STI options in velocity model building. Note that in TTI, the reference velocity is no longer the vertical velocity V_{vert}, but the velocity along the axis of symmetry V_{axis}.

Dehghan and Audebert (2008) discuss the interference between the definition of the axis of symmetry and the estimation of the corresponding velocity in tomography. They use in particular a synthetic model, reproduced in fig. 8.10, which includes an anticline-syncline-shaped STI layer. The velocity along the variable axis of symmetry is constant.

A flat horizon below the STI layer is used to compute the CIGs that are input into a tomography package. With no surprise, isotropic tomography yields velocities that try to compensate anisotropy by increasing velocity and hence depth. The phenomenon is further exaggerated in the syncline. The VTI assumption, with constrained values of ε and δ close to the true ones, gives 'absolute' values of V_{vert} velocities closer to the true ones but still tends to artificially increase them in the syncline and reduce them in the anticline, thus still exaggerating the syncline structure.

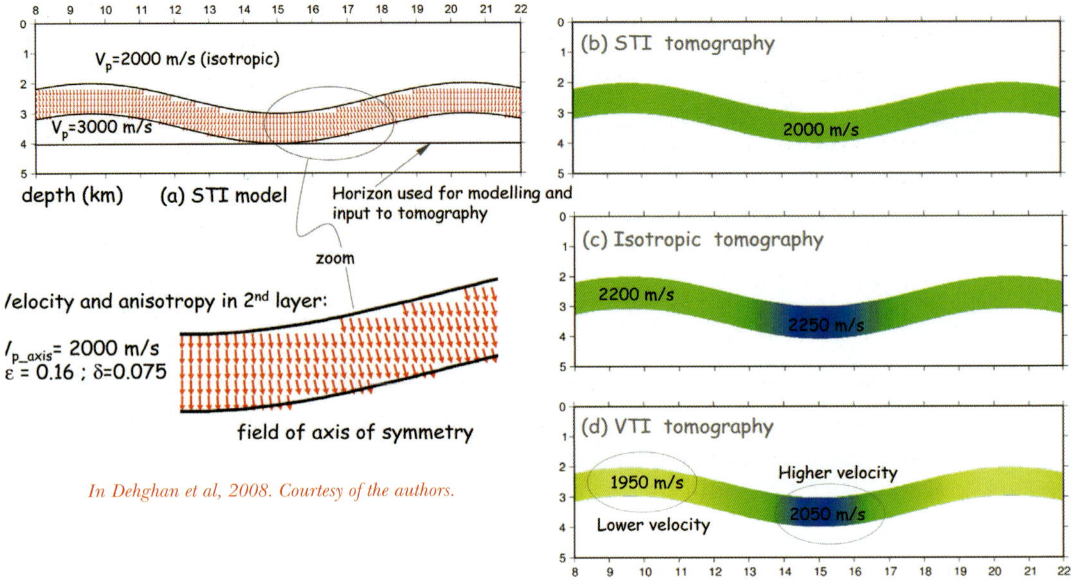

In Dehghan et al, 2008. Courtesy of the authors.

Figure 8.10: Impact of the axis of symmetry on tomography: a synthetic example. The model is made of one homogeneous isotropic layer over one homogeneous STI layer (V_p along the axis of symmetry is constant, but the axis of symmetry follows the structure). STI tomography with the correct axis of symmetry field (b) allows to recover V_{p_axis}. Isotropic tomography (c) oversetimates the velocity. VTI tomography (d) underestimates it on the anticline and overestimates it in the syncline (note that the the V_p colour code is different for the 3 figures).

Without drawing too general conclusions from this synthetic simple exercise, but based on more real data cases, we can deduce as a rule of thumb, firstly that the axis of symmetry is a sensitive parameter and secondly, that 'the best anisotropy option' is the one that produces the 'simplest' corresponding velocity field.

Fig. 8.11 is a real case comparison. In the left-hand side panel, we show the result of PreSDM after VTI tomography and on the right hand side, the same line of this 3D project but now after STI tomography (Dehghan *et al.*, 2008). The improvement in the quality of the image is clear: the STI assumption is probably more 'geologically sound'; it leads to a smoother velocity field and flatter CIGs. To assess the impact of the velocity model on the geological structure, we picked some horizon and faults (in yellow) on the VTI migration and superimposed them in dotted lines on the STI result on the right. The changes in the shape of the structure and the displacement of the fault are noticeable. The STI migration result is confirmed by an improved well-to-seismics tie (not shown).

Anisotropy parameters: how to obtain ε and δ?

We have dealt so far with the axis of symmetry. This is only part of the game: we need the anisotropic parameters too. As mentioned in chapter 4, Thomsen's parameters are used in all projects to characterize anisotropy. The most common practice so far was to have one pair of (ε, δ) per layer and to invert these values in tomographic inversion together with velocities.

With the generalization of high-resolution tomography, we can now envisage to estimate fairly reliably by tomography a field of (ε, δ) pairs discretized over a regular (sparse?) grid, in conjunction with a predefined field of axis of symmetry.

However, anisotropy parameters are poorly constrained by seismic data only. Their inversion consequently requires imposing 'external' constraints. These may have various origins, including:

- General knowledge in a basin.
- Analysis of borehole logs and rock mechanics measurements on cores.
- Well-to-seismics tie while imaging.
- Borehole seismics (multi-offset VSP; walk-above).
- The η parameter derived from previous anisotropic PreSTM projects.

It is generally accepted that shales are an important cause of anisotropy at the seismic scale. However, it is not the only one. Anisotropy in chalk, for instance, is also reported (Hawkins *et al.*, 2002) and should be considered.

Repeated anisotropic migration projects in various zones of a basin helps us to build our general knowledge of a given area. This is an experience that is consolidated in the long term. Geostatistical analysis of the results can give information not only on the absolute value range of anisotropic parameters, but also on their degree of lateral variability, which remains an issue in many instances. As a rule of thumb, we can say that ε and δ depend on lithology and type of shale; they follow a reasonably well-defined compaction trend with depth; ε is generally larger than δ, which is equivalent to saying that η, related to $\varepsilon - \delta$, is generally positive.

Rock-physics measurements on cores are interesting as general pieces of information but difficult to 'upscale' by several orders of magnitude, as is necessary from the plug or core size to the scale appropriate for seismic models.

Attempts have been made to measure ε and δ in the field, such as the P-wave refraction survey presented in Leslie and Lawton (1999). The authors report ε values of around 0.14 on one site and 0.25 on another and δ values close to 0 on both sites. They attribute the variations in ε to "variations in the shale facies between the two sites".

Similar values for ε but unfortunately, also similar ranges of variability, are found in many depth imaging projects, which is satisfactory but not really helpful when starting a project in an unknown basin.

As mentioned earlier, $\delta = 0$ is a possibility but unusual in seismic depth imaging projects.

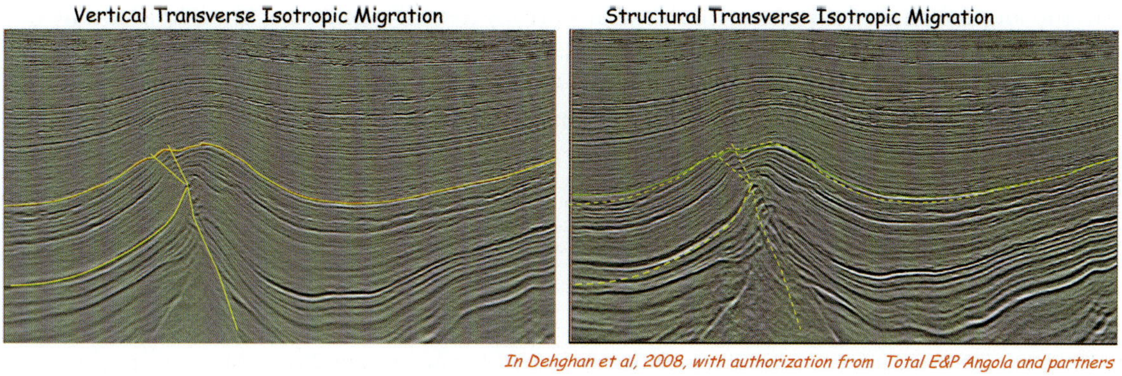

Figure 8.11 : Comparison between VTI and STI anisotropy. In complex geological settings, Structural Transverse Isotropy generally improves the image and the seismic-to-well tie, resulting as well in a more accurate lateral positioning of dipping events such as fault planes. Horizons picked on the VTI result are superimposed onto the STI one to emphasize the differences.

Information derived from *borehole seismics* is local but generally reliable. We discussed in chapter 4 (fig. 4.16) the use of well information to constrain tomographic inversion with the goal of ending up with a seismic image in depth that properly ties well(s).

Let us remind ourselves that forcing markers in wells to match seismic events mainly controls the accuracy with which we can estimate the δ_{int} in the various layers. This comes from the difference between V_{vert}, which controls the vertical time-to-depth relationship and V_{nmo} (equation in fig. 3.10), which includes heterogeneity (V_{rms}) and δ. The comment is true however for markers close to being horizontal. Tying steeply dipping horizons at the wells improves the accuracy of the inversion of both ε and δ, provided that we use the right axis of symmetry!

With the left-hand panel in fig. 8.12 we come back to some issues related to the use of borehole measurements. Firstly, multi-offset VSP or walk-aways can help to constrain the anisotropy parameters around the wells. This was already mentioned by Ball (1995) or Hawkins *et al.* (2002). *Walk-away surveys* provide us with actual traveltimes from a series of points in depth in the well to the surface. Note that this is a direct measurement of Green's functions for the pixels at the receivers' locations (such as the ones marked as 'A' and 'B' in fig. 8.12). The way it may be used in practice is to run a traveltime tomography inversion that consists in the estimation of the layered or gridded model in terms of V_{vert} (or V_{axis}), ε and δ that best explains the measured traveltimes. In contrast to the tomography discussed in chapter 4, there are many more unknowns than measurements. The inversion problem is thus highly under-determined but at least the model resulting from surface data tomography only must be kinematically compatible with the measured traveltimes from the walk-away.

In the case of deviated wells, *walk-above surveys* are shot to estimate the time-depth relationship. Recall however the classical pitfall: the first arrival at the geophone (for instance, point 'C' in the figure) may not correspond to a vertical ray path in the application of Fermat's principle! Interval velocity may still be computed using two close receiver locations. However, note the geometry in the figure: we measure the interval velocity as $\Delta l/\Delta t$ along the actual ray path and not along the vertical. The local ray path direction can be estimated by polarization analysis if a multi-component recording was performed at the receiver in 'C'.

Coming back to anisotropy, the velocity measured in this way is the phase velocity corresponding to the angle β taken with reference to the anisotropy axis of symmetry. As we said, it may not be vertical in the presence of structural dip.

To complete the discussion of this small example, we note that a sonic tool would measure velocities along the well trajectory, consequently yielding a phase velocity but now for the specific angle α taken between the well deviation and the local axis of symmetry, assumed again to be normal to the local structural dip.

With the measurement of phase velocity for two angles and the estimate of the axis of symmetry, we are in principle able to deduce the two local Thomsen's parameters. Needless to say, this is very theoretical and seldom done!

Lastly, some information on anisotropy is provided with anisotropic PreSTM projects if they have been run before or concurrently with the depth imaging project. The right-hand panel in fig. 8.12 is a reminder of fig. 3.24-a. Ideally, PreSTM velocity analysis provides us with measurements of $V_{nmo\text{-}eff}$ and η_{eff} as a function of vertical time at the velocity analysis location. These 'effective' values , i.e., integrating heterogeneity and anisotropy from the surface down to the depth of measurement, can be transformed as we saw, into interval values of V_{nmo_int} and η_{int}. The latter is a first approximation of ε-δ (see equation in fig. 3.10). We know that jointly inverting ε and δ in anisotropic tomographic

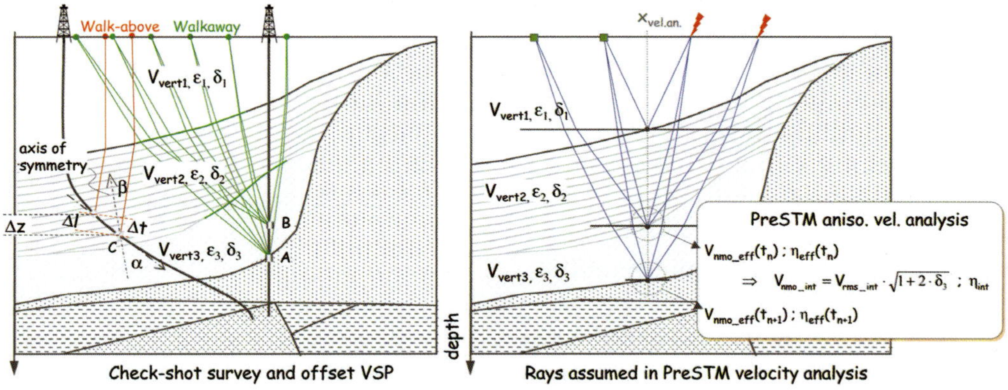

Figure 8.12 : Anisotropy calibration from wells and PreSTM. The anisotropic model around the well used for PreSDM must be consistent with the walkaway travel times. Walk-above data allow us to calibrate V_{phase} (β) (cautious with ray bending !), whilst the sonic log measures $V_{phase}(\alpha)$; angles are taken w/r to the axis of symmetry.
PreSTM yields V_{nmo_eff} and η_{eff}, that can be 'down-scaled' to V_{nmo_int} (a function of V_{rms_int} and δ_{int}) and η_{eff}. ε_{int} is not accessible directly. Moreover PreSTM measurements are biased by the assumptions intrinsic to time processing as discussed in chapter 3.

inversion or migration velocity analysis is an ill-posed problem. Common practice is to assume a fixed ε/δ ratio, typically between 1.5 and 2. The ratio is selected based on regional basin knowledge. An alternative could be to use instead η_{int} measured in PreSTM velocity analysis, to constrain the difference ε-δ in order to reduce the unknowns' dimensionality of the depth tomographic inverse problem. This is seldom done in practice, one of the reasons being the bias in $V_{nmo\text{-}eff}$ and η_{eff} measurements due to intrinsic assumptions made in PreSTM, as discussed at length in chapter 3.

The η_{eff} field we obtain from PreSTM must however be examined before launching a PreSDM project, at least in a qualitative sense. If it mixes information on both heterogeneity and anisotropy, we should nevertheless remember that heterogeneity generally evolves smoothly with depth, unless we hit lithology changes (for instance, from sand-shales layering to chalk). A sharp increase of η_{eff} with depth has thus a good chance of either effectively indicating either a change in lithology or the presence of a more anisotropic layer: the 'δ term' overrides the 'rms term' in the equation of V_{nmo}. This must be taken as a indication of the presence of anisotropy.

Azimuthal velocity anisotropy

The previous section was dedicated to VTI and STI models that assume, as the terms say, 'Transverse Isotropy', i.e., the fact that velocity does not change with the azimuth at which waves propagate in the subsurface.

We know of course that this is again a simplified vision of Nature and in effect, we fully recognize its importance when we try, on the contrary, to use this property to detect fractures in the subsurface.

The measurement of azimuthal anisotropy is made possible only with multi- or wide-azimuth acquisition layouts. As mentioned a few times, this is the case in many current land data in easy access areas (deserts) but also water-bottom (cables or nodes) surveys and nowadays, Towed Streamer Wide Azimuth.

However, it necessitates handling the multi-dimensionality of CIGs: vertical time (for PreSTM) or depth (for PreSDM), offset and azimuth, as shown in fig. 4.15.

The conventional approach consists in sorting the CIGs by sectors: for a given offset range, traces having a shot-receiver falling in a predefined azimuth range are gathered together. Each sector is then processed separately. A typical number of sectors is six (30° range).

Once the Common Offsets CIGs are sorted by sectors, the data in each of them (assuming that they are sufficiently well populated) may undergo a classical velocity analysis with one of the three techniques that we have encountered in this book:

- tomography for depth CIGs after PreSDM;
- inverse NMO followed by one or two-term velocity scanning as in fig. 3.16 for PreSTM CIGs
- Residual Move-Out analysis (as in fig. 3.22).

In doing so, we obtain one velocity field for each azimuth range. Recombining the different fields allows an analysis of velocity as a function of azimuth. This sounds easy but it requires a few comments.

- Performing tomography for different sectors with depth CIGs after PreSDM means that we multiply the number M of 'measurements' and N of 'unknowns' (see chapter 4) by the number of sectors. Moreover, changes of velocity with offset are expected to be subtle. They consequently require high-resolution grid tomography, with the corresponding dense moveout picking, to have a chance to be detected. These considerations make it such that no azimuthal anisotropy after PreSDM has been reported (so far!).
- The method is hence better suited, for the time being, to PreSTM velocity analysis.
- Again, since we increase the number of searched parameters, the usual approach uses one term Residual Move-Out analysis for each sector.

Grechka and Tsvankin (1998) proposed a model that assumes that velocity varies 'elliptically' with azimuth (for horizontal reflectors). Three parameters then define azimuthal anisotropy at a given depth (or vertical time):

- The orientation of the axis of symmetry of the ellipse (denoted α)
- The 'fast' moveout velocity V_{nmo_fast} in direction α.
- The 'slow' moveout velocity V_{nmo_slow} in direction normal to α.

Jenner *et al.* (2001) proposed a method to estimate these three parameters from CIGs sorted by offset and azimuth, hence without 'sectoring' the data. Recently, Calvert *et al.* (2008) and Lecerf *et al.* (2009) suggested working directly in the offset vector domain (see fig. 4.19 or 5.06 for a definition of 'offset vectors'), also with no sectoring. The fitting is now no longer a curve per sector but a surface as a function of h_x and h_y.

As an extension to the parabolic RMO curve, Lecerf *et al.* (2009) propose a parameterization of the type:

$$t(t_0,h_x,h_y,Q,R,S) = t_0 + Q\cdot\left[h_x^2 + h_y^2\right] + R\cdot\left[h_x^2 - h_y^2\right] + S\cdot\left[2\cdot h_x\cdot h_y\right] \quad (8.01)$$

where t is the time of an event on a PreSTM CIG for a trace of vector offset $\{h_x, h_y\}$, t_0 is the time at zero offset and Q, R and S three parameters that generalize the residual move out (hence the correction to the velocity function used to derive the CIG at that location) as a function of both offset and azimuth. Q, R and S are directly related to α, V_{nmo_fast} and V_{nmo_slow}. For instance, Q = R and S=0 would correspond to a major axis in the x direction and a parabolic RMO characterized by R (in this case equivalent to p in fig. 3.22) in that direction.

Q, R and S are estimated by semblance optimization and scanning at each zero-offset time t_0 and for each CIG. The 2D left-hand side and right-hand side panels in fig. 3.22 are now 3D objects (t, h_x, h_y) and the central 2D panel is now a four-dimensional object in which the search is performed (t, p become t, Q, R, S).

Once estimated, Q, R and S can thus be inverted in terms of α, V_{nmo_fast} and V_{nmo_slow}.

Automatic picking of the optimal {Q, R, S} values leads eventually to a 3D field of sets of parameters {α, V_{nmo_fast}, V_{nmo_slow}} that fully characterizes the measured azimuthal anisotropy.

The good point of this technique is that it yields a high-density / high-resolution estimate of the azimuthal behaviour of the velocity field. This permits in particular to perform a stack of the CIG traces corrected with the picked (Q,R,S) field, which is optimized in a 3D sense. This significantly improves the signal/noise of the PreSTM result (azimuthal final RMO).

However, the fact that we use PreSTM rather than PreSDM makes the interpretation of the results dubious in terms, for instance, of the presence and intensity of fracturing. The reason is the impact of structure and shooting azimuth on PreSTM velocities, which were mentioned in fig. 3.27. This warning is given in Lecerf *et al.*, where the estimated field of α, V_{nmo_fast} and V_{nmo_slow} nicely follows the structure of the horizon.

The solution to discriminating between 'structural artefacts' and azimuthal anisotropy related to true rock properties in the subsurface lies probably in azimuthal high-density tomography inversion of depth CIGs. This is at least in theory but practicalities will probably present surprises. As an example, Whitfield *et al.* (2008) report 'apparent azimuthal anisotropy' detected in depth velocity model building coming from short wavelength velocity variations in North Sea chalk. We are still on a learning curve in this domain!

As a side and conclusive comment to this section, note that fig. 4.23 gives an example of depth CIGs sorted by azimuth sectors; however, this is not the classical, but probably a better way to sort data since the parameters are azimuth and scattering angles at the reflection point in depth, instead of the more traditional surface offset and azimuth. It may be used all the same for azimuthal velocity analysis in depth.

Is there an optimum algorithm for imaging?

This is clearly 'the' question to ask when embarking on an imaging project. It would be nice if we could conclude this course with a definite answer. Unfortunately, the present review of the main methods available to date should make it clear that there is no general answer: no single method is well adapted to all cases. In the most recent literature, we find authors presenting good results with Controlled Beam Migration (Ben-Brahim *et al.*, 2009; Pham *et al.*, 2008); in different parts of the world others will advocate the benefit of Common Reflection Angle and Kirchhoff migrations (Winbow, 2009); others again show how RTM (Farmer *et al.*, 2006; Buur and Kühnel, 2008;) can help in salt tectonics settings, just to cite a few examples.

It was precisely one of the ambitions of this course to guide readers in the selection of the method from among the many available that could be best adapted to their own imaging problem. We also emphasized the importance of the choice of parameters, and of course, the migration velocity building issues.

Just as a final example of the benefits and limitations of various methods, we present some last comparisons in fig. 8.13 and the following ones.

In fig. 8.13, we show the result of Kirchhoff migration in a halokinetic environment. The method yields good resolution of the sediments but presents some migration 'smiles' in low signal areas, for reasons explained in chapter 4.

Fig. 8.14 now compares it with Gaussian beam. Similar to Kirchhoff, the resolution is good in good signal/noise areas but the migration noise is reduced with respect to Kirchhoff, as discussed in chapter 5 (see also fig. 5.14).

Fig. 8.15 shows the results of Controlled Beam Migration. It performs well for structural interpretation. The base of salt bodies, for instance, is well defined. The migration noise is limited but the character of good signal to noise ratio areas has deteriorated with respect to Gaussian Beam or Kirchhoff.

Finally, fig. 8.16 is an example of RTM and we find what we would expect from the method: the best image of salt flanks and salt base, probably some resulting from turning waves. However for computing cost reasons that were addressed in chapter 6, the maximum frequency is limited, resulting in limited resolution that shows up in the shallow sediment basins.

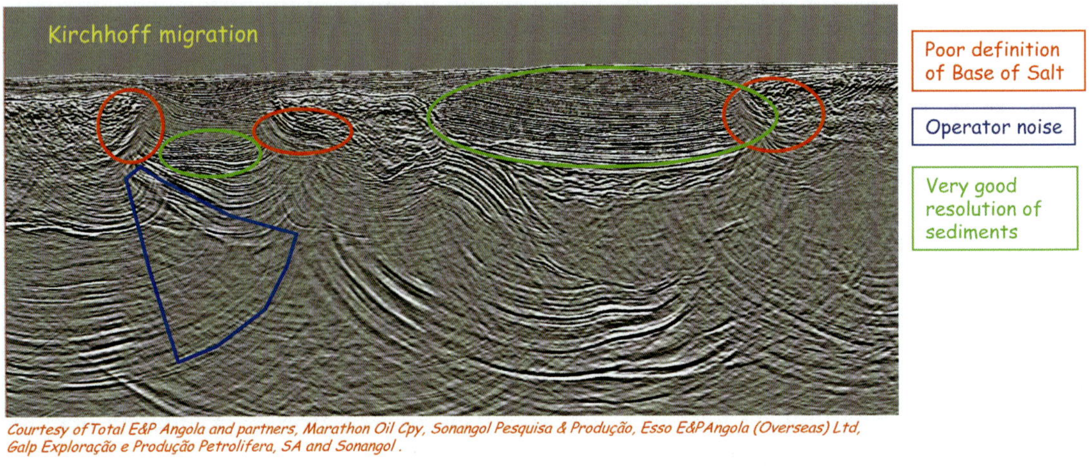

Figure 8.13 : In complex tectonics environment, there is no 'ideal' imaging tool. Each one may bring its own piece of information ... and noise. Here 3D Kirchhoff migration described in chapter 4.

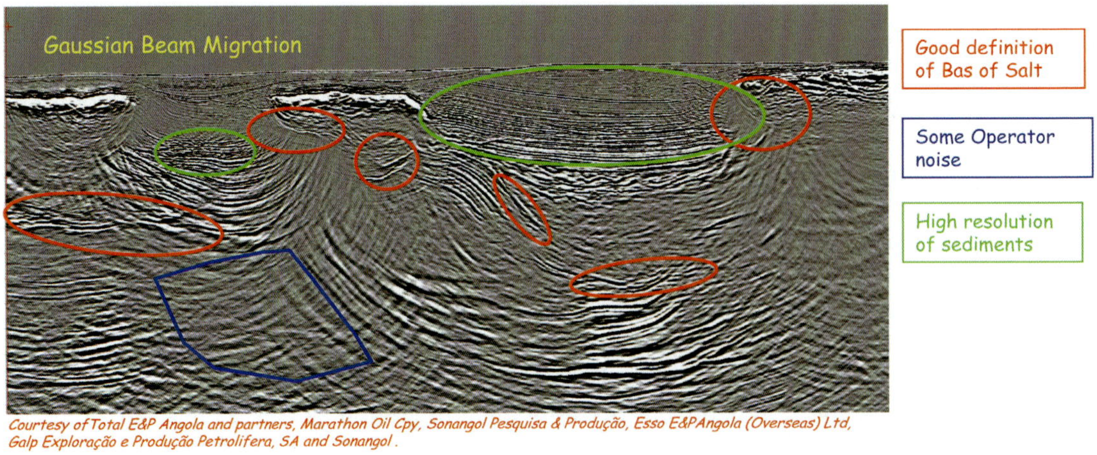

Figure 8.14 : Same line as in fig. 8.13, but now processed with Gaussian Beam Migration (in chapter 5).

With this set of comparisons we wanted to illustrate once again that the selection of an imaging method depends on many parameters, some of them technical (the maximum dip to migrate, possibility to image overhangs, presence of well calibration), some related to the geological setting (salt tectonics, foothills, fast subsiding basins, structurally quiet areas, clastics versus carbonates, fractured reservoirs, ...), some to geophysics (requested resolution, depth of target, need for amplitude preservation, level of accuracy of the reflectors location, severe multiple issues or not, ...), some to project management (turnaround time, budget, code and computer power available), etc.

Imaging is team work and geological input is of particular importance

The selection of the migration method was the main object of this course but we recognize that it is only one aspect of any imaging project. Each step is crucial from survey design, quality control

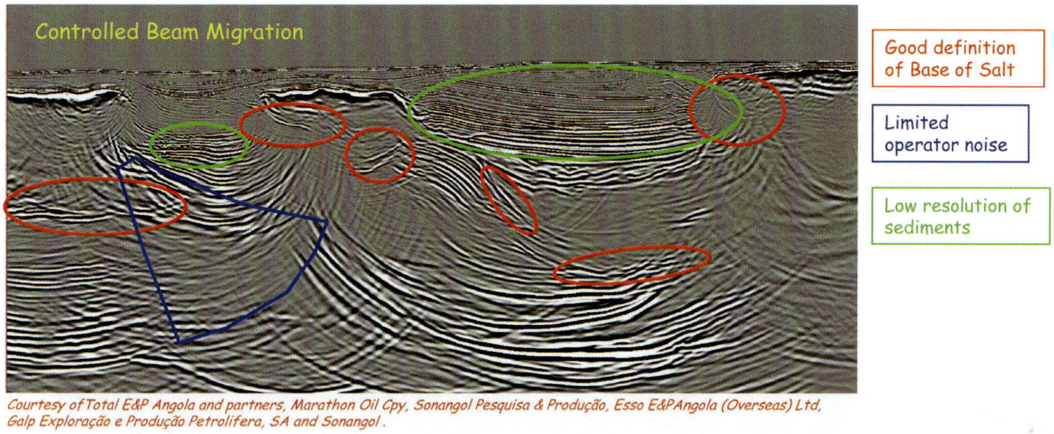

Figure 8.15 : Same as fig. 8.13, but now with Controlled Beam Migration described in chapter 5.

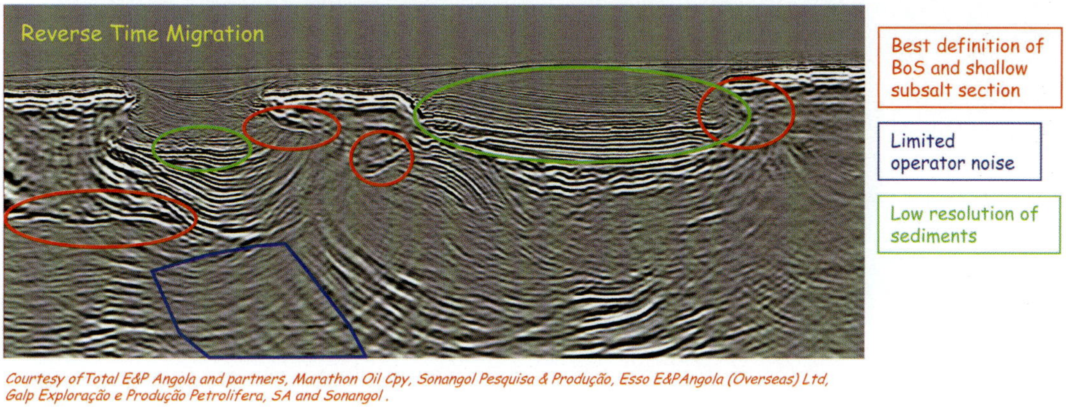

Figure 8.16 : Same as fig. 8.13, but now with Reverse Time Migration described in chapter 6.

during acquisition, pre-processing steps (recall fig. 8.08), extensive parameters testing, velocity model building, to the final migration followed by adequate post-processing. At each step the expertise of a specialist is requested but his or her decisions always interact with other issues of the project.

Further to 'classical' seismics and geology, we note also more and more the benefit of 'thinking in geomechanical terms'. Geomechanics has appeared in the field of seismics first for repeat 4D seismics. Hatchell *et al.* (2003), for instance, showed how changes during production in the stress and strain in and as importantly around the reservoir, impacts 4D measurements. This has changed the way we now interpret the results of repeat surveys in some areas (Taylor *et al.*, 2007).

On the other hand, geomechanics is also considered in 'conventional' 3D imaging projects. More specifically, it can help us to constrain or update anisotropic velocity models (Birdus, 2008; Sengupta and Bachrach, 2008). Geomechanics is a new discipline for many geophysicists. It deals however with the elastic rock properties, hence with propagation velocity and anisotropy and we foresee a development of these methods in the future.

Being very orientated towards wave propagation and seismics techniques, we would like to conclude this book by acknowledging the key role and the necessary expertise of interpreters, sedimentologists and structural geologists in any imaging project.

We take the example in fig. 8.17 for this. It presents a velocity model that cannot be the result uniquely of tomography or Full Waveform Inversion. Geologists were involved from the beginning of the project, adding their input at each step of tomography, helping to interpret all tests, amalgamating any piece of information available at all scales to eventually build a model that is compatible with the recorded data, certainly not unique but 'geologically sound'.

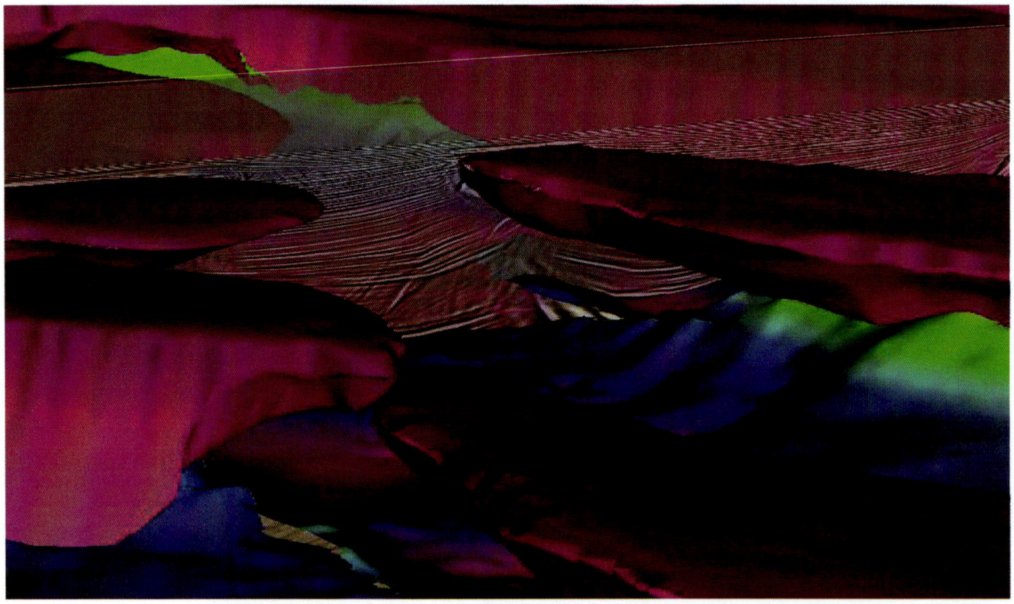

Courtesy of Total E&P Angola and partners, Marathon Oil Cpy, Sonangol Pesquisa & Produção, Esso E&PAngola (Overseas) Ltd, Galp Exploração e Produção Petrolifera, SA and Sonangol.

Figure. 8.17 : A 3D vision of complex velocity models. This type of model is the result of several iterations of tomography, testing of various imaging tools, but also inclusion of a lot of geological knowledge to ensure it is not only 'seismically compatible' with the data, but also 'geologically sound'.

As a conclusion ...

Dramatic improvements have been made in the domain of seismic imaging in recent years. Despite these improvements and all the efforts made, the images that we obtain today from a given data set cannot be considered as the ultimate ones. Experience in seismic imaging since its inception in the late seventies shows that this technology has never stopped providing clearer, sharper and more accurate images. In this book we showed examples of spectacular improvements with vintage data sets using new imaging tools and there are no reasons why this trend should stop in the future.

Acquisition technology, survey design, geological concepts, structural modelling and 3D section balancing, computer power and architecture, powerful workstations, fewer approximations in the use of the full elastic wave equation, more complex, higher-resolution anisotropic models that are geomechanically constrained, all are moving forward to give us good reasons to foresee that the trend of progress in seismic imaging will continue.

9 Acknowledgements

First, I would like to thank the management of Total E&P and particularly, M. Lepoutre, M. Blaizot and B. Seiller, for allowing me to dedicate time for the preparation of the course and writing this course book.

I am particularly indebted to the many colleagues who gave me their time and advice through helpful discussions and thorough review of the course book: Samuel Gray, Philippe Herrmann, Gilles Lambaré, Risto Siliqi from CGGVeritas, Prof. Stéphane Operto and Prof. Jean Virieux leaders of the Seiscope project, Reda Baina from Opera and my colleagues in Total's Processing, Depth Imaging and Research groups Jean Arnaud, François Audebert, Henri Calandra, Thierry Castex, Bertrand Duquet, Eric Dussaud, Weihong Fei, Pierre Hardy, Huimin Guan, Ronan Petton, Cyril Saint-André, Paul Sexton, Laurent Sirgue, Paul Williamson and Enrico Zamboni.

I am especially grateful to those who provided me with many illustrative examples: Paul LeCoq, John and Kevin Sherwood and Steve Pitman from PGS, Ian Jones and Alexander Calvert from ION GXtech, Jerry Kapoor, Irina Marin, Paul Taylor, Christophe Rimlinger and Hasbi Lubis from WesterGeco, Keith Hawkins, Vetle Vinje and Gareth Williams from CGGVeritas, Zwi Koren from Paradigm, Nathan Sharff and Tamir Tal from Geomage, Mike Hebert from TGS-Nopec, Prof. Changsoo Shin from Seoul National University, Prof. Biondo Biondi from Stanford University, Hafedh Ben-HadjAli and Romain Brossier from Université Sciences Azur, Frederic Billette and Olav-Inge Barkfeld from bp, Ted Bakamjian from SEG and my colleagues in Total's Head Quarters and affiliates: Lotfi Ben-Brahim, Florent Bertini, Jean-Luc Bouroullec, Jean-Luc Debroux, Magali Collin, Bernard Geiss, Jérôme Guilbot, Pierre Jousselin, Paul Matthews, Antony McAteer and Alain Riou.

Finally, thank you also to the EAGE staff, in particular Salima Gader, Anne-Claire Hoenson, Karin Merkens and Linda Molenaar, for their help in editing the course book and organizing the lectures tour.

10 References

Text books and general interest papers

Bancroft, J.C. (1997) *A practical understanding of pre and poststack migrations, 2 volumes*, SEG Course Notes Series N° 7.

Bednar J.B. (2005) A brief history of seismic migration. *Geophysics*, **70**, 3MJ-20MJ.

Berkhout, A. (1982) *Imaging of Acoustic Energy by Wavefield Extrapolation*. Elsevier.

Biondi, B. (2006) *Seismic Imaging*. SEG.

Bleinstein, N., Cohen, K. and Stockwell, A. (2001) *Mathematics of Multi-Dimensional Seismic Inversion*. Springer-Verlag, New York.

Claerbout, J.F. (1976) *Fundamentals of Geophysical Data Processing*. McGraw-Hill.

Claerbout, J.F. (1985) *Imaging the earth interior*. Blackwell Scientific Publications, Boston.

Grechka, V. (2008) *Application of Seismic Anisotropy in the Oil and Gas Industry*. EAGE Publications bv.

Jones, I.F., Bloor, R.I., Biondi, B.L. and Etgen, J.T. (2008) *Prestack Depth Migration and Velocity Model Building*. SEG, Geophysics reprint series N°25.

Landa, E. (2007) *Beyond Conventional Seismic Imaging*. EAGE Publications bv.

Robein, E. (2003) *Velocities, Time-imaging and Depth-imaging in reflection seismics. Principles and methods*. EAGE Publications bv.

Stolt, R.H. and Benson, A.K. (1986) *Seismic migration: theory and practice*. Pergamon.

Tarantola, A. (1987) *Inverse problem theory: methods for data fitting and model parameter estimation*. Elsevier.

Thomsen, L. (1986) Weak elastic anisotropy. *Geophysics*, **59**, 1954-1966.

Tsvankin, Y. (2006) *Seismic signatures and analysis of Reflection Data in Anisotropic Media*, Klaus Helbig & Sven Trietel Editors, Vol.29. Elsevier.

Verschuur, D.J. (2006) *Seismic multiple removal techniques. Past, present and future*. EAGE Publications bv.

Yilmaz, Ö. (1987) *Seismic data analysis*. SEG investigations in Geophysics, Series N°2.

Yilmaz, Ö. (2001) *Seismic data analysis; processing, inversion and interpretation of seismic data, 2 volumes*. SEG investigations in Geophysics, Series N°10.

Chapter 3

Adler, F. and Brandwood, S. (1999) Robust estimation of dense 3D stacking velocities from automated picking. *69th SEG Annual International Meeting*, Expanded Abstracts, 1162–1165.

Al-Chalabi, M. (1994) Seismic velocities: a critique. *First Break*, **12**, 589-595.

Alkhalifah, T.and Tsvankin, I. (1995) Velocity analysis of transversely isotropic media. *Geophysics*, **60**, 1550-1566.

Alkhalifah, T. (1996) Transformation to zero offset in transversely isotropic media. *Geophysics*, **61**, 947-963.

Alkhalifah, T (1997) Velocity analysis using non-hyperbolic moveout in transversely isotropic media. *Geophysics*, **62**, 1839-1854.

Alkhalifah, T. (2003) Tau migration and velocity analysis: Theory and synthetic examples. *Geophysics*, **68**, 1331-1339.

Black, J. L. and Brzostowski, M.A. (1994) Systematics of time-migration errors. *Geophysics*, **59**, 1419-1434.

Castle, R. (1994) A theory of normal moveout. *Geophysics*, **59**, 983-999.

De Bazelaire, E. (1988) Normal moveout revisited: Inhomogeneous media and curved interfaces. *Geophysics*, **53**, 143-157.

De Bazelaire, E. and Viallix, J. (1994) Normal moveout in focus. *Geophysical Prospecting*, **42**, 477-499.

Dix, C.H. (1955) Seismic velocities from surface measurements. *Geophysics*, **20**, 68-86.

Gancarski, S. and Kaculini, S. (1998) Where is the well? A time migrated case study. *68th SEG Annual International Meeting*, Expanded Abstracts.

Hubral, P. (1977) Time migration – Some ray theoretical aspects. *Geophysical Prospecting*, **25**, 738-745.

Kaculini, S. and Arnaud, J. (1997) Integration of well data with a 3D time migrated seismic. *67th SEG Annual International Meeting*, Expanded Abstracts. INT 1.9.

Lambaré, G. et al (2008) Velocity update for pre-stack time migration. *78th SEG Annual International Meeting*, Expanded Abstratcs.

Lambaré, G., Herrmann, P., Touré, J.P., Suaudeau, E. and Lecerf D. (2008) Computation of kinematic attributes for pre-stack time migration. *78th SEG Annual International Meeting*, Expanded Abstracts.

Le Meur, D., Siliqi, R., Gamar, F., Touré, J.P. and Herrmann, P. (2003) High density moveout parameter fields V and η. Part two: Simultaneous geostatistical filtering. *73rd SEG Annual International Meeting*, Expanded Abstracts.

Robein, E. and Hanitzsch, C. (2001) Benefits of pre-stack time migration in model building: a case history in the Souh Caspian Sea. *First Break*, **19** (4), 183-189.

Robein, E. (2002) On the meaning of Kirchhoff Pres-Stack time migration velocities. *64th EAGE Conference & Exhibition*, Extended Abstracts, P005.

Schleicher, J., Costa, J.C. and Novais, A. (2008) Time migration velocity analysis by image-wave propagation of common-image gathers. *78th SEG Annual International Meeting*, Expanded Abstracts.

Siliqi, R and Bousquié, N. (2000) Anelliptic time processing based on a shifted hyperbola approach. *70th SEG Annual International Meeting*, Expanded Abstracts.

Siliqi, R. (2001) Technological leap in time processing focuses the data throughout anisotropic media. *First Break*, **19** (11), 612-618.

Siliqi, R., Le Meur, D., Gamar, F., Smith, L., Touré, J.P. and Herrmann, P. (2003) High density moveout parameter fields V and η. Part one: Simultaneous automatic picking. *73rd SEG Annual International Meeting*, Expanded Abstracts.

Siliqi, R. and Le Meur, D. (2004) How to estimate uncorrelated dense V-η fields. *66th EAGE Conference & Exhibition*, Extended Abstracts, P066.

Siliqi, R., Herrmann, P., Prescott, A. and Capar, L. (2007) Automatic dense High Order RMO Picking. *67th EAGE Conference & Exhibition*, Extended Abstracts, P037.

Stovas, A. (2009) Direct Dix-type inversion in a layered VTI medium. *79th SEG Annual International Meeting*, Expanded Abstracts, 2506-2510.

Taner, M. and Koelher, F. (1969) Velocity spectra – digital computer derivation and application of velocity functions. *Geophysics*, **34**, 859-881.

Ursin, B. and Stovas, A. (2006) Travel time approximation for a layered transversely isotropic medium. *Geophysics*, **71**, D23-D33.

Chapter 4

Adler, F. (2004) Anisotropic velocity model building for time-to-depth conversion after non-hyperbolic Kirchhoff PreSTM. *74th SEG Annual International Meeting*, Expanded Abstracts, 2235-2338.

Adler, F., Baina, R., Soudani, M.A., Cardon, P. and Richard, J.B. (2008) Non-linear tomographic least-square inversion of migrated depth in Kirchhoff PreSDM common-image gathers. *Geophysics*, **73**, VE13-VE23.

Audebert, F., Froidevaux, P., Rakotoarisoa, H. and Svay-Lucas, J. (2002) Insights into migration in the angle domain. *72nd SEG Annual International Meeting*, Expanded Abstracts, MIG 3.5

Audebert, F., Jousselin, P., Duquet, B., Dehghan, K. and Le Stunff, Y. (2008) Velocity model building in complex settings: combining the strength of tomography and wave-equation PreSDM. *78th SEG Annual International Meeting*, Expanded Abstracts.

Bleinstein, N. (1987) On the imaging of reflectors in the earth, *Geophysics*, **52**, 931 – 942.

Brandsberg-Dahl, S., de Hoop, M. and Ursin, B. (1999) Velocity analysis in the common scattering-angle/azimuth domain. *69th SEG Annual International Meeting*, Expanded Abstracts, STHRY 1.3.

Brittan, J., Yuan, J., Long, A. (2006) Resolving small scale heterogeneities with dense non-hyperbolic seismic tomography. *20thASEG International Geophysical Conference and Exhibition*.

Claerbout, J. (1994) Applications of two-and three-dimensional filtering. *64th SEG Annual International Meeting*, Expanded Abstracts, 1572-1577.

Duquet, B., Marfurt, K. and Dellinger, J. (2000) Kirchhoff modelling, inversion for reflectivity, and subsurface illumination. *Geophysics*, **65**, 1195 – 1209.

Duquet, B., Lemaistre, L., Le Stunff, Y., Gancarsky, S. and Camez, J.C. (2008) Multi-domain and multi-scale tomography for precise reservoir imaging. *78th SEG Annual International Meeting*, Expanded Abstracts.

Etgen, J., Billette, F., Sandshaper, R. and Rietfeld, E. (2002) The key practical aspects of 3D tomography: data picking and model representation. *72nd SEG Annual International Meeting*, Expanded Abstracts.

Fagin, S. (1999) *Model based depth imaging*. Course notes series N°10 (Edition RA Young). SEG.

Fruhen, J., Jones, I., Valler, V., Sangvai, P., Biswal, A. and Mathur, M. (2008) Resolving Near-Seabed Velocity Anomalies: Deep Water Offshore South East India. *Geophysics* (Special Issue), **73**, N°5, VE235 – VE241.

Gray, S. and Marfurt, K. (1995) Migration from topography: Improving the near surface image. *Canadian Journal of Exploration Geophysics*, **31**, 18-24.

Guillaume, P. *et al.* (2008) Kinematics invariants: an efficient and flexible approach for velocity model building. *78th SEG Annual International Meeting*, Expanded Abstracts.

Gulunay, N., Magesan, M. and Roende, H. (2007) Gather flattening. *The Leading Edge*, **26**, 1538 – 1543.

Hanitzsch, C. (1997) Comparison of weights in prestack amplitude-preserving Kirchhoff depth migration. *Geophysics*, **62**, 1812 -1816.

Hardy, P. and Jeannot, J-P. (1999) 3D reflection tomography in time-migrated space. *69th SEG Annual International Meeting*, Expanded Abstracts, 1287-1290.

Hardy, P. (2004) High resolution tomographic MVA with automation. *69th EAGE Conference & Exhibition*, Extended Abstracts, C019.

Hardy, P. (2007) Improved RMO picking and velocity model building with prestack event skeletons. *77th SEG Annual International Meeting*, Expanded Abstracts.

Fomel, S., Landa, E. and Turner, M. (2007) Poststack velocity analysis by separation and imaging of seismic diffractions. *Geophysics*, **72**, U89-U94.

Jiao, J., Lowrey, D.R., Willis, J.F. and Martinez, R.D. (2008) Practical approaches for subsalt velocity model building, *Geophysics*, **73**, VE183 –VE194.

Jones, I.F., Baud, H., Ibbotson, K., and Audebert, F. (2000) Continuous 3-D preSDM velocity analysis. *The Leading Edge*, 263-269.

Jones, I.F. and Baud, H. (2001) Velocity as an attribute: continuous velocity estimation from 3D preS-DM CRP gathers. *First Break*, **19** (1), 19-23.

Jones, I.F. (2003) A review of 3D preSDM velocity model buiding techniques. *First Break*, **21**, 45-48.

Jones, I.F., Sugrue, M. and Hardy, P. (2007) Hybrid gridded tomography, *First Break*, **25**, 15-21.

Jones, I.F. (2010) A practical review of migration issues and solutions. In Preparation, EAGE Publications bv.

Koren, Z., Sheng, X. and Kosloff, D. (2002) Target oriented common-reflection angle migration. *72nd SEG Annual International Meeting*, Expanded Abstracts.

Koren, Z., Ravve, I., Bartana, A., and Kosloff, D. (2007) Local angle domain in seismic imaging. *69th EAGE Conference & Exhibition*, Extended Abstracts, P287.

Koren, Z., Ravve, I., Gonzalez, G. and Kosloff, D. (2009) Anisotropic local tomography. *Geophysics*, **73**, VE75-VE92.

Lambaré, G. (2008) Stereotomography. *Geophysics*, **73**, VE25-VE34.

Landa, E., Shtivelman, V. and Gelchinsky, B. (1987) A method for detection of diffracted waves on common-offset sections. *Geophysical Prospecting*, **35**, 359-373.

Landa, E. and Keydar, S. (1998) Seismic monitoring of diffraction images for detection of local heterogeneities. *Geophysics*, **63**, 1093-1100.

Lui, J., Bear, L., Krebs, J., Montelli, R. and Palacharla, G. (2008) Tomographic inversion by matrix transformation. *Geophysics*, **73**, VE35-VE38.

Reshef, M. and Rueger, A. (2008) Influence of structural dip angles on interval velocity analysis. *Geophysics*, **73**, U13-U18.

Reshef, M. and Landa, E. (2009) Post-stack velocity analysis in the dip-angle domain using diffractions. *Geophysical Prospecting*, **57**, 811-821.

Reynaud, B. and Robein, E. (1998) *Method for modelling kinematic seismic data processed with at least one motion operator*. US patent 5 764514.

Robein, E., Jones, I.F. and De Bazelaire, E. (2003) High-density high resolution seismic velocity (review article). *First Break*, **21**.

Rousseau, V., Svay-Lucas, J., Nicoletis, L. and Rakotoarisoa, H. (2000) 3D true-amplitude migration by regularization in angle domain. *62nd EAGE Conference & Exhibition*, Extended Abstracts.

Schleicher, J., Tygel, M. and Hubral, P. (1993) Parabolic and hyperbolic paraxial two-point traveltimes in 3D media. *Geophysical Prospecting*, **41**, 495-513.

Schultz, Ph. (1999) *The seismic velocity model as an interpretation asset*. SEG, Distinguished Instructor Series, N°2.

Shumaker, N., Lindsay, R. and Ogilvie, J. (2007) Depth-calibrating seismic data in the presence of allochtonous salt. *The Leading Edge*, **26**, 1442 – 1453.

Taner, T., Fomel, S. and Landa, E. (2006) Prestack separation of seismic diffractions using plane wave decomposition. *76th SEG Annual International Meeting*, Expanded Abstracts, 2401-2404.

Ursin, B. (2004) Parameter inversion and angle migration in anisotropic media. *Geophysics*, **69**, 1125 -1142.

Versteeg, R. (1994) The Marmousi experience: velocity model determination on a synthetic complex data set. *The Leading Edge*, **18**, 86 – 91.

Whitfield, P., Dazley, M., Santos-Luis, B., Nieuwland, F. and Lemaistre, L. (2008) Building velocity models for depth imaging in the presence of short-wavelength velocity variations – a case study from the North Sea. *First Break*, **26**, 45-50.

Woodward, M., Nichols, D., Zdraveva, O., Whitfield, P. and Johns, T. (2008) A decade of tomography. *Geophysics*, **73**, VE5-VE11.

Xu, S., Chauris, H., Lambaré, G. and Noble, M. (1998) Common angle image gather: a strategy for imaging complex media. *68th SEG Annual International Meeting*, Expanded Abstracts, ST4.3.

Chapter 5

Alkhallifah, T. (1995) Gaussian beam depth migration for anisotropic media. *Geophysics*, **60**, 1474-1484.

Ben-Brahim, L., Coombes, R., Cooper, R., Fredonnet, J., Taylor, D. and Valery, S. (2009) Re-Evaluation of a Complex Mature Field (Dunbar, UKCS): a Geosciences Integrated Work. *SPE 123624*.

Beydoun, W. and Keho, T. (1987) The paraxial ray method. *Geophysics*, **52**, 1639-1653.

Cerveny, V., Popov, M. and Psencik, I. (1982) Computation of wave fields in inhomogeneous media – Gaussian beam approach. *Geophysical Journal of the Royal Astronomy Society*, **70**, 109-128.

Pham, D., Sun, J., Tang, Q., Bone, G. and Giang, N.T. (2008) Imaging of fractures and faults inside granite basement using controlled beam migration. *Paper presented at the Asian Pacific Workshop.* Abstracts.

Gao, F., Zhang, P., Wang, B. and Dirks, V. (2006) Fast Beam migration – a step toward interactive imaging. *76th SEG Annual International Meeting, Expanded Abstracts, 2470-2473*

Gray, S. (2005) Gaussian beam migration of common shot records. *Geophysics*, **70**, S71-S77.

Gray, S., Xie, Yi, Notfors, C., Zhu, T., Wang, D. and Ting, C.O. (2009) Taking apart beam migration. *The Leading Edge*, 1098-1108.

Gray, S. and Bleistein, N. (2009) True-amplitude Gaussian-beam migration. *Geophysics*, **74**, S11-S23.

Hill, N.R. (1990) Gaussian-beam migration. *Geophysics*, **55**, 1416 -1428.

Hill, N.R. (2001) Prestack Gaussian-beam depth migration. *Geophysics*, **66**, 1240-1250.

Peng, C. (2007) Diplet-based seismic processing. *WIPO patent WO 2007/143355 A2*, Nexus Geosciences Inc applicant.

Sherwood, J., Sherwood, K., Tieman, H. and Schleicher, K. (2008) 3D beam prestack migration with examples from around the world. *78th SEG Annual International Meeting*, Expanded Abstracts, 438-442. Also in: The Leading Edge, Sept 2009, 1120-1127.

Ting, C. and Wang, D. (2008) Controlled Beam Migration application in Gulf of Mexico. *68th SEG Annual International Meeting*, Expanded Abstracts.

Virgilio, M., Re, S. and Colombo, D. (2008) Gaussian beam migration: prestack, common-shot, TTI, true amplitude in angular domain. *78th SEG Annual International Meeting*, Expanded Abstracts.

Wang, B., Ji, J., Mason, C., Gajawada, S. and Kim, Y. (2008) Beam-based Interactive Imaging for Salt Interpretation and Salt Model Building. *68th SEG Annual International Meeting*, Expanded Abstracts.

Chapter 6

Berkhout, A. (1992) Areal shot record technology. *Journal of Seis. Expl.*, **1**, 254-264.

Duquet, B., Lailly, P. and Ehinger, A. (2001) 3D plane wave migration of streamer data. *71st SEG Annual International Meeting*, Expanded Abstracts, 1033-1036.

Baysal, E., Kosloff, D.D. and Sherwood, J.W.C. (1983) Reverse-time migration. *Geophysics*, **48**, 1514-1524.

Baysal, E., Kosloff, D.D. and Sherwood, J.W.C. (1984) A two-way non-reflecting wave equation. *Geophysics*, **49**, 132-141.

Biondi, B. (2003) Equivalence of source-receiver migration and shot-profile migration. Short note. *Geophysics*, **68**, 1340-1347.

Biondi, B. and Symes, W.W. (2004) Angle domain common-image gathers for migration velocity analysis by wavefield-continuation imaging. *Geophysics*, **69**, 1283-1298.

Chang, W. and McMechan, G. (1989) 3-D acoustic reverse time migration. *Geophysical prospecting*, **37**, 243-256.

Clapp, R. (2009) Reverse time migration with random boundaries. *79th SEG Annual International Meeting*, Expanded Abstracts, 2809-2813.

Claerbout, J.F. (1971) Toward a unified theory of reflector mapping. *Geophysics*, **36**, 467 – 481.

Dussaud, E., Symes, W.W., Williamson, P., Lemaistre, L., Singer, P., Denel, B. and Cherrett, A. (2008) Computational strategies for reverse time migration. *78th SEG Annual International Meeting*, Expanded Abstracts.

Etgen, J.T. (1986) Prestack reverse time migration of shot profiles, *SEP-50*, 151-169.

Etgen, J.T. (2005) How many angles do we really need for delayed-shot migration? *75th SEG Annual International Meeting*, Expanded Abstracts, 1985-1988.

Etgen, J.T. and O'Brien, M. (2006) Computational methods for large-scale 3D acoustic finite-difference modeling: A tutorial. *Geophysics*, **72**, SM223-SM230.

Farmer, A., Jones, I.F., Zhou, H., Bloor, R.I. and Goodwin, M.C. (2006) Application of reverse time migration to complex imaging problems, *First Break*, **24**, 65-73.

Fink, M. (2006) Time-reversal acoustics in complex environments. *Geophysics*, **71**, SI151-SI164.

Gazdag, J. (1978) Wave-equation migration with the phase shift method. *Geophysics*, **43**, 1342-1351.

Gazdag, J. and Sguazzero, P. (1984) Migration of seismic data by phase shift plus interpolation. *Geophysics*, **49**, 124-131.

Guan, H., Kim, Y., Ji, J., Yoon, K., Wang, B., Xu, W. and Li, Z. (2009) Multi-Step Reverse-Time Migration. *The Leading Edge*, 442-447.

Guitton, A., Kaelin, B. and Biondi, B. (2007) Annual meeting selection: Lesat-squares attenuation of reverse-time-migration artifacts. *Geophysics*, **72**, S19-S23.

Kaelin, B. and Guitton, A. (2006) Imaging condition for reverse time migration. *76th SEG Annual International Meeting*, Expanded Abstracts, 2594-2598.

Hua, B., Williamson, P., Zhang, L. and Calandra, H. (2007) 3-D prestack depth migration with survey sinking operator in source-geophone and mid-point-offset hybrid domain. *77th SEG Annual International Meeting*, Expanded Abstracts.

Liu, F., Hanson, D., Whitmore, N., Day, R. and Stolt, R. (2006) Toward a unified analysis for source plane-wave migration. *Geophysics*, **71**, S129-S139.

Liu, F, Zhang, G., Morton, S. and Leveille, J. (2007) Reverse-time migration using one-way wavefield imaging condition. *77th SEG Annual International Meeting*, Expanded Abstracts, 2170-2173.

Jones, I., Goodwin, M.C., Berranger, I.D., Zhou, H. and Farme, P.A. (2007) Application of anisotropic 3D reverse-time migration to complex North Sea imaging. *77th SEG Annual International Meeting*, Expanded Abstracts, 2140-2143.

Jones, I. (2008) Effects of Pre-Processing on Reverse Time Migration: a North Sea study. *First Break*, **26** (6), 73-80.

Karazincir, M. and Gerrard, C. (2006) An efficient 3D reverse-time pre-stack depth migration. *68th EAGE Conference & Exhibition*, Extended abstracts, P261.

McMechan, G.A. (1983) *Geophysical Prospecting*, **31**, 413-420.

Rietfeld, W. (1995) Controlled illumination of prestack seismic migration, *Ph.D. thesis*, Delft University of Technology.

Romero, L., Ghiglia, D., Ober, C. and Morton, S. (2000) Phase encoding of shot records in prestack migration. *Geophysics*, **65**, 426-436.

Sava, P., Hill, S.J. (2009) Overview and classification of wavefield seismic imaging methods. *The Leading Edge*, **28**, 170 – 182.

Schultz, P.S. and Sherwood, J.W.C. (1980) Depth migration before stack. *Geophysics*, **45**, 376-393.

Soubaras, R. (2006) Modulated-shot migration. *76th SEG Annual International Meeting*, Expanded Abstracts, 2430-2433.

Symes, W. and Corazzon, J. (1991) Velocity inversion by differential semblance optimization. *Geophysics*, **56**, 654-663.

Symes, W. (2007) Reverse time migration with optimal checkpointing. *Geophysics*, **72**, SM213-SM221.

Vigh, D. and Starr, W. (2006) Comparisons of shot-profile vs. plane-wave reverse time migration. *76th SEG Annual International Meeting*, Expanded Abstracts, 2358 -2361.

Wang, B., Mason, C., Guo, M., Guan, H., Yoon, K. and Li, Z. (2008) Subsalt velocity update using RTM-based DT scan. *78th SEG Annual International Meeting, Expanded Abstracts, 27, 3265.*

Whitmore, N.D. (1983) Iterative depth migration by backward time propagation. *53rd SEG Annual International Meeting*, Expanded Abstracts, S10.1.

Whitmore, N. (1995) An imaging hierarchy for common angle plane seismogram. *Ph.D. Thesis*, University of Tulsa.

Zhang, Y., Sun, J., Notfors, C., Gray, S., Chernis, L. and Young, J. (2005) Delayed-shot 3D prestack depth migration. *Geophysics*, **70**, E21-E28.

Zhang, Y., Sun, J., Gray, S. and Young, J. (2006) Sampling issues in delayed-shot migration and common-shot migration. 68th EAGE Conference & Exhibition, Extended Abstracts, G012.

Zhang, Y., Sun, J. and Gray, S. (2007) Reverse-time migration: amplitude and implementation issues. *77th SEG Annual International Meeting*, Expanded Abstracts.

Zhang, Y. and Sun, J. (2009) Practical issues in reverse time migration: true amplitude gathers, noise removal and harmonic source encoding. *First Break*, **27**, 53-59.

Chapter 7

Alkhallifah, T. (2009) Angle gathers in wave-equation imaging for VTI media. *79th SEG Annual International Meeting*, Expanded Abstracts, 2899-2903.

Biondi, B. and Symes, W. (2004) Angle-domian common-image gathers for migration velocity analysis by wavefield-continuation imaging. *Geophysics*, **69**, 1283-1298.

Brossier, R. (2009) Two-dimensional frequency-domain visco-elastic full waveform inversion: parallel algorithms, optimization and performance. *Computers & Geosciences*. Elsevier. In-press.

Bunks, C., Saleck, F., Zaleski, S. and Chavent, G. (1995) Multiscale seismic waveform inversion. *Geophysics*, **60**, 1457-1473.

De Bruin, C., Wapenaar, C. and Berkhout, J. (1990) Angle-dependent reflectivity by means of prestack migration. *Geophysics*, **55**, 1223-1234.

Fei, W., Williamson, P. and Khoury, A. (2009) 3-D common-azimuth wave-equation migration velocity anlysis. *79th SEG Annual International Meeting*, Expanded Abstracts, 2283-2287.

Fomel, S. (2004) Theory of 3-D angle gathers in wave-equation imaging. *74th SEG Annual International Meeting*, Expanded Abstracts.

Gao, F. and Symes, W. (2009) Differential semblance velocity analysis by Reverse Time Migration: Image gathers and theory. *79th SEG Annual International Meeting*, Expanded Abstracts, 2317-2321.

Geckenbach, M., Reynolds, D., Shen, P. and Symes, W. (2002) Efficient and automatic implementation of the adjoint state method. *ACM Transactions on Mathematical Software (TOMS)*, **Volume 28**, 22 - 441.

Jannane, M., Beydoun, W., Crase, E., Cao, D., Koren, Z., Landa, E., Menses, M., Pica, A., Noble, M, Roeth, G., Singh, S. Snieder, R., Tarantola, A., Trezeguet, D. and Xie, M. (1989) Short Note: Wavelengths of earth structures that can be resolved from seismic reflection data. *Geophysics*, **54**, 906-910.

Jeannot, JP., Faye, JP. and Denelle, E. (1986) Prestack migration velocities from depth focusing analysis. *56th SEG Annual International Meeting*, Expanded Abstracts, 438-440.

Hu, J., Schuster, G. and Valaseck, P. (2001) Post-stack migration deconvolution. *Geophysics*, **66**, 939-952.

Kühl, H. and Sacchi, M. (2003) Least-squares wave-equation migration for AVP/AVA inversion. *Geophysics*, **68**, 262-273.

Lailly, P. (1983) The seismic inverse problem as a sequence of before stack migrations. *Conference on Inverse Scattering, theory and application*. Society for Industrial and Applied Mathematics, Philadelphia, Expanded abstracts, 206-220.

Landa, E., Kosloff, D., Keidra, S., Koren, Z. and Resheff, M. (1988) A method for determination of velocity and depth from seismic reflection data. *Geophysical Prospecting*, **36**, 223-243.

Landa, E., Beydoun, W. and Tarantola, A. (1989) Reference velocity model estimation from prestack waveforms: Coherency optimisation by simulated annealing. *Geophysics*, **54**, 984-990.

Luo, Y., Zhu, H., Nissen-Meyer, T., Morency, C. and Tromp, J. (2009) Seismic modelling and imaging based upon spectral-element and adjoint methods. *The Leading Edge*, 568-574.

Nemeth, T., Wu, C. and Schuster, G. (1999) Least-square migration of incomplete reflection data. *Geophysics*, **64**, 208-221.

Operto, S., Ravaut, C., Impotra, L., Virieux, J., Herrero, A. and Dell'Aversana, P. (2004) Qunatitative imaging of complex structures from dense wide-aperture seismic data by multiscale traveltime and waveform inversions: a case study. *Geophysical Prospecting*, **52**, 625-651.

Operto, S., Virieux, J., Amestoy, P, L'Excellent, JY., Giraud, L. and Ben-Hadj-Ali, H. (2007) 3D finite-difference frequency-domain modelling of visco-acoustic wave propagation using a massively parallel direct solver: a feasibility study. *Geophysics*, **72**, SM195-SM211.

Plessix, RE. (2006) A review of the adjoint-state method for computing the gradient of a functional with geophysical applications. *Geophysical Journal International*, **Volume 167**, Issue 3, pp. 495-503.

Pratt, R.G. (1999) Seismic waveform inversion in the frequency domain, part 1: theory and verification in a physical scale model. *Geophysics*, **64**, 888-901.

Pratt, R.G. and Symes, W. (2002) Semblance and differential semblance optimisation for waveform tomography: a frequency domain implementation. *Subsalt imaging: Journal of Conference abstracts*, **7**(2), 183-184, Cambridge University press.

Rickett, J. and Sava, P. (2002) Offset and angle-domain common image-point gathers for shot-profile migration. *Geophysics*, **67**, 883_889.

Sava, P. and Fomel, S. (2003) Angle-domain common-image gathers by wavefield continuation methods. *Geophysics*, **68**, 1065-1074.

Sava, P. and Biondi, B. (2004a) Wave-equation migration velocity analysis – 1: Theory. *Geophysical Prospecting*, **52**, 593-606.

Sava, P. and Biondi, B. (2004b) Wave-equation migration velocity analysis – 2: Sub salt imaging example. *Geophysical Prospecting*, **52**, 607-623.

Sava, P. and Vasconcelos, I. (2009) Efficient computation of extended images by wavefield-based migration. *79th SEG Annual International Meeting*, Expanded Abstracts, 2824-2828.

Shen, P., Symes, W. and Stolk, C. (2003) Differential semblance velocity analysis by wave-equation migration. *73rd SEG Annual International Meeting*, Expanded Abstracts 2135-2139.

Shen, P., Symes, W. (2008) Automatic velocity analysis via shot profile migration. *Geophysics*, **73**, VE49-VE59.

Shen, P., Symes, W., Morton, S. and Calandra, H. (2005) Differential semblance velocity analysis via shot profile migration. *75th SEG Annual International Meeting*, Expanded Abstracts, 2249-2252.

Shin, C., Jang, S. and Min, D. (2001) Improved amplitude preservation for prestack depth migration by inverse scattering theory. *Geophysical Prospecting*, **49**, 592-606.

Shin, C. and Cha, Y.H. (2008) Waveform inversion in the Laplace domain. *Geophysical Journal International*, **Volume 173**, Issue 3, 922-931.

Shin, C. and Ha, W. (2008) A comparison between the behaviour of objective functions for waveform inversion in the frequency and Laplace domain. *Geophysics*, **73**, Issue 5, Pages: VE119-VE133.

Shin, C., and Cha, Y.H. (2009) Waveform inversion in the Laplace–Fourier domain. *Geophysical Journal International*, **Volume 177**, Issue 3, 1067-1079.

Sirgue, L. and Pratt, R.G. (2003) Waveform inversion under realistic conditions : migration of non-linearity. *73rd SEG Annual International Meeting*, Expanded Abstracts, 694-697.

Sirgue, L. and Pratt, R.G. (2004) Efficient waveform inversion and imaging: a strategy for selecting temporal frequencies. *Geophysics*, **69**, 231-248.

Sirgue, L. (2006) The importance of low frequency and large offset in waveform inversion. *68th EAGE Conference & Exhibition*, Extended abstracts, A037.

Sirgue, L. (2008) 3D Full Waveform Inversion: a complex recipe for success. 70th *EAGE Conference & Exhibition,* Extended abstracts, 41-44.

Sourbier, F., Operto, S., Virieux, J., Amestoy, P. and L'Excellent, JY. (2009) FWT2D: A massively parallel program for frequency-domainfull-waveform tomopgraphy of wide-aperture seismic data- Part1: Algorithm. *Computer & Geosciences*, **35**, 487-495.

_____ (2009) FWT2D: A massively parallel program for frequency-domain full-waveform tomography of wide-aperture seismic data- Part2: Numerical examples and scalability analysis. *Computer & Geosciences*, **35**, 487-495.

Stelk, I., Warner, M. and Umpleby, A. (2007) 3D frequency domain waveform inversion – synthetic shallow channel example. *69th EAGE Conference & Exhibition,* Extended Abstracts, C026.

Symes, W. (1993) A differential semblance criterion for inversion of multi-offset seismic reflection data. *Journal of Geophysical Research*, **98**, 2061-2073.

Symes, W. and Versteeg, R. (1993) Velocity model determination using differential semblance optimization. *63rd SEG Annual International Meeting*, Expanded Abstracts, 696-699.

Sun, H. and Schuster, G. (2001) 2-D wavepath migration. *Geophysics*, **66**, 1528-1537.

Tarantola, A. (1984) Inversion of seismic reflection data in the acoustic approximation. *Geophysics,* **49**, 1259 – 1266.

Toldi, J. (1989) Velocity analysis without picking. *Geophysics*, **54**, 191-199.

Ursin, B. (2004) Tutorial: parameter inversion and angle migration in anisotropic elastic media. *Geophysics*, **69**, 1125-1142.

Vigh, D. and Starr, E.W. (2008) 3D plane wave full waveform inversion. *Geophysics*, **73**, VE135 –VE144.

Vigh, D. and Starr, E.W. (2008) Comparison for Waveform Inversion, Time domain or Frequency domain? *78th SEG Annual International Meeting*, Expanded Abstracts, 1890-1894.

Vigh, D., Starr, W. and Kapoor, J. (2009) Developing Earth model with full waveform inversion. *The Leading Edge*, **28**(4), 432-435.

Virieux, J., Operto, S., Ben-Hadj-Ali, H., Brossier, R., Etienne, V., Sourbier, F., Giraud, L. and Haidar, A. (2009) Seismic wave modelling for seismic imaging. *The Leading Edge*, **28**, 538-544.

Virieux, J.and Operto, S. (2009) An overview of full waveform inversion in exploration geophysics. *Geophysics*, in-press.

Warner, M, Stekl, I. and Umpleby, A. (2007) Full wavefield seismic tomography – iterative forward modelling in 3D. *69th EAGE Conference & Exhibition*, Extended Abstracts, C025.

Zhou, H., Liao, Q. and Ortigosa, F. (2009) Migration Velocity Inversion with Semblance Analysis (mVISA). *71st EAGE Conference & Exhibition*, Extended Abstracts, U022.

Chapter 8

Adler, F. (2004) Anisotropic velocity model building for time-to-depth conversion after non-hyperbolic Kirchhoff PreSTM. *74th SEG Annual International Meeting*, Expanded Abstracts.

Audebert, F., Dirks, V. and Pettenati, A. (2006) TTI anisotropic depth migration: what tilt estimate should we use? *76th SEG Annual International Meeting*, Expanded Abstracts.

Anquelle, L., Arnaud, J., Beele, M., Esquier, P. and Le Stunff, Y. (2006) Anisotropic prestack time migration and prestack depth migration for reservoir evaluation in offshore West Africa. *76th SEG Annual International Meeting*, Expanded Abstracts, 510-513.

Baldock, S., Miller, B., Reta-Tang, C., Rodriguez, G. and Specht, J. (2008) Challenges in Time and Depth: imaging the Eastern Delta OBC survey, a case history. *78th SEG Annual International Meeting*, Expanded Abstracts.

Ball, G. (1995) Estimation of anisotropy and anisotropic 3-D prestack depth migration, offshore Zaïre. *Geophysics*, **60**, 1495-1513.

Baykulov, M. and Gajewski, D. (2008) Seismic data enhancement with Common Reflection Surface (CRS) stack method. *78th SEG Annual International Meeting*, Expanded Abstracts, 2596-2600.

Ben-Brahim, L., Debec, P., Bachtanik, C., Joncour, F. and Rahmanov, O. (2008) How to keep alive an aging field with seismic technology and integrated studies: Alwyn Field case study. PETEX Annual Meeting, Extended Abstracts.

Ben-Brahim, L., Varley, S., Michaud, O., Buffet, P., Hollingworth, S, Woodcock, A. and Purcell, C. (2009) Application of Beam migration to the evaluation of an exploration prospect (Balvenie): a case history from the UKCS Northern North Sea. *71st EAGE Conference & Exhibition*, Extended Abstracts.

Bernitsas, N., Sun, J. and Sicking, C. (1997) Prism waves – an explanation for curved seismic horizons below the edge of salt bodies. *59th EAGE Conference & Exhibition*, Extended Abstracts, E038.

Berkovitch, A., Gelchinsky, B. and Keydar, S. (1994) Basic formulae for multifocusing stack. *56th EAGE Conference & Exhibition*, Extended Abstracts, P140.

Berkovitch, A., Belfer, I. and Landa, E. (2008) Multifocusing as a method of improving subsurface imaging. *The Leading Edge*, 250-256.

Beydoun, W. and Mendes, M. (1989) Elastic ray-Born L^2 migration/inversion. *Geophys. J. Internat.*, **97**, 151-160.

Birdus, S. (2008) Restoring velocity variations below seafloor with complex topography by Geomechanical modelling. *78th SEG Annual International Meeting*, Expanded Abstracts.

Boudou, F., Le Stunff, Y., Arnaud, J. Esquier, P., Kenworthy, A., Whitfield, P. and Soufleris, C. (2008) Benefits of tilted transverse isotropy prestack depth migration for reservoir evaluation offshore West Africa, *First Break*, **26**, 65-69.

Borrini, D., Cristini, A., Follino, P., Marchetti, P. and Zamboni, E. (2005) 3D CRS Processing: a better use of Pre-stack data. *75th SEG Annual International Meeting*, Expanded Abstracts, 2233-2236.

Buur, J. and Kühnel, T. (2008) Salt interpretation enabled by reverse-time migration. *Geophysics*, **73**, VE211-VE216.

Calvert, A., *et al.* (2008) Preserving azimuthal velocity information: experience with cross-spread noise attenuation and offset vector tile PreSTM. *78th SEG Annual International Meeting*, Expanded Abstracts.

Cavalca, M. and Lailly, P. (2005) Prismatic reflections for the delineation of salt bodies. *75th SEG Annual International Meeting*, Expanded Abstracts, 2550-2553.

Cestari-Cuenca, J., Arnaud, J. and Elias, M. (2008) Depth imaging over Structural Transverse Isotropy (STI) media in the presence of complex salt bodies, offshore West Africa. *78th SEG Annual International Meeting*, Expanded Abstracts.

Charles, S., Mitchell, D., Holt, R., Lin, J. and Mathewson, J (2008) Data-driven tomographic velocity analysis in tilted transversely isotropic media: a 3D case history from the Canadian Foothills. *Geophysics*, **73**, VE261-VE268.

De Gennaro, S., Grandi, A., Escobar, I., Onaisi, A., Ben-Brahim, L., Joffroy, G., Tindle, C. and Neillo, V. (2008) Integrating 4D seismic, geomechanics and reservoir simulations in Elgin and Franklin. *70th EAGE Conference & Exhibition*, Extended Abstracts, E019.

Dehghan, K., Audebert, F., and Cestari-Cuenca, J. (2008) STI media in practice, why and when to use it. *78th SEG Annual International Meeting*, Expanded Abstracts, 227-231.

Dehghan, K., Audebert, F. (2009) STI vs VTI velocity model building- More plausible velocity, better tie to the well. *71st EAGE Conference & Exhibition*, Extended Abstracts, U012.

Duquet, B., Marfurt, K. and Dellinger, J. (2000) Kirchhoff modelling, inversion for reflectivity and subsurface illumination. *Geophysics*, **65**, 1195-1209.

Etgen, J., and Regone, C. (1998) Strike shooting, dip shooting, widepatch shooting – Does prestack depth migration care? *68th SEG Annual International Meeting*, Expanded Abstracts, 66-69.

Farmer, P., Jones, I, Zhou, H., Bloor, R. and Goodwin, M. (2006) Application of Reverse Time Migration to complex imaging problems. *First Break*, **24**, 65-73.

Grechka, V. and Tsvankin, I. (1998) 3-D description of normal-moveout in anisotropic inhomogeneous media. *Geophysics*, **63**, 1079-1092.

Hanitzsch, C., Schleicher, J. and Hubral, P. (1994) True-amplitude migration of 2D synthetic data. *Geophysical Prospecting*, **42**, 445-462.

Hatchell, P. and Bourne, S. (2005) Rocks under strain: Strain-induced time-lapse time shifts are observed for depleting reservoirs. *The Leading Edge*, 1222-1225.

Hatchell, P., van den Beukel, A., Molenaar, M., Maron, K., Kenter, C., Stammeijer, J., van der Velde, J. and Sayers, C. (2003) Whole earth 4D: reservoir monitoring geomechanics. *73rd SEG Annual International Meeting*, Expanded Abstracts, 1330-1333.

Hawkins, K., Legott, R., Williams, G. and Kat, H. (2001) Addressing anisotropy in 3-D prestack depth migration: a case study from the Southern North Sea. *The Leading Edge*, 528-533.

Hawkins, K., Leggot, R. and Williams, G. (2002) An integrated geoscience approach to reservoir imaging using anisotropic pres-SDM. *72nd SEG Annual International Meeting*, Expanded Abstracts.

Hawkins, K. (2008) Defining the extent of the compacting Elgin reservoir by measuring stress-induced anisotropy. *First Break*, **26**, 81-88.

Holt, R.A., Campbell, D. (2008) Estimating lateral positioning uncertainty after anisotropic depth migration: a thrust belt case history. *78th SEG Annual International Meeting*, Expanded Abstracts, 232-236.

Hu, J., Schuster, G. and Valasek, P. (2001) Poststack migration deconvolution. *Geophysics*, **66**, 939-952.

Isaac, H. and Lawton, D. (2004) A practical method for estimating effective parameters of anisotropy from reflection seismic data. *Geophysics*, **69**, 681-689.

Isaac, H. and Lawton, D. (2008) Seismic velocity model building in an area of complex geology, southern Alberta, Canada. *Geophysics*, **73**, VE255-VE260.

Jenner, E., Williams, M. and Davis, T. (2001) A new method for azimuthal velocity analysis and applicaton to a 3D survey, Weyburn field, Saskatchewan, Canada. *71st SEG Annual International Meeting*, Expanded Abstracts, 102-105.

Jones, I.F., Bridson, M.L. and Bernitsas, N. (2003) Anisotropic ambiguities in TI media. *First Break*, **21**.

Jones, I.F. (2008) Effects of pre-processing on reverse time migration: a North Sea study. *First Break*, **26**, 73-80.

Landro, M. and Stammeijer, J. (2004) Quantitative estimation of compaction and velocity changes. *Geophysics*, **69**, 949-957.

Lecerf, D., Navion, S., Boelle, J.L., Belmokhtar, A. and Ladmek, A. (2009) Azimuthal residual velocity analysis in offset vector for WAZ Imaging. *71st EAGE Conference & Exhibition*, Extended Abstracts, V013.

Leslie, J. and Lawton, D. (1999) A refraction-seismic field study to determine the anisotropic parameters of shales. *Geophysics*, **64**, 1247-1252.

Müller, T., Jäger, R. and Höcht, G. (1998) Common reflection surface stacking method – Imaging with an unknown velocity model. *68th SEG Annual International Meeting*, Expanded Abstracts.

Nemeth, T., Wu, C. and Schuster, G. (1999) Least squares migration of incomplete reflection data. *Geophysics*, **64**, 208-221.

Roende, H. et al (2008) Estimating subsurface stress direction and intensity from surface full azimuth land data. *78th SEG Annual International Meeting*, Expanded Abstracts.

Sexton, P. and Williamson, P. (1998) 3D anisotropic velocity estimation by model-based inversion of Pre-stack traveltimes. *68th SEG Annual International Meeting*, Expanded Abstracts.

Suiter, J., Romani, R., Arnaud, J., Hollingworth, S. and Hawkins, K. (2004) Reducing structural uncertainties through anisotropic prestack depth imaging: examples from the Elgin/Franklin/ Glenelg HP/HT fields area, central North Sea. *Proceedings of the 6th Petroleum Geology Conference*. Doré, A. G. & Vining, B. (eds). Geological Society, London.

Sengubta, M. and Bachrach, R. (2008) Velocity updating around salt bodies using stress modelling solutions and non-linear elasticity. *78th SEG Annual International Meeting*, Expanded Abstracts.

Taylor, N., Ben-Brahim, L., Tindle, C., Joffroy, G., Hubans, C. and Onaisi, A. (2007) Elgin-Franklin 4D seismic: encouraging results for reservoir monitoring & development planning. *69th EAGE Conference & Exhibition*, Extended Abstracts.

Vestrum, R.W., Lawton, D.C. and Schmid, R. (1999) Imaging structures below dipping TI media. *Geophysics*, **64**, 1239-1246.

Wang, J. and Sacchi, M. (2009) Structure constrained least-squares migration. *79th SEG Annual International Meeting*, Expanded Abstracts, 2763-2767.

Winbow, G., Clee, T. and Rainwater, M. (2007) Multipath imaging into common-reflection-angle at a GOM site. *The Leading Edge*, **26**, 1430-1433.

Winbow, G. (2009) CRAM and Kirchhoff images of a West Africa salt structure. *The Leading Edge*, **28**, 146-155.

11 Index